Handbook of Single Molecule Fluorescence Spectroscopy

Handbook of Single Molecule Fluorescence Spectroscopy

Chris Gell

Laser facilities Manager, Institute of Molecular Biophysics, University of Leeds

David Brockwell

Lecturer, School of Biochemistry and Microbiology, University of Leeds

Alastair Smith

Director, Institute of Molecular Biophysics, University of Leeds

OXFORD
UNIVERSITY PRESS

OXFORD
UNIVERSITY PRESS

Great Clarendon Street, Oxford, OX2 6DP,
United Kingdom

Oxford University Press is a department of the University of Oxford.
It furthers the University's objective of excellence in research, scholarship,
and education by publishing worldwide. Oxford is a registered trade mark of
Oxford University Press in the UK and in certain other countries

First Edition published in 2006
First published in paperback 2013

Impression: 1

Published in the United States of America by Oxford University Press
198 Madison Avenue, New York, NY 10016, United States of America

British Library Cataloguing in Publication Data
Data available

ISBN 978–0–19–852942–2 (hbk.)
ISBN 978–0–19–967384–1 (pbk.)

Printed and bound by CPI Group (UK) Ltd, Croydon, CR0 4YY

"There is nothing, Sir, too little for so little a creature as man. It is by studying little things that we attain the great art of having as little misery and as much happiness as possible."

Samuel Johnson, 1763

Preface

The development of techniques capable of studying the properties of an individual molecule have been strongly driven by new applications in drug discovery and quantum information processing for example, and by an interest in the heterogeneity in the physical, chemical, and biological properties within an ensemble of molecules. Tools for studying the structure and photochemistry of single molecules are now well established and becoming available to a broad range of non-specialist scientists. The most widely used spectroscopic probe at the single molecule level is fluorescence, due to the relatively high quantum efficiency of the process compared with other possible probes. In molecular biology in particular, we have recently seen much wider use of single molecule techniques such as fluorescence resonance energy transfer and fluorescence correlation spectroscopy that have revealed new and interesting kinetic and structural information about a variety of macromolecules.

This book is aimed at experimental scientists with a physical chemistry or biochemistry background who wish to enter this new and exciting field of research and to apply single molecule fluorescence techniques to studies of macromolecular structure and function. The book is designed to present a complete introduction, from the motivation for single molecule experiments to their implementation and the analysis of results.

In Chapter 1 the motivation for single molecule experiments is discussed. Experiments capable of resolving individual molecules are described as a probe of heterogeneity and identification of rare states that are lost within the average signal obtained from conventional ensemble measurements. Then the core experiments and techniques are outlined along with an overview of the information content of the resulting data. In Chapter 2, detailed phenomenological and mathematical descriptions of three principle single molecule fluorescence methods are given (fluorescence correlation spectroscopy, fluorescence resonance energy transfer, and the photon counting histogram). These powerful techniques are discussed in detail along with the methodologies and special considerations needed to collect and analyse the data. In Chapter 3, a thorough description of the implementation of these techniques is presented including many aspects of optical design. In particular, apparatus for far field confocal and total internal reflection type geometries is described in detail. The aim of this chapter is to give the reader a complete practical insight into the realization of single molecule fluorescence experiments. With the framework of motivation, technique, and instrumentation firmly established, Chapter 4 discusses a number of practical considerations. These include selection of chromophores, both intrinsic to the molecule and extrinsic dye molecules, suitable as fluorescent reporters of structure or function. The practicalities of labelling large macromolecules, that is, the chemical attachment of extrinsic dyes to (principally) biological molecules, will also be described in detail since it presents a significant barrier to be overcome in single molecule spectroscopy. The immobilization of molecules on surfaces and within matrices, as well as purification and other related issues for sample preparation will also be discussed. Chapters 5 and 6 will provide a review of the applications of single molecule fluorescence spectroscopy, and discuss these with relation to the practical problems that have been encountered and overcome and the potential for new experiments that are exposed. The corollary in Chapter 7 highlights the exciting outlook for the analysis of individual molecules, with particular attention paid to fundamental studies of biomolecular structure and conformational dynamics.

The authors intend that this volume will give the reader a complete guide to the practical implementation of single molecule experiments and stimulate the same excitement they feel for this growing field.

Chris Gell
David Brockwell
Alastair Smith

Contents

Acknowledgements

The authors thank everyone who has had some input into the compilation and editing of this text. In particular we acknowledge collaborators Sheena Radford, Peter Stockley, and Nicola Stonehouse, all at Leeds University, without whom much of the work that provided the motivation for this text would not have been performed. The ongoing projects conceived with them, and the questions they asked, led to our need to implement many of the techniques we describe. The lessons we learnt (and are still learning) provided the basis to enable us to write this text—hopefully it will help anyone else with similar questions.

Much of the data that we present was measured (unless otherwise referenced) by scientists working in the Institute of Molecular Biophysics in Leeds. In particular we thank Tomoko Tezuka-Kawakami, Tara Sabir, Rob Leach, Sara Pugh, Jennifer Clark, and Mark Robinson. We are also very grateful to Clive Bagshaw for critical reading of the manuscript. For informal, but precise, editorial input we are indebted to Claire Friel and Kurt Baldwin. We also thank Andrea Rawse for her efforts in obtaining reprint permissions.

Chris Gell
David Brockwell
Alastair Smith

Glossary of terms and symbols

Here we list a number of mathematical symbols, specialist terms, and acronyms used throughout this text. Where possible we have used the commonly accepted convention, although some duplication and repetition is present as symbols and acronyms are not always used consistently in the literature. In places we have indeed used different symbols to identify the same parameters. Generally, this is in order to maintain consistency with the nomenclature found in the original work that we describe or review.

Å	Angstroms
AFM	atomic force microscopy
ALEX	alternating laser excitation
APD	avalanche photodiode
Apo	apochromatic
BFP	back focal plane (the focal plane typically on the illumination side of the lens)
bp	base pair/s
BSA	bovine serum albumin
CCD	charge coupled device
CEF	collection efficiency function (see PSF)
CMOS	complimentary metal oxide semiconductor
Corr{x,y}	mathematical correlation of x and y
cpspm	counts per second per molecule
D	penetration depth, beam diameter, aperture diameter, translational diffusion coefficient
DC	value to which the autocorrelation function decays—the baseline (typically 1)
DLL	dynamic linked library
DTT	dithiothreitol
e	base of the natural logarithm function ($e = \exp(1) = 2.71828$)
E_{FRET}	fluorescence resonance energy transfer efficiency
η	viscosity
η_I	proportionality constant between the amount of light falling on a detector and detected signal
η_x	detection efficiency of x

ϵ molecular brightness (see cpspm)
EMCCD electron multiplying charge coupled device
ESI-MS electro-spray ionization mass spectrometry
f focal length
F_T FCS triplet fraction (molecule in the triplet state)
FAM carboxyfluorescein
FAMS fluorescence aided molecular sorting
FCS fluorescence correlation spectroscopy
FIDA fluorescence intensity distribution analysis
FITC reactive isothiocyanate form of fluorescein
Fluor fluorite
FRET fluorescence resonance energy transfer
γ constant depending on the PSF (in FCS), constant accounting for differential detection efficiencies, and quantum yields of the donor and acceptor in spFRET detection
$g_{3DG}(\tau)$ diffusion only part of the analytical description of the autocorrelation function with the sample volume (PSF) approximated to a three-dimensional Gaussian
$G(\tau)$ autocorrelation function with lag-time (see τ)
GdnCl guanidine chloride
GdnHCl guanidine hydrochloride
GFP green fluorescent protein
HPLC high-performance liquid chromatography
Hz hertz
I_A number of acceptor photon counts
I_D number of donor photon counts, light intensity at the detector
I(f) Fourier transform of the intensity signal I(t)
I^*(f) complex conjugate of the Fourier transform of the intensity signal I(t)
IA iodoacetamide
IC internal conversion (photophysics)
ICCD intensified charge coupled device
ISC inter-system crossing (photophysics)
ISF instrument spread function
J spectral overlap integral (spFRET)
κ electric dipole orientation factor (spFRET)
k photon counts per unit time interval, Boltzmann constant
k_B Boltzmann constant
k^2 PSF shape parameter in FCS ($= \omega_0/z_0$)
K_d disassociation constant
k_x rate constant for the process x

λ	wavelength
μ	micro-, mean or moment
μ_m	mth moment of μ
n	refractive index
M	molarity
M^2	laser beam quality parameter
MAFID	moment analysis of the fluorescence intensity distribution
MCP	micro channel plate
MCS	multi channel scalar card
N_A	Avogadro's number
N	time averaged number of molecules in the PSF
NA	numerical aperture
Nd:YAG	neodymium-doped yttrium aluminum garnet, (a typical solid laser gain medium)
Nsm^{-2}	Newton seconds per square meter
ω_0	radius of the point spread function perpendicular to the optic axis
OD	optical density
OPSL	optically pumped semiconductor laser
$p^{(k)}$	probability distribution or PCH of k
$p^{(1)}(k,x)$	PCH for a 1 particle system with photon counts k and involving parameters x
$p^{(\text{fixed})}(k,x)$	PCH for a particle fixed at the origin with photon counts k and involving parameters x
P	proximity ratio (in spFRET)
P_{FRET}	proximity ratio (in spFRET)
PCH	photon counting histogram
PCI	peripheral component interconnect
PCR	polymerase chain reaction
PEG	polyethylene glycol
Poi(k,x)	a Poisson distribution of the independent variable k, involving parameters x
PMMA	polymethyl-methacrylate
PMT	photomultiplier tube
PSF	point spread function
PVA	polyvinyl alcohol
r	anisotropy
r_0	intrinsic molecular anisotropy
RNA	ribonucleic acid
σ	standard deviation
S	stoichiometry based ratio in ALEX

S_x	singlet level x (photophysics)
SCM	scanning confocal microscopy
SDF	spatial detectivity function (see PSF)
SH	sulphydryl group
smMFD	single molecule multiparameter fluorescence detection
spFRET	single pair fluorescence resonance energy transfer
τ	lifetime, lag-time (correlation time or delay time)
τ_D	FCS diffusion time
τ_R	reaction rate for a reversible two-state process manifesting in the FCS autocorrelation function
τ_T	FCS triplet correlation time
TCSPC	time correlated single photon counting
TMR	tetramethylrhodamine
ϕ	rotational correlation time (in anisotropy)
ϑ	angle (degrees)
ϑ_T	critical angle for total internal reflection
Q	quantum yield
Q_D	quantum yield of the donor dye (spFRET)
QE	quantum efficiency
R	scalar dye separation in spFRET
R_0	Förster distance (spFRET, R for 50% transfer efficiency)
Φ_x	quantum yield of x
T	threshold (in spFRET), temperature
$T_{A/D}$	threshold for the donor or acceptor detection channel (in spFRET)
TCEP	tris(2-carboxyethyl)phosphine hydrochloride
TEM	transverse electromagnetic mode
TIR	total internal reflection
TIRF	total internal reflection fluorescence
TIRFM	total internal reflection fluorescence microscope/microscopy
TTL	transistor transistor logic
UV	ultra-violet part of the electromagnetic spectrum
V_x	vibrational energy level
V_0	volume (typically of the PSF)
z_0	radius of the point spread function in the direction of the optic axis
$<k>$	mean of k
$<\Delta k^2>$	variance of k
$\otimes$	mathematical convolution operation

ONE
Introduction

1.1 Motivation

The measurement of single fluorescent molecules is, in principle, straightforward as suitable detectors, light sources, and sampling optics are readily available. The major hurdle lies in the interference from the massive excess of other molecules that contribute to the background, along with photophysical artifacts from the dye labels, that both reduce the signal-to-noise of the measurement. So what is the motivation for facing the challenges of making measurements with single molecule resolution? The simplest reason is the need to achieve very high sensitivity. Single molecule measurements represent the ultimate level of sensitivity—the ability to detect 1.66×10^{-24} mole of the object of interest (1.66 yoctomole), that is, the inverse of Avogadro's number, which has also been referred to as a 'guacamole'. Sensitivity is clearly a strong driving force in applications such as pathology and diagnostic medicine in which one would ideally like to be able to detect one copy of a protein or gene, perhaps within a cell, that is indicative of disease. Measurement sensitivity will also be key to overcoming the contemporary challenges of studying and developing nanoscale devices and subsequently interfacing with them. Nanotechnology is creating new requirements for optical and photonic probes for which single molecule techniques can provide solutions. As well as high sensitivity, single molecule measurements also provide information about the environment local to the probe fluorophore with extremely high spatial resolution. A conventional microspectroscopy measurement typically samples a volume of 10^5 nm^3 and even near-field optical probes, which avoid the diffraction limit to spatial resolution (which is of the order of half the wavelength of light used in conventional optical microscopies), sample several hundreds of cubic nanometres. However, an individual molecule samples its surroundings within a much smaller volume, possibly only a few cubic nanometres and can therefore relay chemical information with very high spatial resolution and probes the local environment with similar resolution, providing a valuable interface with the macroscopic world.

Importantly, single molecule spectroscopy has other merits in addition to sensitivity and high resolution. Measurements of concentrated samples yield only an ensemble average of the properties of interest and provide no means of assessing the heterogeneity of complex systems. Single molecule spectroscopy on the other hand provides an insight into the behaviour of each individual molecule and therefore allows the detail of subpopulations in structure or dynamics within an ensemble to be delineated. In addition, single molecule methods provide a way of probing fluctuating systems under equilibrium conditions, allowing kinetic pathways to be studied without the need for synchronization. For example, in a typical ensemble experiment designed to observe protein folding kinetics, the folding of many molecules must be synchronized by some starting event such as a rapid temperature increase [1], pH jump [2] or change in the chemical conditions by mixing two solutions [3]. If the kinetics of each molecule are not synchronized in an ensemble experiment then the kinetic rate constants cannot be measured. Such initiating events have finite duration; for example, mixing two solutions requires hundreds of microseconds or milliseconds [3] depending on the experimental design. Therefore, the synchronization event in ensemble experiments creates a dead time for observation that can make it impossible to observe early kinetic events that may determine the route taken through the folding energy landscape. Single molecule measurements intrinsically require no such synchronization and therefore reaction or folding kinetics can, in principle, be studied with shorter dead time. The ability of single molecule measurements to observe heterogeneity in kinetic pathways by a series of measurements allows the experimentalist to dissect the ensemble average. This procedure may allow rare intermediates to be observed that would be swamped by the signal from more abundantly populated states in an ensemble experiment. This is perhaps the key motivation for many researchers adopting single molecule techniques.

Finally, it should be noted that single molecule measurements can also provide an important direct comparison with theory and the results of computer simulations. Many theoretical approaches and computer simulations inherently deal with the properties of an individual molecule; a comparison with the average properties of the system yielded by ensemble experiments may therefore be far from ideal. Single molecule measurements require no assumption about how molecular properties scale to the bulk and these studies thus allow direct comparison with the results of theory and simulations.

1.2 A historical perspective

Arguably the first single molecule spectroscopic measurement was made by Rotman in the 1960s [4] in which a single enzyme was detected indirectly through its reaction products. However, Hirschfeld [5, 6] was probably the first to make direct measurements with single molecule sensitivity when he demonstrated the detection of an individual antibody molecule, albeit labelled with ~100 fluorophores! One can argue that his is not the seminal work but

Hirschfeld's contributions were significant since he recognized the need for reduced excitation and collection volumes and discussed photobleaching as one of the essential limitations in single molecule spectroscopy. Moerner and Kador [7,8] clearly demonstrated the detection of an individual molecule using absorption measurements at low temperature in 1989 and since then there has been a rapid growth in the number of reports of single molecule spectroscopy focusing mainly on fluorescence studies, the first of which was made by Orrit [9]. Keller's group at Los Alamos [10–12] was one of the major contributors to the development of room temperature single molecule spectroscopy in fluids. Their work had a great influence on the growth in interest of single molecule fluorescence measurements of biological molecules under physiological conditions. Betzig [13] made significant contributions to the field in the early nineties by using near-field fiber optical probes to detect single fluorophores immobilized on surfaces and showed that the orientation of their transition dipole moments could be mapped using the technique. However, the field really began to gain momentum when a simple confocal optical microscope arrangement was shown to be capable of making single molecule measurements [14–16]. The simplicity and relatively low cost of this approach has, over the last 10 years, resulted in an explosion in the number of publications applying single molecule techniques in chemistry, physics, and biophysics. A recent step was made when wide field microscopy was used to image single molecules immobilized on surfaces using a total internal reflection illumination geometry and an intensified Charge Coupled Device camera, and this arrangement has since proved the method of choice for studying immobilized biological systems and even individual molecules within cells [17,18]. Along with the development of optical arrangements came the methods for analyzing the data. In the case of freely diffusing molecules, the fluorescence bursts can be analysed in terms of their brightness, duration, polarization, and fluorescence lifetime or wavelength using correlation spectroscopy or other statistical methods, which will be discussed in some detail in Chapter 2. In the case of immobilized molecules similar statistical approaches can be employed to the time series of fluorescence photons emitted by a single molecule to report on dynamical processes such as protein folding or the action of molecular motors.

1.3 This book

Although absorption and Raman spectroscopy [19,20] have also been shown to provide single molecule sensitivity, we shall concentrate in this book on fluorescence techniques and their applications. This is by no means a limitation; fluorescence spectroscopy and microscopy provides a vast range of opportunities for experiments in physics, chemistry and biology owing to the availability of new detectors and light sources, fluorescent dyes and labelling chemistries coupled to the never ending supply of fascinating problems. In the remaining sections of this chapter we provide a simple phenomenological introduction of the two core groups of single molecule fluorescence experiments: measurements of diffusing single molecules and measurements of immobilized single fluorescent molecules, in order to provide the reader with a basic understanding of the experiments and the information content of the data. Subsequent chapters then greatly expand upon this brief overview: in Chapter 2, detailed phenomenological and mathematical descriptions of three major single molecule fluorescence methods are given (fluorescence correlation spectroscopy, fluorescence resonance energy transfer and the photon counting histogram) along with a discussion of the simpler analysis of data resulting from studies on immobilized molecules. In Chapter 3, a thorough description of the implementation of these techniques is presented including simplified aspects of optical design and data collection. In particular, basic apparatus for far-field confocal, far-field multi-photon and total internal reflection geometries are described in detail. The aim of this chapter is to give the reader a practical insight into the implementation of single molecule fluorescence measurements. With the framework of technique and instrumentation firmly established, Chapter 4 discusses a number of practical considerations. These include selection of chromophores, both intrinsic molecular fluorophores and extrinsic dye molecules that are suitable as fluorescent reporters of structure or function, along with a basic introduction to dye photophysics relevant to single molecule work. The practicalities of labelling large macromolecules, that is, the chemical attachment of extrinsic dyes to (principally) biological molecules, will also be described since it represents a significant challenge that has to be overcome prior to making single molecule measurements. The immobilization of molecules on surfaces and within matrices as well as purification and other related issues for sample preparation will also be discussed. Chapters 5 and 6 will provide a non-exhaustive review of existing applications of single molecule fluorescence spectroscopy, and discuss these in relation to the practical problems that have been encountered and overcome. The potential for new experiments that are exposed as a result of these studies are also highlighted. The corollary in Chapter 7 highlights the exciting outlook for single molecule fluorescence analysis.

This book is aimed at experimental scientists with a physical chemistry or biochemistry background who wish to enter this new and exciting field of research and to apply single molecule fluorescence techniques to studies of macromolecular structure and function. The book is designed to present an introduction to the topic, from the practical implementation of single molecule fluorescence experiments, through methods of data analysis to a description of a range of current and future applications. We hope that we have provided enough useful information to make starting such experiments straightforward and rewarding.

1.4 **Single molecule measurements**

In this section we outline the two main types of single molecule measurement that we have chosen to discuss in detail in this text; measurements on diffusing fluorescent single molecules and measurements on immobilized single fluorescent molecules. We introduce the basic concepts of these experiments, which we then expand upon in both a phenomenological and rigorous mathematical way in subsequent chapters.

1.4.1 **Diffusion studies**

The basic concept of a single molecule diffusion fluorescence experiment is illustrated in Figure 1.1. The labelled analyte (Figure 1.1(a)), in this case a nucleotide stem – loop structure with a single fluorescent dye label at one terminus and a quencher for this dye at the other terminus, is allowed to diffuse freely in solution. The molecule can undergo a reversible conformational transition between the folded (stem – loop) and unfolded (denatured random coil) conformations with some rate constants. The fluorescence from a small sample volume (<0.1 femtolitre, Figure 1.1 (b)) in this solution is then monitored as a function of time. When the analyte diffuses into the volume a transient burst of fluorescence is observed above the background level (Figure 1.1 (c)). The temporal persistence of this burst is a function of a number of variables including: solvent viscosity, molecule size, path through the volume, quantum yield of the dye and the size of the volume. If the molecule is in a dynamic equilibrium, where the energy barrier between the conformations is of the order of the energy available to the system (~$k_B T$, the Boltzman constant multiplied with the temperature), then the molecule may, at any point, undergo a reversible transition to the other conformation. If the rate of the conformational fluctuations (the observed rate constant) is faster than the time taken to diffuse through the volume, then the quenching or recovery of fluorescence for the folded and unfolded states, respectively, will modulate the burst between high and low signal levels (Figure 1.1(d)).

The amount of information contained within these bursts is large. Techniques such as fluorescence correlation spectroscopy (FCS, Chapter 2) are able to extract the average width (the amount of time) that fluctuations in the signal last for and so can measure the width of the transients (i.e. diffusion coefficients) or the width of the features within the modulated transients (i.e. the observed rate constant for the dynamics). Further, by measuring for several minutes, sufficient statistics can be built up to allow the analysis of the heights (intensities) of the transient bursts through the fitting of histograms of the number of counts observed in each counting interval (PCH, Chapter 2). These data can then be used to explore heterogeneity in the sample, if that heterogeneity is marked by species with differential mean intensities. If the analyte is labelled with two-dye molecules selected so that inter-dye, distance-dependant energy transfer can occur, then measurements of the signals from each dye simultaneously can revel both structural and dynamic information by using single-pair fluorescence resonance energy transfer analysis techniques (spFRET, Chapter 2).

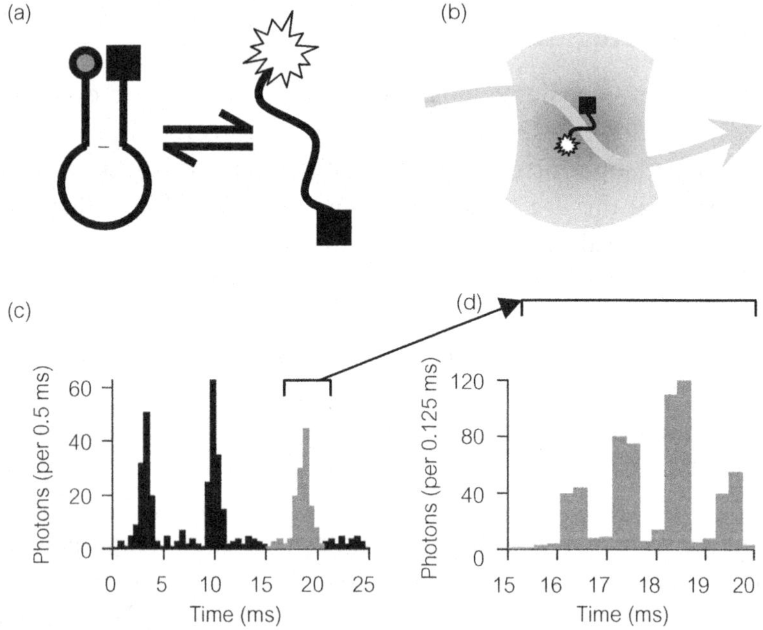

Figure 1.1 Illustration of the concept of measuring the fluorescence from an individual single molecule diffusing in solution. (a) A molecule, which can undergo a reversible transition between a folded and unfolded conformation, is labelled with a dye at one terminus and a quencher at the other. In the folded (native) conformation the fluorescence from the dye is quenched. In the unfolded (denatured) conformation the fluorescence is enhanced. (b) The molecule is allowed to diffuse freely in solution. Passage through the sample volume is detected by fluorescence emission from the single dye label. (c) The resulting fluorescence signal in a counting interval (0.5 ms), measured as a function of time, reveals transient bursts above a background. Many bursts can occur, each from a different single molecule. (d) If the rate of reversible conformational fluctuations is faster than the diffusion time through the volume then the individual transient bursts are themselves modulated by the conformational dynamics and contain important information (assuming that the time resolution is increased in order to follow the fluctuations).

1.4.2 **Immobilization studies**

One disadvantage of the measurement of freely diffusing single molecules is the transient nature of the detected fluorescence, which means that in order to build up statistics the measurements of many different single molecules are required. Whilst such an experiment allows tremendous insight, such as resolution of folded and unfolded protein species in a solution, some sensitivity to determine heterogeneity may be lost due to the averaging over molecules within each state. Further, it may not be possible to differentiate static or dynamic heterogeneity—does each member of an ensemble (each folded molecule for example) provide the same, but different for different molecules, time-invariant signal; or does each molecule contribute an unsynchronized time-dependant signal.

Studies of immobilized single molecules can help to answer some of these questions. In these experiments the time available to monitor the molecule in the small observation volume is extended by immobilizing the molecule of interest, either tethered to a substrate (generally a glass slide, Figure 1.2(b)) or by immobilization in a gel or tethered liposome. In this way the observation time is only limited by the stability of the instrument used, the signal to noise ratio of the

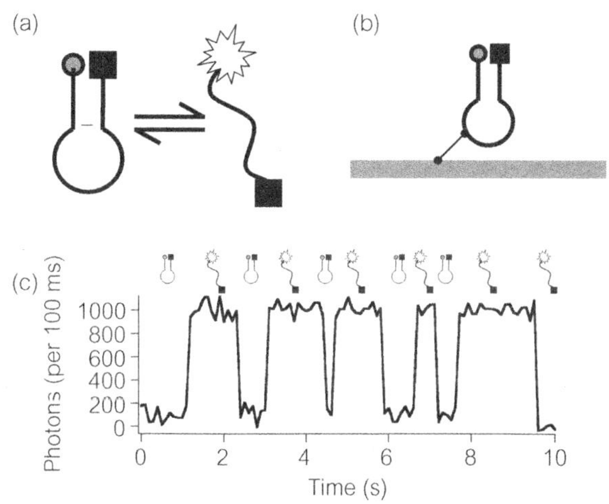

Figure 1.2 Illustration of the concept of measuring the fluorescence from an immobilized single molecule. (a) A molecule, which can undergo a reversible transition between folded and unfolded conformations, is labelled with a dye at one terminus and a quencher at the other. In the folded (native) conformation the fluorescence from the dye is quenched. In the unfolded (denatured) conformation the fluorescence is enhanced. (b) The molecule is immobilized (tethered) onto a solid substrate and the fluorescence signal from a small volume near the surface monitored as a function of time. (c) The same molecule can be monitored for a considerable length of time and the stochastic transitions (the number of which depend on the height of the energy barrier for the transition) can be observed. Eventually (at around 9.5 s in this simulated example), photobleaching of the dye occurs to a non-fluorescent state, at which point no more information can be extracted from this molecule.

experiment and irreversible photobleaching of the dye to some non-fluorescent state. In this way it is possible to monitor the fluorescence of a single molecule as a function of time (an intensity trajectory) for up to several minutes (Figure 1.2(c)). If sufficient time resolution and signal to noise is available in the experiment then one can directly extract kinetic information for individual molecules. As for diffusion techniques these studies can be extended with spFRET to provide a powerful probe of structure and dynamics for complex systems. Again, it may be necessary to combine the results of many individual molecules' trajectories, but unlike the diffusion case little information is lost in this way.

1.4.3 **Interpretation of single molecule data**

In the examples described earlier much of the detail has been omitted. For example, detailed statistical analysis of data is often necessary to determine the reliability of any findings, as the data are often dominated by random contributions (for example, the path taken to diffuse through the volume and shot noise from the detection of small numbers of photons per molecule). Of particular concern is the photo-physics of the labels used: transient dark states, photobleaching, and quenching—all can be mistakenly interpreted as reporting on the behaviour of the host molecule if care is not taken. Further, dye labelling and immobilization must be shown not to perturb the molecule being probed in any significant manner. In the remainder of this text we hope to give the reader an introduction to many of these topics and hope that it enables the application of these techniques in exciting new ways.

References

[1] **Dimitriadis, G, Drysdale, A, Myers, JK, Arora, P, Radford, SE, Oas, TG,** *et al.*, Microsecond folding dynamics of the F13W G29A mutant of the B domain of staphylococcal protein A by laser-induced temperature jump. *Proceedings of the National Academy of Sciences of the United States of America* **101** (2004) 3809–3814.

[2] **Rami, BR** and **Udgaonkar, JB**, pH-Jump-induced folding and unfolding studies of barstar: Evidence for multiple folding and unfolding pathways. *Biochemistry* **40** (2001) 15267–15279.

[3] **Roder, H, Maki, K, Cheng, H,** and **Shastry, MCR**, Rapid mixing methods for exploring the kinetics of protein folding. *Methods* **34** (2004) 15–27.

[4] **Rotman, B**, Measurement of Activity of single molecules of beta-D- galactosidase. *Proceedings of the National Academy of Sciences of the United States of America* **47** (1961) 1981–1991.

[5] **Hirschfeld, T**, Optical microscopic observation of single small molecules. *Applied Optics* **15** (1976) 2965–2966.

[6] **Hirschfeld, T**, Quantum efficiency independence of time integrated emission from a fluorescent molecule. *Applied Optics* **15** (1976) 3135–3139.

[7] **Moerner, WE** and **Kador, L**, Finding a single molecule in a haystack—optical-detection and spectroscopy of single absorbers in solids. *Analytical Chemistry* **61** (1989) A1217–A1223.

[8] **Moerner, WE** and **Kador, L**, Optical-detection and spectroscopy of single molecules in a solid. *Physical Review Letters* **62** (1989) 2535–2538.

[9] **Orrit, M** and **Bernard, J**, Single pentacene molecules detected by fluorescence excitation in a para-terphenyl crystal. *Physical Review Letters* **65** (1990) 2716–2719.

[10] **Dovichi, NJ**, **Martin, JC**, **Jett, JH**, and **Keller, RA**, Attogram detection limit for aqueous dye samples by laser- induced fluorescence. *Science* **219** (1983) 845–847.

[11] **Nguyen, DC**, **Keller, RA**, and **Trkula, M**, Ultrasensitive laser-induced fluorescence detection in hydrodynamically focused flows. *Journal of the Optical Society of America B-Optical Physics* **4** (1987) 138–143.

[12] **Shera, EB**, **Seitzinger, NK**, **Davis, LM**, **Keller, RA**, and **Soper, SA**, Detection of single fluorescent molecules. *Chemical Physics Letters* **174** (1990) 553–557.

[13] **Betzig, E** and **Chichester, RJ**, Single molecules observed by near-field scanning optical microscopy. *Science* **262** (1993) 1422–1425.

[14] **Bian, RX**, **Dunn, RC** and **Xie, XS**, Single molecule emission characteristics in near-field microscopy. *Physical Review Letters* **75** (1995) 4772–4775.

[15] **Macklin, JJ**, **Trautman, JK**, **Harris, TD**, **Brus, LE**, Imaging and time-resolved spectroscopy of single molecules at an interface. *Science* **272** (1996) 255–258.

[16] **Rigler, R** and **Mets, U**, Diffusion of single molecules through a Gaussian laser beam. *Laser Spectroscopy of Biomolecules* 1921 (1992) 239.

[17] **Mashanov, GI**, **Tacon, D**, **Knight, AE**, **Peckham, M**, and **Molloy, JE**, Visualizing single molecules inside living cells using total internal reflection fluorescence microscopy. *Methods* **29** (2003) 142–152.

[18] **Mashanov, GI**, **Tacon, D**, **Peckham, M**, and **Molloy, JE**, The spatial and temporal dynamics of pleckstrin homology domain binding at the plasma membrane measured by imaging single molecules in live mouse myoblasts. *Journal of Biological Chemistry* **279** (2004) 15274–15280.

[19] **Kneipp, K**, **Wang, Y**, **Kneipp, H**, **Perelman, LT**, **Itzkan, I**, **Dasari, R**, *et al.*, Single molecule detection using surface-enhanced Raman scattering (SERS). *Physical Review Letters* **78** (1997) 1667–1670.

[20] **Nie, SM** and **Emery, SR**, Probing single molecules and single nanoparticles by surface-enhanced Raman scattering. *Science* **275** (1997) 1102–1106.

TWO
Single molecule fluorescence techniques

2.1 Introduction

The information that can be obtained from single molecule fluorescence data depends on the sophistication and reliability of the analysis employed. In order to understand what information single molecule fluorescence experiments can provide, we will take the approach of first understanding the methods of data analysis. For this purpose we will define two classes of single molecule fluorescence experiment; those involving freely diffusing molecules in solution and those in which molecules are immobilized at a surface. We shall describe the concepts of some of the data analysis techniques appropriate to these two classes of experiment including burst analysis, fluorescence correlation spectroscopy (FCS), single pair fluorescence resonance energy transfer (spFRET) and photon counting histograms (PCH). A short discussion will also be given of more advanced methodologies that follow naturally from these basic areas, including multi-parameter analysis, moment analysis and higher order autocorrelation analysis. These methodologies form the basis of single molecule fluorescence spectroscopy, but the reader should be aware that new approaches are constantly being developed. For all of the experiments that we discuss, we assume a basic understanding of the theories of the interaction of light with matter, geometrical optics, polarization and luminescence. For a grounding in these topics we refer the reader elsewhere [1–3].

2.2 Burst analysis

If the fluorescent analyte is allowed to flow or diffuse into and out of a small excitation/collection volume defined, in part, by a focused laser beam, then this gives rise to a stochastic series of short-lived fluorescence bursts detected above the background noise level (see Figure 2.1(a)). This type of experiment was one of the first used to demonstrate the feasibility of fluorescence detection of single

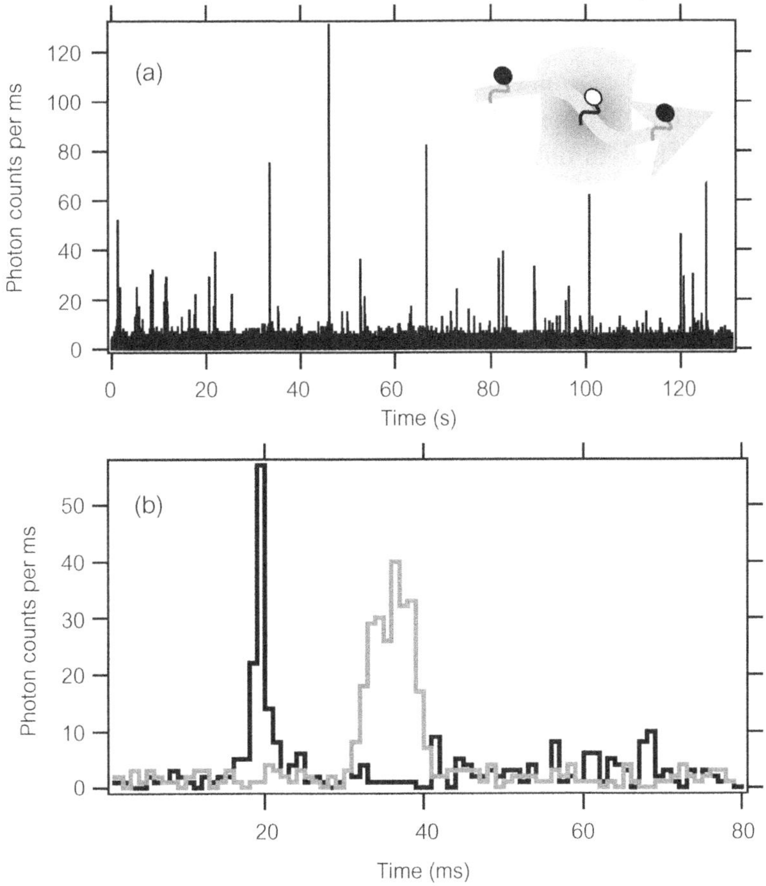

Figure 2.1 (a) Typical burst trace for a 100 pM sample of the fluorescent dye fluorescein in water. As molecules diffuse into and out of the small excitation/collection volume (cartoon inset) they lead to bursts of fluorescence observed above a background signal generated by Rayleigh and Raman scattering from the solvent, fluorescence from impurities and noise from detectors and other electronics. (b) A close up of two typical bursts from differing data sets. Black shows an individual burst from fluorescein diffusing in water, showing the transient nature of the burst due to the short transit time through the <0.1 fl volume. Grey shows the same molecule but in a solution containing 50% glycerol to increase solvent viscosity. In this case the rate of diffusion of the molecule is reduced and so the width of the burst is increased. Burst widths may also vary because molecules may take a long or a short path through the excitation/collection volume. The burst intensity also depends on the path through the excitation/collection volume and shot noise.

molecules in solution at room temperature [4]. However, despite the simplicity of the approach, the stochastic nature of the data requires sophisticated analyses. Bursts often consist of <100 photons and the data are therefore dominated by shot noise (see later), in addition each molecule is able to take any path through the excitation/collection volume which has a spatially dependent excitation intensity and collection efficiency (see Chapter 3, Section 3.2.2), resulting in a range of burst widths and intensities.

The simplest approach to analyse such transient signals is often referred to as burst analysis. Burst analysis involves the straightforward counting of bursts, the quantification of the number of photons in a burst, the length of the burst or the time between bursts (recurrence time). Burst analysis has been used quite widely, for example in high throughput screening and medical diagnostics applications. Ferris *et al.* [5] used burst analysis to treat the data from experiments using flowing sample streams containing fluorescently labelled respiratory viruses. After careful instrument calibration they showed that the number of viruses present could be obtained rapidly and accurately by simply 'counting' the fluorescence bursts. Furthermore, they demonstrated that under conditions when the relative contribution of shot noise in the data is low, when multiple dye-labelling of the viral complexes was employed, the fluorescence intensity of the bursts can even be used to estimate the size of each individual complex. An interesting observation was made by Osborne and colleagues [6] who carried out a similar simple statistical analysis on the fluorescence burst traces of a number of different fluorescent molecules and demonstrated that the distribution of recurrence times was non-random despite the stochastic nature of the experiment. They explained this by suggesting that there was a biasing potential, probably due to the electric field of the focused laser beam, which increases the probability of a molecule diffusing back into the volume after it has just diffused out. This leads to a bunching of burst events, which has important consequences for burst analysis for all applications.

The reliability of screening or identification assays using simple forms of burst analysis has been improved by developing methods for the coincident detection of two dye labels attached to a target molecule. In this way the properties of the particular fluorescence bursts are of somewhat less concern as coincident bursts can be detected with significantly more confidence, facilitating the discrimination of signals from uncorrelated background events. For example Li and co-workers [7] used coincidence detection to distinguish double labelled DNA in a solution in the presence of 1000 fold excess of single labelled DNA. This has obvious applications in detecting bound complexes over unbound monomers in high throughput screening/drug discovery applications, for example.

2.3 Photon counting histograms

A higher level of sophistication in analysis is achieved by using photon counting histograms (PCH). PCH are formed by a thorough statistical analysis of the distribution of the number of detected photons in each burst (or the distribution of the fluorescence intensity measured in each counting interval). PCH is mainly used to measure sample heterogeneity by determining the concentration of each species in the sample and the brightness of its fluorescence. Thorough but manageable theoretical treatments of the distribution of photon counts from single molecules were developed independently by two groups. PCH was introduced by Enrico Gratton's group [8,9] and fluorescence intensity distribution analysis (FIDA) was developed in Kirsten Gall's laboratory [10]. Both methods have the same physical origins, that is, the statistical analysis of the number of fluorescence photons detected as single molecules diffuse freely into and out of a small excitation/collection volume, and both have their origins in the moment analysis of fluorescence intensity distribution (MAFID) introduced by Elson in 1990 [11]. There the higher order moments (see Section 2.7) of fluorescence fluctuation data are calculated and the mean number of fluorescent molecules in the excitation/collection volume is recovered. FIDA and PCH rely on the calculation of the probability of observing k photons during an integration time T (referred to as $p(k)$) which is dependent on the fluorescence brightness and the concentration of the molecules (taking into account all stochastic contributions). PCH and FIDA differ in their mathematical methodology; FIDA incorporates a more sophisticated algorithm with an empirical description of the excitation/collection volume (see Chapter 3) rather than the theoretical approximation used in PCH [12]. However, others have extended PCH by incorporating semi-empirical parameters into the model to account for the non-ideality of the excitation/collection volume [13–15]. A detailed description of PCH along with a rigorous mathematical derivation can be found elsewhere [8,9,16]. Here we will provide a simplified description as a practical introduction to the technique.

PCH characterizes fluorescence fluctuation data using two parameters, the average number of molecules present $\overline{N}$ in the excitation/collection volume of the instrument (see Chapter 3) and the molecular brightness ε. The brightness is defined as the mean number of photon counts detected per molecule per sampling interval but is often expressed as the number of counts per second per molecule (cpspm). The goal is to develop an expression that can be used to fit an experimentally determined photon counting histogram taking into account the stochastic nature of the experiment (and thereby recover information about the system under study). This goal is achieved via a number of stages (see Figure 2.2):

1. Consider the distribution of counts arising from the intrinsic stochastic nature of photon detection by first assuming that the light intensity arriving at the detector is constant; for example from a steady fluorescence source, fixed in space (Figure 2.2(a)).

2. Consider fluctuations in the light intensity falling on the detector caused by the diffusion of a single fluorescent particle around the excitation/collection

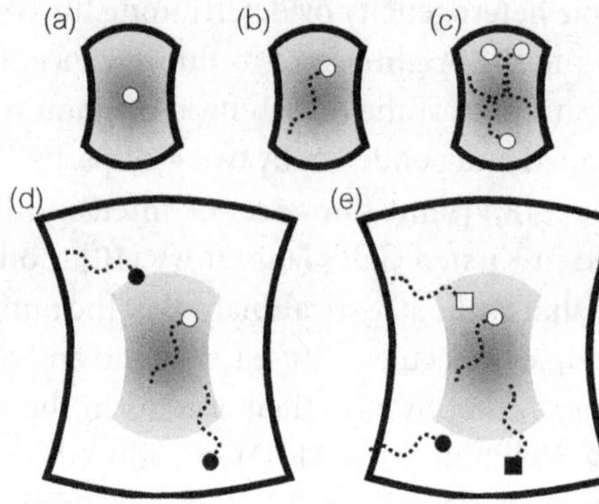

Figure 2.2 Schematic illustration of the conceptual stages in the development of a model to fit photon counting histograms. (a) The case of a non-fluctuating fluorescent particle fixed at the centre of a closed excitation/detection volume (V_0). (b) The case when fluctuations are created by diffusion of the fluorescent molecule around a closed volume with spatially varying excitation/detection efficiency. (c) The case of multiple diffusing molecules in the closed volume. (d) The case when molecules can enter and leave the volume. (e) The case when molecules with different molecular brightness can enter and leave the volume.

volume (Figure 2.2(b)), which has a spatially varying excitation/collection efficiency (e.g. a focused laser beam and confocal detection optics).

3. Extend the model to the case of many fluorescent particles diffusing around but unable to leave the excitation/collection volume (Figure 2.2(c)).
4. Incorporate the concept of a sample volume that is larger than the excitation/collection volume, which therefore implies that the fluorescent particles may enter and leave the volume (Figure 2.2(d)).
5. Consider the case of two or more distinct species, defined by differing molecular brightnesses, able to diffuse into and out of the volume (Figure 2.2(e)).

With these concepts in mind we may now proceed to place them within a mathematical framework.

2.3.1 **Photon detection statistics**

First let us consider the process of detecting a single photon. If light from a source with constant output intensity, such as a 'perfect' fluorescent particle fixed at the centre of an excitation/collection volume (Figure 2.2(a)), is incident on a detector, then the output of the detector (photon counts per time interval) will not be steady but will contain fluctuations. This is because the quantum mechanical nature of the interaction of a photon with the detector material leads to a probability that the arrival of the photon results in an output count. These fluctuations in output signal are referred to as shot noise. A semi-classical treatment of this has been carried out by Mandel [17]. The probability of observing k photon counts is given by,

$$p(k)=\int_0^\infty \frac{(\eta_I I_D)^k e^{-\eta_I I_D}}{k!} p(I_D)\mathrm{d}I_D = \int_0^\infty \mathrm{Poi}(k,\eta_I I_D)p(I_D)\mathrm{d}I_D. \quad (2.1)$$

The notation $\mathrm{Poi}(k,\eta_I I_D)$ is used to denote a Poisson distribution in k with mean $\eta_I I_D$. $p(k)$ is a function of the intensity at the detector and the probability distribution of the intensity $p(I_D)$. The constant η_I is proportional to the detection efficiency (the proportionality constant between the amount of light falling on the detector and the average number of photon counts $<k>$ detected) and incorporates the sampling time. This distribution is thus the Poisson transform of the light intensity distribution at the detector. Thus, for a perfectly steady light source (i.e. one in which $p(I_D)$ is a delta function), the distribution $p(k)$ will be Poissonian [8,9] and can therefore be described in terms of just the mean count number $<k> = \eta_I I_D$;

$$p(k)=\mathrm{Poi}(k,<k>)=\frac{(\eta_I I_D)^k e^{-\eta_I I_D}}{k!}. \quad (2.2)$$

2.3.2 **Photon counting statistics incorporating fluctuations: PCH for a single diffusing particle**

In the case where fluctuations are present in the light intensity falling on the detector, the probability distribution $p(I_D)$ is no longer a delta function, and $p(k)$ is given by equation 2.1 with the appropriate form of $p(I_D)$. A Poisson distribution (equation 2.2) is defined as having its variance equal to its mean,

$$\langle \Delta k^2 \rangle = \langle k \rangle. \quad (2.3)$$

Any additional fluctuations in light intensity described by $p(I_D)$ causes broadening of the distribution which results in a variance greater than the mean

$$\langle \Delta k^2 \rangle > \langle k \rangle. \quad (2.4)$$

Such fluctuations can be generated by the single fluorescent particle diffusing inside a closed, excitation/collection volume which has varying illumination intensity or collection efficiency or both (Figure 2.2 (b)). The fluorescence intensity at the detector I_D due to a fluorescent particle within the sample volume at a point $\vec{r}_0$ then is related to the excitation intensity at that location (assuming one-photon excitation) according to [9],

$$I_D = I_0 \beta \overline{\mathrm{PSF}}(\vec{r}_0) \quad (2.5)$$

where the spatial distribution of the excitation/collection efficiency is given by the point spread function[1] (PSF) of the particular instrument, which is normalized to be equal to unity at the origin. I_D is thus normalized to the intensity at the centre of the PSF, I_0. The coefficient β contains such factors as the transmittance of the microscope and quantum yield of the detector.

Equation 2.1 is written in terms of the probability distribution of intensities falling on the detector. From equation 2.5 it is clear that the probability distribution of intensities is thus connected to the probability distribution of position of the particle. It can be shown that equation 2.1 may be rewritten [9]:

$$p^{(1)}(k; V_0, \varepsilon) = \int \mathrm{Poi}(k, \varepsilon \overline{\mathrm{PSF}(\vec{r})}) p(\vec{r}) d\vec{r}, \tag{2.6}$$

where $\varepsilon = I_0 \beta \eta_I$ and the notation $p^{(1)}(k; V_0, \epsilon)$ has been used to indicate the PCH for a single diffusing particle with a distribution in counts k, confined in the volume V_0 with a molecular brightness ε. Essentially then, each position sampled by the diffusing molecule can be considered to contribute a Poisson distribution to the total distribution whose mean is related to the position of the molecule, as given by equation 2.5.

Further, if the fluorescent particle is confined within the volume V_0 defined by the PSF (Figure 2.2 (b)), then $p(\vec{r})$, the probability of finding the fluorescent particle at some point $\vec{r}$, is given by

$$p(\vec{r}) = \begin{cases} 1/V_0, & \text{for } \vec{r} \in V_0, \\ 0, & \text{for } \vec{r} \notin V_0. \end{cases} \tag{2.7}$$

and so we can further rewrite equation 2.6 as

$$p^{(1)}(k; V_0, \varepsilon) = \int \mathrm{Poi}(k, \varepsilon \overline{\mathrm{PSF}(\vec{r})}) p(\vec{r}) d\vec{r} = \frac{1}{V_0} \int_{V_0} \mathrm{Poi}(k, \varepsilon\, \overline{\mathrm{PSF}(\vec{r})}) d\vec{r}. \tag{2.8}$$

Conceptually, each point within the sample volume contributes a Poisson distribution in photon counts as if there were a particle at each of those points (we must assume that all points are sampled equally during a finite time period). Therefore equation 2.8 describes the weighted average of all these Poisson distributions, for all possible positions of a fluorescent particle within the volume V_0, each distribution having a mean value of $\varepsilon\, \overline{\mathrm{PSF}(\vec{r})}$. If there are no fluctuations caused by diffusion of the particle, (that is, the fluorescent particle is fixed at the origin, $\vec{r}_0$ (see Figure 2.2(a)), then the case of a single Poisson function is recovered [9].

$$p^{\mathrm{fixed}}(k; \vec{r}_0) = \mathrm{Poi}(k, \varepsilon \overline{\mathrm{PSF}(\vec{r}_0)}). \tag{2.9}$$

[1] Commonly the PSF is also known as the instrument spread function (ISF), the collection efficiency function (CEF) or the spatial detectivity function (SDF). The PSF is a convolution of the intensity profile of the excitation light with the volume from which fluorescence is collected (in the one-photon case) see [9] and Chapter 3, Section 3.2.2.

An analytical description of the PSF is now needed to insert into equation 2.8 (see Chapter 3, Section 3.2.2) to provide a practical expression for the PCH of a single particle valid for $k > 0$. (Note that the case of $k = 0$ is not valid [9] because the integral diverges in this limit.) The solution of the integral in equation 2.8 for $k > 0$ can be evaluated numerically thus enabling the calculation of the form of the PCH for the physical situation modelled by the PSF.

This discussion highlights two important concepts. Even if strong intensity fluctuations are present, then in the limit of a very large integration time, all such fluctuations will be averaged out (fluctuations from diffusion are lost). It is therefore essential to make sure that the integration time chosen in an experiment is sufficiently short to follow the fluctuations that the model describes. Further, it should be noted that in PCH analysis the assumption has been made that the coordinates of a fluorescent particle do not to change significantly during the integration time interval [8–10]. In a typical experiment the integration time used might be of the order 10–30 μs, during which some molecular motion is likely to occur (based on the typical diffusion coefficients for small molecules in solution). However, it has been shown that even with an integration time of 40 μs PCH analysis still holds [10]. In addition we assume that sufficient data is collected such that all spatial positions within the volume have been sampled and that no additional photophysics, such as triplet crossing, occurs. All of these assumptions are violated to a certain extent in real experiments and it is essential therefore that controls be performed (see later) to ensure the validity of the results.

2.3.3 The PCH for multiple diffusing particles

The model can now be extended to describe the case of many particles in the closed volume (Figure 2.2(c)). The PCH for two independent particles is given by the Poisson function of the combined intensity of the particles averaged over all possible spatial configurations of the two particles. Thus, re-writing equation 2.8 [9],

$$p^{(2)}(k; V_0, \varepsilon) = \int \mathrm{Poi}(k, [\varepsilon\overline{\mathrm{PSF}}(\vec{r}_1) + \varepsilon\overline{\mathrm{PSF}}(\vec{r}_2)])p(\vec{r}_1)p(\vec{r}_2)\mathrm{d}\vec{r}_1\mathrm{d}\vec{r}_2. \tag{2.10}$$

For a system of N particles the PCH is simply the Poisson function of the combined intensity of all the particles averaged over all space within the sample volume V_0 [9] and is given by,

$$p^{(N)}(k; V_0, \varepsilon) = \int \cdots \int \mathrm{Poi}\left(k, \sum_{i=1}^{N} \varepsilon_i \overline{\mathrm{PSF}}(\vec{r}_i)\right) p(\vec{r}_1) \cdots p(\vec{r}_N)\mathrm{d}\vec{r}_1 \cdots \mathrm{d}\vec{r}_N. \tag{2.11}$$

It is not necessary to evaluate this large integral [9], which would be computationally intensive, because the probability distribution for a sum of statistically independent variables is the convolution of the probability distribution of the individual variables [18]. Equation 2.11 may therefore be re-written [9] as

$$p^{(N)}(k; V_0, \varepsilon) = \underbrace{\left\{ p^{(1)}(k; V_0, \varepsilon) \otimes \cdots \otimes p^{(1)}(k; V_0, \varepsilon) \right\}}_{N-\text{times}}, \tag{2.12}$$

which can be evaluated numerically quite straightforwardly.

2.3.4 The PCH for an open volume with Poisson number fluctuations

Equation 2.12 describes the theoretical form of the PCH for a system of N particles diffusing around a closed volume that is spatially identical to the PSF of the microscope. However, the more usual experimental situation consists of a microscopic excitation/collection volume defined by the PSF, which is part of a larger sample volume. Molecules can therefore diffuse into and out of the PSF (Figure 2.2(d)), generating fluctuations in the measured signal in addition to those due to diffusion within the inhomogeneous excitation profile, as was discussed earlier. The distribution of particles inside such a sub-volume [6] is described by Poisson statistics [9, 19]. Thus, the PCH for an open system is given by the expectation value of the N-particle PCH for some average number of molecules $\overline{N}$ [9];

$$p(k; V_0, \overline{N}, \varepsilon) = \langle p^{(N)}(k; V_0, \varepsilon) \rangle_N. \tag{2.13}$$

That is, the PCH is the convolution of the *average* number of single particle PCH. Following the convention generally used, $\overline{N}$ is chosen to be the average number of molecules inside the PSF volume V_0, although the choice of sample volume can be shown to be arbitrary [9]. In equation 2.13, a complex system has been reduced to a function of two variables, the brightness of the molecule and the average number of molecules inside the PSF, which can be recovered from the fit to the experimental data.

We have seen that fluctuations can be cancelled out by using integration times tending to infinity leaving a shot noise limited Poisson distribution and the same effect occurs as N is increased. As the concentration is increased, the fluctuations in the fluorescence signal are lost. It naturally follows that the best conditions are those of strongly fluorescent molecules and single molecule occupancy of the PSF.

2.3.5 PCH for multiple independent species in an open volume

Thus far we have considered the case of multiple diffusing particles of the *same* species but it is possible that multiple particles of different species with different fluorescence characteristics could be present (Figure 2.2(e)). One way of dealing with this eventuality is to absorb variations in the quantum yield of the different particles, along with differences in the detection efficiency of the microscope at different wavelengths, into the molecular brightness parameter ε. In a slight modification of equations 2.12 and 2.13, the PCH of the mixture of particles is given by the convolution of the many particle PCH for each species (which are themselves the convolution of the single particle PCH for each species) [9]. Thus, in the case of two different species, equation 2.13 is extended to [9],

$$p(k) = \langle p^{(N_1)}(k; V_0, \varepsilon_1)\rangle_{N_1} \otimes \langle p^{(N_2)}(k; V_0, \varepsilon_2)\rangle_{N_2}. \tag{2.14}$$

2.3.6 **Implementing PCH analysis**

Data collection for PCH is straightforward; all that is required is a method of detecting and recording a time trace (Figure 2.1) of photon counts at a detector with sufficient time resolution such that the assumptions discussed, such as particles being stationary during each measurement window, hold. Often it is necessary to combine many separate time traces because the number of 'bins' that commercial acquisition cards (typically multi-channel scalar cards, see Chapter 3, Section 3.8) provide is often limited to either 64 or 128 K. Concatenation of multiple data sets provides reasonable length time traces in which each bin contains the number of photon counts in a ~20–40 μs time window. The total amount of data necessary for a reliable PCH analysis is a function of the sample concentration (which affects the number of photons per unit time), the signal-to-noise of the measurements and the integration time. It has been suggested that it is necessary to collect between 10^5 and 10^6 bins of data [8, 9] when the sample concentrations are in the single molecule regime, that is, ~0.1 nM or less. The histograms can easily be constructed using any common data analysis package simply by plotting the occurrence (on the ordinate) against the number of photon counts k in each time bin (on the abscissa) for all values of k observed. Often the occurrence will vary over many orders of magnitude over the range of k values; high k values corresponding to large numbers of photons detected in a bin will be rare, but low numbers of photons corresponding to one or no molecule being present will occur frequently and so PCH are often presented on a semi-logarithmic graph.

Once the experimental data has been plotted as a PCH, the next stage is to fit the data with the model (equation 2.13 or 2.14) to recover parameters such as the molecular brightness, sample concentration and to reveal whether any heterogeneity is present. A simple approach to this is to use the PCH modelling and fitting algorithms that have been incorporated into a commercial software package called Globals WE, produced by and available from Enrico Gratton [20]. We shall not go into the detail of how to implement this fitting procedure here but those who are sufficiently mathematically aware will be able to carry out the modelling in any suitable computational analysis package.

An excellent test of instrumentation and analysis procedures is to measure the PCH of an ideal scatterer. This should not introduce any super-Poissonian fluctuations into the measurement, assuming the scatterer does not degrade and the instrumentation does not introduce any other fluctuations. Such a test therefore provides a way of examining the stability of the light source and the instrumentation prior to any further experiments. A concentrated emulsion made from powdered milk in water provides an ideal non-fluorescent (in the visible region of the spectrum) scattering sample. The emission filters should be removed from the detection path (see Chapter 3, Section 3.9) and replaced with neutral density filters [2] to reduce the scattered light intensity to similar mean photon count rates as in a single molecule fluorescence experiment (1–10 KHz). Shown in Figure 2.3 is a typical PCH for a scattering sample. The experimental data has been fit with a Poisson function. The residuals[3] of the fit indicate that these data are described very well by the Poisson distribution and therefore the instrumentation and light source used appear to be stable and suitable for single molecule fluorescence fluctuation studies. Rather than using an ideal scatterer it is also possible to use a fluorescent dye solution at very high concentration and a lower excitation power. Clearly, however, in order to ensure no fluctuations in the fluorescence it is essential that the concentration is high enough that, on the timescale of interest, no variation in the number of fluorescence molecules in PSF volume occurs by diffusion or photobleaching. As a daily check of instrument stability this approach is useful since it does not require filters to be removed from the optical path, a process requiring time consuming alignment. Figure 2.4 shows the PCH for a labelled protein sample at high concentration (E Colicin immunity protein Im9 [22] labelled with the dye BODIPY FL IA (Molecular Probes, USA)). The distribution is well described by a Poisson function and the average count rate is high, as expected.

When it is known that the instrument is only shot noise limited, one can proceed with single molecule experiments. The PCH for a dilute dye solution shows super-Poissonian characteristics (Figure 2.5), with the measured distribution (grey) being much broader than a Poissonian distribution (black) with the same average number of counts. The tail on the right hand side of the distribution arises because there are more counting periods (bins) containing a higher number of

[2] Neutral density filters are either reflecting or absorbing light attenuators, often simply smoked or semi-silvered glass.

[3] Many aspects of single molecule analysis require a good understanding of statistics and the evaluation of fits using chi-squared minimization, weighting and residual analysis. The reader is referred to the following text for a treatment of these methods [21].

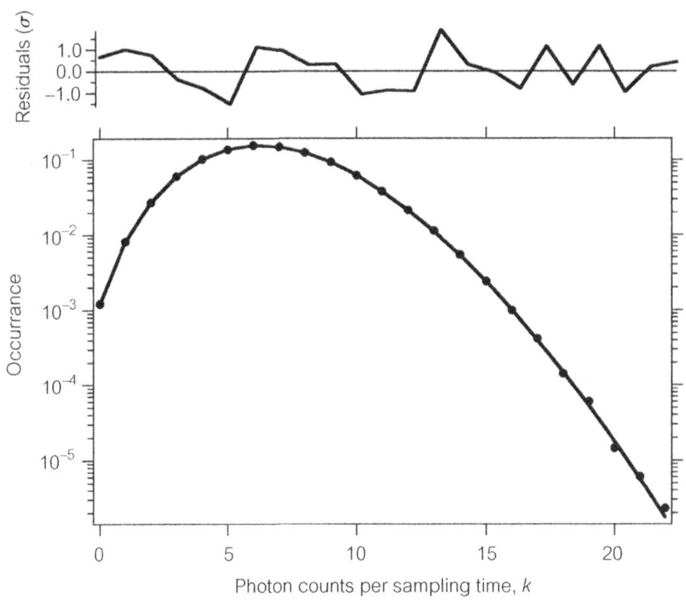

Figure 2.3 PCH for an ideal scatterer placed at the laser focus. The experimental photon-count distribution (solid circles) is exactly fit by a Poisson function (solid line) with an average number of photon counts $<k> = 6.7$. Residuals are shown in units of standard deviations. The fit gives $\chi^2 = 0.91$. A total of 131072 data points were collected.

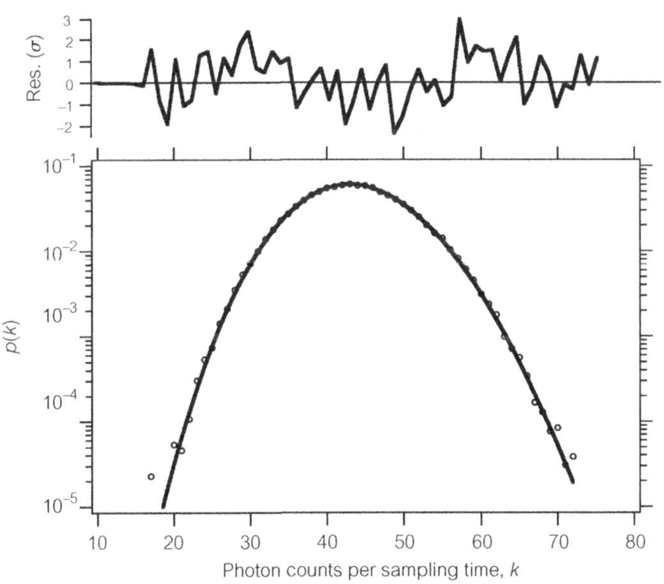

Figure 2.4 PCH (circles) from a concentrated ($\approx$100 nM) solution of the protein Im9 [22] labelled with the dye BODIPY Fl IA in 4 M urea in 50 mM sodium phosphate buffer, pH 7.0. The PCH is well described (see residuals above and $\chi^2 = 1.1$) by a Poisson distribution (solid line) with average $<k> = 43.5$.

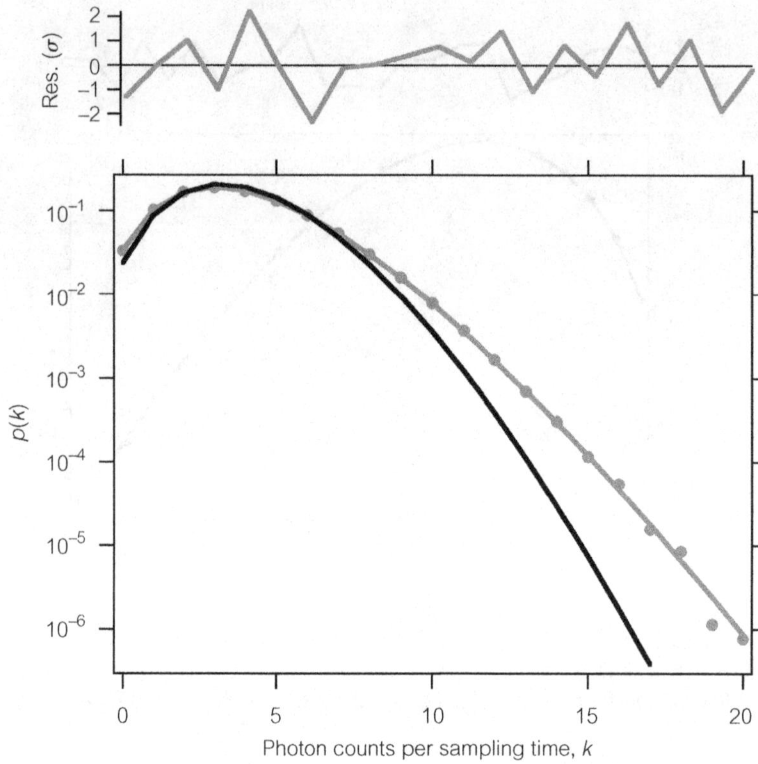

Figure 2.5 PCH from a photon burst trace of Fluorescein 27 in 50 mM sodium phosphate buffer at pH 7.0 (circles). A simple Poisson function, with an average of $<k> = 0.82$ (black line, equal to the mean number of photon counts in the recorded dataset) does not describe the data. The data are fit with a single species PCH (solid grey line) with $N = 0.13$ and $\varepsilon = 123800$ cpspm with $\chi^2 = 2.4$. A total of 131072 data points were collected. Such data (grey) are referred to as super-Poissonian.

counts than can be explained by shot noise variations of the average signal alone. We have occasional relatively intense bursts above the background (see Figure 2.1). The data are now well described by a single species PCH (of the form of equation 2.14) which yields the molecular brightness and the concentration of the analyte.

Figure 2.6 shows the PCH for a mixture of the two dyes R110 and F27 that have molecular brightnesses differing by a factor of >3 and a single- and a two-species PCH fit are shown. The PCH analysis is able to resolve the heterogeneity in this sample and would be able to do so in the absence of the *a priori* knowledge of the sample composition. Generally, we have found that fitting PCH of single species containing solutions with single species models produce a reduced $\chi^2 \leq 3$; χ^2 values significantly higher than this, accompanied by large fluctuations in the fit residuals, suggest that the sample contains more than one species. PCH can

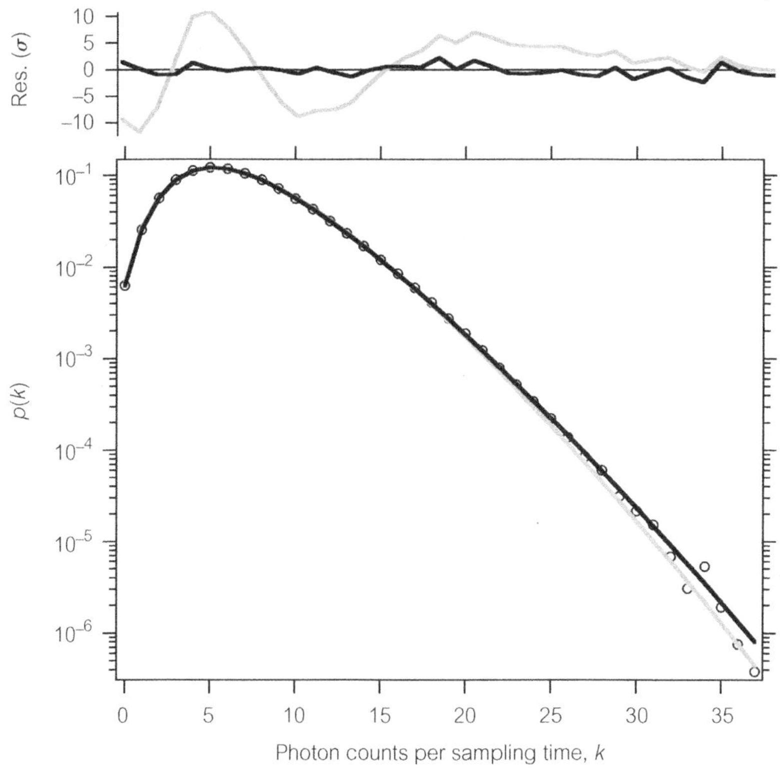

Figure 2.6 PCH for a mixture of the two fluorescent dyes R110 and F27 in 50 mM sodium phosphate buffer at pH 7.0 (open black circles). The molecular brightness of these two dyes differs by a factor of ~3. The fit with a single species PCH function (grey line; fit parameters are $\varepsilon = 263000$ cpspm and $N = 52.0$ is poor (see residuals) and $\chi^2 = 32.1$. The fit to a two species PCH model (black line; fit parameters are $N_1 = 43.01$, $N_2 = 33.0$, $\varepsilon_1 = 69200$ cpspm and $\varepsilon_2 = 321200$ cpspm) describes the data well with $\chi^2 = 1.1$.

provide quantitative values for molecular brightness of the species and their relative and absolute concentrations [16, 23]. Generally, it is easier to accurately detect the presence of two species that have only small differences in molecular brightness if lower concentrations are used because this makes the relative amplitude of the fluctuations greatest. For example, two species with a relative difference in brightness as low as 1.5 can be resolved if the absolute concentration is reduced sufficiently. A further consideration is that the 'quality' of single species fits is dependent on the PSF model chosen [13, 14] and this will be dependent on optical alignment and other parameters such as the refractive index of the sample or solvent (see Chapter 3, Section 3.2.2 for a discussion of this). The choice of PSF model can be supported by an experiment on a homogeneous single species sample and confirming that the measured distribution is well described by a single species PCH.

PCH can be used to extract useful parameters from photon count traces for a number of applications. Chirico *et al.* [24] used PCH analysis to compliment other fluorescence techniques in order to identify heterogeneity in the structure/function of the enzyme o-acetylserine sulfhydrylase. Palo *et al.* [25] extended PCH analysis by recording photon count traces as a function of integration time and showed how information about molecular brightness and diffusion rates can be obtained concurrently. Some of the same authors [26] also showed how PCH can be combined with fluorescence lifetime measurements to resolve heterogeneity with greater sensitivity. In Chapter 5, we present a detailed review of the work of Schaertl and colleagues [27] demonstrating how PCH can be applied to high-throughput screening.

2.4 Fluorescence correlation spectroscopy

PCH analysis is based on the statistical treatment of fluctuations in intensity in single molecule burst data. Fluorescence correlation spectroscopy (FCS) employs a statistical analysis of the time dependence of such fluctuations. A method of interrogating the microscopic molecular properties of a sample through what are essentially concentration fluctuations (the presence or absence of a molecule in a defined volume) was first suggested by Magde, Elson and Webb [28] in a seminal paper in 1972. An excellent review of the technique they developed, which was to become more widely known as fluctuation correlation spectroscopy, along with a historical perspective can be found elsewhere [29]. Briefly, Elliot Elson at Cornell approached Watt Webb with the problem of understanding how DNA became denatured for transcription. With the help of Douglas Magde [28–30] they developed a new experimental and theoretical framework that took advantage of the fluorescence intensity enhancement experienced by the dye ethidium bromide upon intercalation with DNA. They showed that by observing the effective concentration fluctuations of the dye they could probe the dynamics of association with DNA and thus the dynamics of dissociation as the DNA duplex is denatured.

The fluctuations in the fluorescence signal from a sample are used in FCS to probe the processes that cause them. In terms of a diffusing sample, the effective number (concentration) fluctuations of a fluorescent species in an open sample volume defined by a fluorescence microscope are monitored (see Figure 2.1). The number fluctuations of the fluorescent species can be due to random diffusion of molecules into and out of the sample volume, chemical reactions, or structural changes—indeed any effect that generates, extinguishes, enhances or modulates a fluorescence signal from a species in the sample volume. Measurements of the fluctuations are thus measurements of deviations from the equilibrium 'concentration'. Probing the timescale of the fluctuations elegantly probes the timescales of the mechanisms of the fluctuations. FCS has undergone extensive development over the last 30 years and here we begin with a phenomenological description of the theory to avoid some of the mathematical complexity.

The theoretical treatment of FCS considers the persistence of temporal information in a fluctuation trace through the construction of a *temporal autocorrelation function*, which can then be fit using models based on physical processes which may be present in the sample and which cause the fluctuations. For a more detailed review the reader is referred elsewhere [28–34].

2.4.1 The autocorrelation function for fluorescence fluctuations

Time domain autocorrelation analysis provides a measure of the self-similarity of a time series signal[4] and the decay of the autocorrelation function describes the temporal persistence of information carried by it. The normalized fluorescence correlation function of a fluctuating signal $F(t)$ can be written as [29],

$$G(\tau)=\frac{\langle \delta F(t)\delta F(t+\tau)\rangle}{\langle F(t)\rangle^2}, \tag{2.15}$$

where $\delta F(t)$ and $\delta F(t+\tau)$ are the amplitudes of fluctuations from the mean at time t and $t+\tau$ respectively, and $<F(t)>$ is the mean value of the signal (see Figure 2.7). Thus, the autocorrelation is the normalized average product of the fluctuation of a signal from the mean at some time, t, with the fluctuation from the mean at some later time, $t+\tau$. The time, τ, is known as the delay time (sometimes also called the correlation or lag time) and is the time delay over which the fluctuations are compared. The autocorrelation function calculated for a signal $F(t)$ is the value of this normalized product as a function of the delay time. The amplitude of the autocorrelation function at any given delay time τ is therefore related to the relative persistence of fluctuations in the measured data on the timescale τ. This fundamental description of autocorrelation is represented schematically in Figure 2.8, which shows how such a calculation might provide information on the temporal persistence of fluctuations that might be present in a single molecule experiment. Autocorrelation curves are generally presented on semi-logarithmic plots as the range of delay times is likely to span many orders of magnitude: the shape of the autocorrelation function is approximately exponential in many cases and it is possible to read off the approximate lifetime of the

[4] Autocorrelation is often applied to any signal or data set and used to show up trends that may exist. For example, autocorrelation of residuals from fitting can give information on trends in a fit to data, which may reflect systematic errors in the experiment or inappropriateness of the model fit.

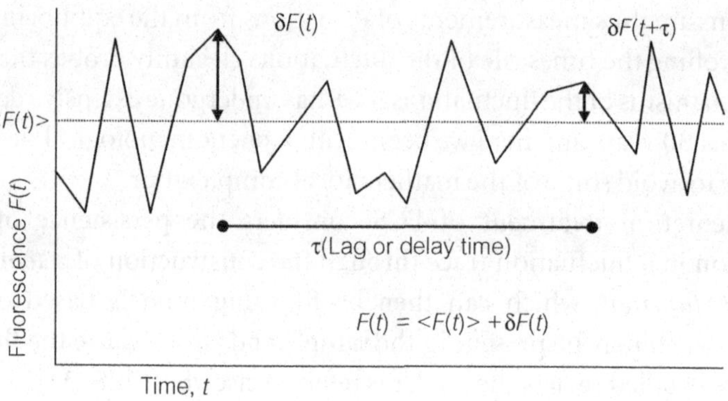

Figure 2.7 Illustration showing a generalized representative fluctuation trace with the definitions of parameters used in equation 2.15 shown.

decay to get an indication of the timescale of the fluctuations present. Thus the autocorrelation function for the type of raw single molecule diffusion data shown in Figure 2.8 can be used to give information about the timescale of the fluctuation events, in this case the widths of the fluctuations (bursts) produced by diffusion into and out of the volume. Note that the autocorrelation is therefore built up from the temporaly similar signals of many single molecules. Further, the autocorrelation function may contain additional information on any other processes which cause fluctuations on a time scale faster than the occupation times of molecules in the volume.

Figure 2.9 shows a 'real' autocorrelation function calculated for a small 19 nucleotide RNA hairpin, labelled with the dye fluorescein, diffusing in water. In this particular case two components are seen in the autocorrelation function occurring on different timescales. There is a diffusion component caused by fluctuations from diffusion into and out of the PSF and a faster component caused by blinking (alternating periods of fluorescence and darkness) of the dye as it enters and leaves a triplet state (see Chapter 4, Section 4.2.1). As is indicated in Figure 2.9, a number of parameters can be extracted from the autocorrelation function regardless of the mechanism of the fluctuations. The amplitudes of the decay components give information about the relative strength of the fluctuations; in the case of triplet crossing the fraction of time spent in the triplet state and in the case of diffusion the amplitude provides a measure of the average number of molecules in the small excitation/collection volume (proportional to the concentration). Additionally, the decay rate of the processes gives an indication of the timescale of the processes that cause the fluctuation. In the case of triplet crossing the $1/e$ point provides an estimate of the crossing rate and for diffusional processes, the half-life gives the average time taken for a molecule to transit the excitation/collection volume. A simple inspection such as this provides only approximate values and in the following sections we will show how the experimental autocorrelation

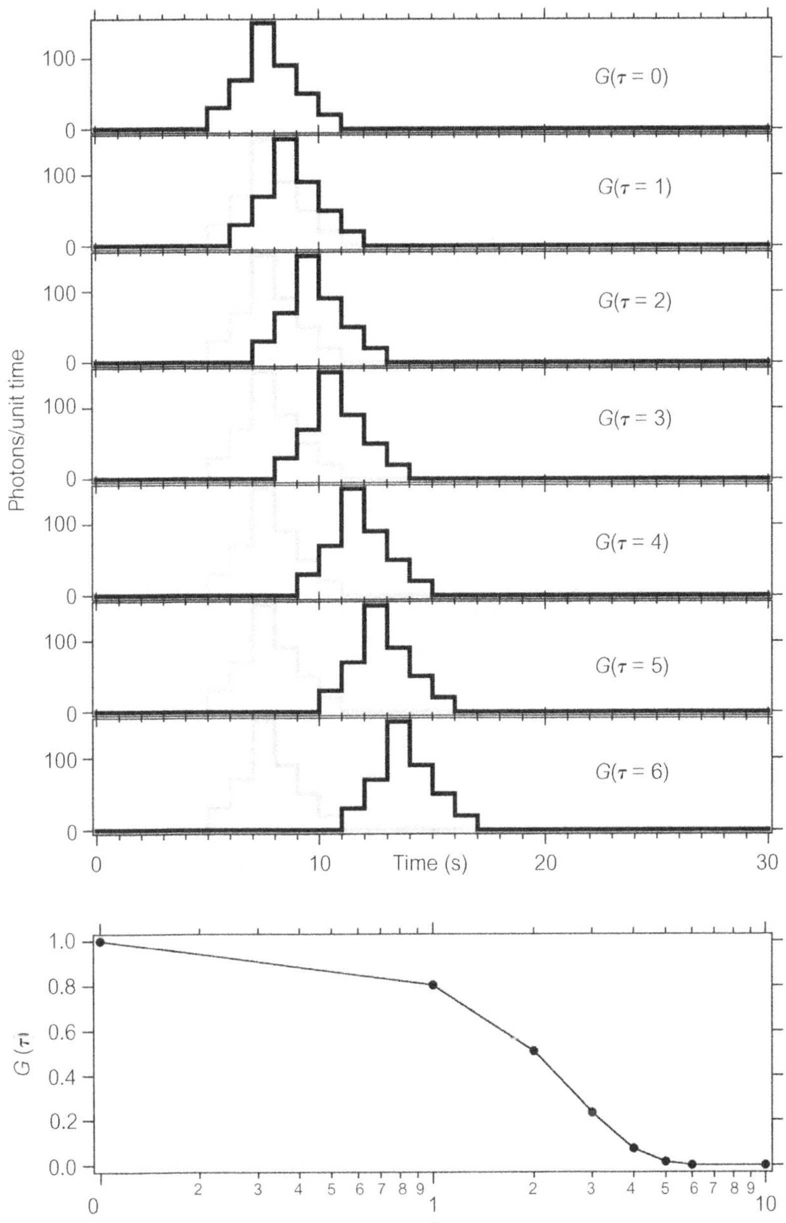

Figure 2.8 Schematic representation of the principle of an autocorrelation calculation on a single molecule data set (top). A fluorescence burst ($F(t)$) is shifted by the integration time (lag time) τ. The original and shifted traces are then multiplied together, $F(t) * F(t + \tau)$, and the integrated area is stored as the value of the autocorrelation function at lag time τ. The values of the autocorrelation function ($G(\tau)$) are then plotted on a logarithmic lag timescale (bottom). The points shown are the actual overlap integrals (normalized) from the data shown (top). At short lag times the overlap integral is large whereas at longer lag times the overlap integral diminishes to zero. In this way the autocorrelation function contains information on the width of the feature in the data set.

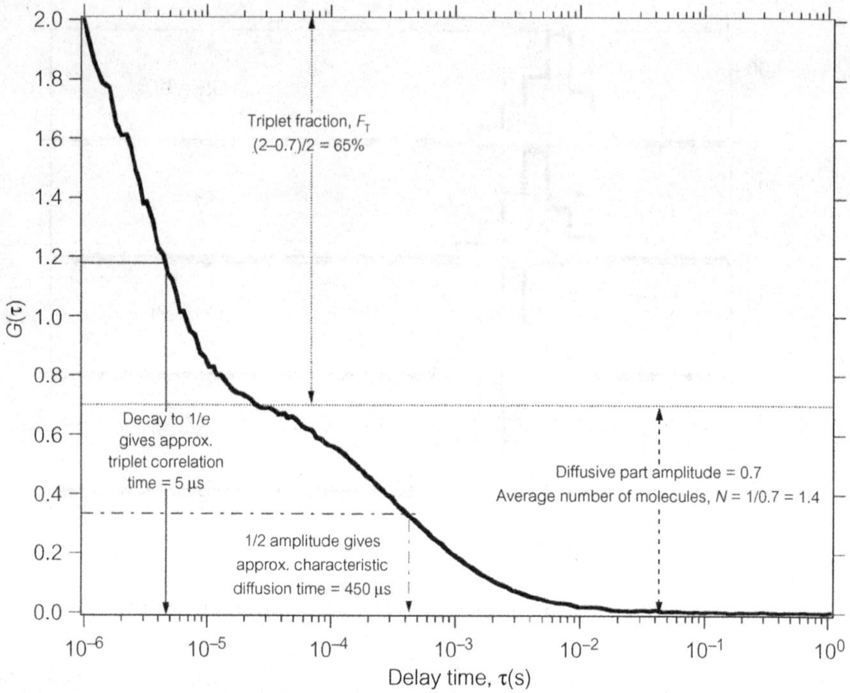

Figure 2.9 An example of the autocorrelation of a real single molecule fluorescence data set. This autocorrelation function was measured for a short 19 nucleotide RNA molecule labelled with the dye fluorescein diffusing in a suitable buffer. The autocorrelation function shows two components: a diffusive component in the lower portion of the graph, and a component at shorter delay times that is assigned to the strong triplet crossing behaviour of this dye. The lifetime of the triplet decay gives the rate of triplet crossing and its amplitude gives the proportion of time spent in the triplet state. The reciprocal of the amplitude of the diffusive part provides an estimate of the average number of molecules being observed at any instant while the decay at 1/2 maximum gives the typical time taken to diffuse through the excitation/collection volume.

function is calculated, what fluctuations can manifest themselves in this analysis and the timescale on which they appear as well as describing some common models that are applied to extract physical parameters with greater certainty.

In FCS of freely diffusing particles the primary fluctuations occur due to the presence or absence of a fluctuating species within the excitation/collection volume. However, in typical FCS instrumentation a spatial inhomogeneity also exists in the excitation/collection volume that leads to fluctuations without concentration changes. Thus the amplitude of a given fluctuation is modulated by its position in the volume. Fluctuations are therefore expressed as spatially weighted concentration changes [29] according to,

$$\delta F(\vec{r}, t) = \phi(\vec{r})\delta C(\vec{r}, t), \tag{2.16}$$

where $\delta C(\vec{r}, t)$ is the concentration fluctuation and $\phi(\vec{r})$ is the point spread function (PSF, see Chapter 3, Section 3.2.2). Integration over all space (the entire sample volume) gives the total fluorescence fluctuation signal and, assuming the existence of a single fluorescent species, the amplitude of the fluorescence fluctuations is given by [29],

$$\delta F(t)=\int \phi(\vec{r})\delta C(\vec{r},t)\,\mathrm{d}^3\mathrm{r}. \tag{2.17}$$

The total fluorescence signal can be shown to be given by [29],

$$F(t)=\int \phi(\vec{r})C(\vec{r},t)\,\mathrm{d}^3\mathrm{r}. \tag{2.18}$$

and the average fluorescence signal is thus [29],

$$<F(t)> = <C(t)>\int \phi(\vec{r})\,\mathrm{d}^3\mathrm{r}. \tag{2.19}$$

Combining equations 2.15, 2.17, and 2.19 yields the fluorescence fluctuation autocorrelation function [34],

$$G(\tau)=\frac{\iint \phi(\vec{r})\phi(\vec{r}')<\delta C(\vec{r},t)\delta C(\vec{r}',t+\tau)>\mathrm{d}^3\mathrm{r}\mathrm{d}^3\mathrm{r}'}{\left[<C(t)>\int \phi(\vec{r})\,\mathrm{d}^3\mathrm{r}\right]^2}, \tag{2.20}$$

where $<\delta C(\bar{r},t)\delta C(\bar{r}',t+\tau)>$ is referred to as the correlation function of a concentration fluctuation at some point $\vec{r}$ at time t with the concentration fluctuation at a point $\vec{r}'$ at some later time $t+\tau$ [34]. Equation 2.20 can be extended to a solution containing several different chemical species by representing the fluorescence signal as the sum of a series of different signals [34].

The particular case of $G(0)$ represents the correlation of a molecule at $\vec{r}'$ with a molecule at $\vec{r}$ at the same instant. In a sample in which there are no long-range interactions, no spatial correlations such as this exist and therefore fluctuations are only correlated at the same instant at the same position (and all positions are equivalent). In this limit it can then be shown that equation 2.20 reduces to [34],

$$G(0)=\gamma\frac{<\delta C(t)^2>}{<C(t)>^2}, \tag{2.21}$$

where γ is a constant depending only on the PSF [29]. Equation 2.21 then, is the relative mean square amplitude of fluctuations, which for independent random molecular processes can be shown to be inversely proportional to the average number of processes $\bar{N}$ [34]. Thus,

$$G(0)=\gamma\frac{1}{\bar{N}}. \tag{2.22}$$

a typical value of γ for common experimental geometries is ~0.5 and depends only on $\phi(r)$ with a weak dependence on sample volume shape [34]. Thus $G(0)$ depends strongly on the number of fluorescent molecules in the sample volume, and so FCS like PCH can probe sample concentrations directly which has been exploited in a number of studies [35].

An interesting result of fluctuation analysis of this type is that, as with PCH, it is not necessary to have only a single molecule in the PSF. If, on average, small numbers of molecules are present in the PSF then temporal fluctuations in the fluorescence signal will still be detected when one molecule enters or leaves the volume; the fluctuations caused by a single molecule are still being probed. Single molecule sensitivity is only lost entirely if, when one molecule leaves the volume (by diffusion or chemical reaction) it is immediately replaced by another, in which case the fluctuations tend to zero. FCS is therefore, in principle, sensitive to single molecule fluctuations over quite a broad range of concentration. Experiments are, however, best performed in conditions where fluctuations are maximized, that is, at or near single molecule concentrations.

2.4.2 **Experimental determination of the autocorrelation of a signal**

The autocorrelation function can be calculated in real time using a hardware correlator or in software after the collection of a photon count trace using a multichannel scalar card. Details of the instrumentation will be discussed in Chapter 3. Formally, the un-normalized autocorrelation of time series data is given as [6]:

$$G(\tau) = \mathrm{Corr}\{I(t), I(t+\tau)\} = \frac{1}{T}\int_0^T I(t)I(t+\tau)\,\mathrm{d}t, \tag{2.23}$$

where T is the total length of the time trace. One may proceed with this calculation giving consideration to the integration times ('bin' widths) used and the timescale of the fluctuations of interest (see next section), however, the autocorrelation function is most easily (computationally efficiently) evaluated in Fourier space (the frequency domain) [36] as follows:

$$\mathrm{Corr}\{I(t), I(t+\tau)\} \Leftrightarrow I(f)I^*(f), \tag{2.24}$$

where $I(f)$ represents the Fourier transform of $I(t)$ and * indicates the complex conjugate. A fast Fourier transform routine (available in numerous scientific data analysis packages) can easily be used to obtain $I(f)$ and its complex conjugate. Thus the autocorrelation is computed as follows (see equation 2.15). The data set first has its mean value subtracted from all points (because we wish to determine the autocorrelation of the fluctuations about the mean). The data set is then Fourier transformed to yield $I(f)$ and its complex conjugate computed ($I^*(f)$). The two data sets in Fourier space are then multiplied and the inverse Fourier transform of this result computed. This results in the un-normalized autocorrelation function at different lag times ($0, \tau, 2\tau, \ldots$) where τ is equal to the integration time (bin width) of the original, uniformly spaced data. The function is now normalized (in order to account for the different number of possible calculations at different lag times) and finally if the function is to be fitted with a physical model (see Section 2.4.4), then the first point $G(0)$ should be discarded as this is a special case which is not allowed for in these models.

Using this method, referred to as a software autocorrelation, the data is generally a time trace of photons detected in a given integration time. The main disadvantage with this 'linear' time base correlation is the fixed integration time and the amount of data that can be collected. Clearly, in order to calculate the autocorrelation at long times it is necessary to measure for at least as long as the maximum lag time. Similarly, the integration time needs to be short enough to be able to follow the fluctuations of interest. Many common autocorrelation measurements are concerned with diffusion and triplet activity which typically require large numbers of correlations to be calculated across the time range 1 μs–0.1 s. Thus with a linear correlation configuration one must generally measure for at least several tens of seconds (to build up a statistically representative number of long lag times) and so the number of measurements (integration times) is very large. This is not impossible to achieve but an alternative to this post-processing software-based methodology is to use a hardware correlator (see Chapter 3 and [37]). These work in the same fundamental way as the software method:- intensity fluctuations are recorded, stored, multiplied with fluctuations at later times and summed and normalized to create the autocorrelation function. However, sophisticated electronics allow the use of varying integration times and are able to display the autocorrelation function in real time for those delay times that are possible at that stage of the measurement. In this way autocorrelation functions can be calculated in several minutes that span delay times ranging from 200 ns to several seconds (see Figure 2.10). One disadvantage of hardware correlation is that many of these cards do not preserve the raw photon count time trace data, the unavailability of which, combined with the use of different integration times at different delay times, complicates the statistical analysis. However, the advantage of being able to observe the autocorrelation function essentially instantaneously (and the small data file sizes) cannot be underestimated.

2.4.3 **Processes which can be monitored by FCS**

A number of common physical phenomena can affect and influence the autocorrelation function of a diffusion single molecule fluorescence experiment (Figure 2.1) and they are summarized in Figure 2.11. The principle component which generally dominates the autocorrelation function is diffusion (Figure 2.11(a)) [38]. The Stokes–Einstein relation describes the translational diffusion coefficient of a particle in a viscous medium,

$$D = \frac{kT}{f}, \tag{2.25}$$

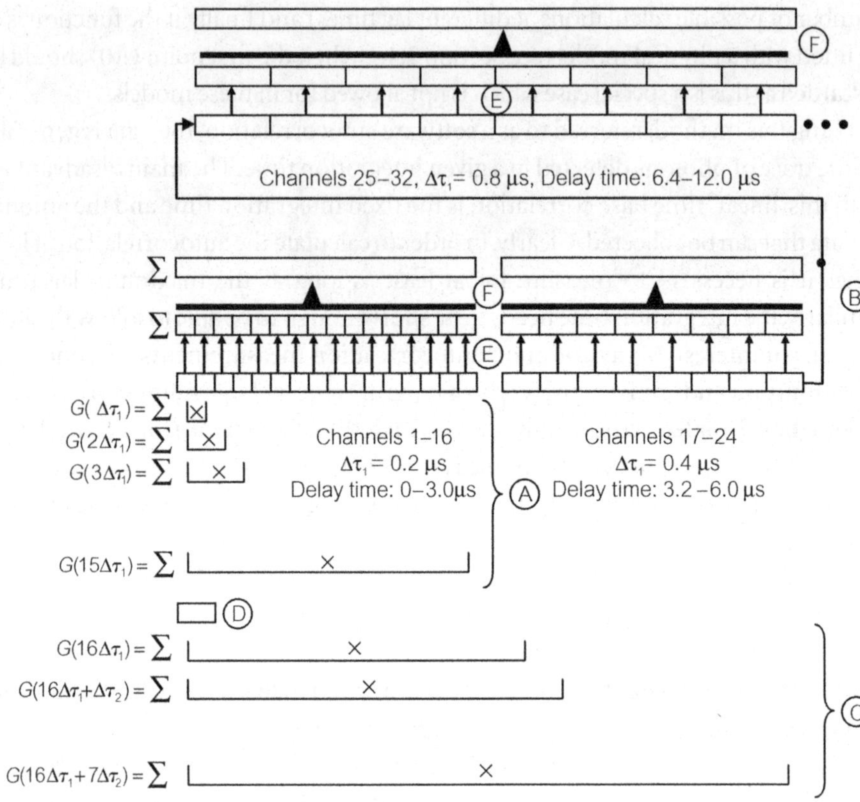

Figure 2.10 Schematic representation of the channel structure and calculation processes for a hardware digital correlator. This design is that used in the ALV5000 multiple tau digital hardware correlator (ALV GmbH, Germany), however, it is typical of this type of hardware. The measured fluorescence signal is recorded with a 200 ns integration time and stored in channel 1. After each measurement three processes are executed. First, the signal stored in each channel (1–15) is moved to the right (B), and the new measurement is stored in the first channel. Channels 15 and 16 are added together and shifted to channel 17, giving an integration time of 400 ns and so on. Second, the products between channels, as indicated, are continually carried out in real time and added to the displayed output correlation function at the appropriate delay time. Channels 1–16 have products formed in the manner shown (A). Products between longer delay times (channels > 16), with larger integration times, are formed by multiplying the channel with the appropriate number of summed channels, using earlier shorter integration channels (C). The summed channel for an integration time of 400 ns, for correlation with channels 17–24 is shown (D). Thus, for a given measurement duration, fewer products are summed at longer delay times. Finally, while all other processes are carried out, a 'delayed monitor' sums up all the counts that pass through each and every channel (E) and the counts passing through groups of channels with the same integration time are summed into the 'direct monitor' (F) and used, in combination with the number of measurements to provide normalization. A more detailed mathematical and quantitative description of the operation can be found elsewhere [37].

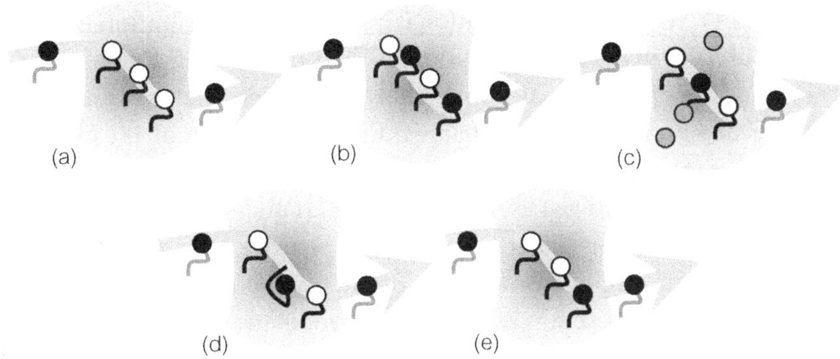

Figure 2.11 Schematic showing of some of the processes that lead to fluctuations in a diffusion single molecule fluorescence experiment. White circles represent active fluorescent molecules (a) diffusion of a single labelled molecule in the inhomogeneous excitation volume, (b) triplet crossing causing intermittent fluorescence, (c) reversible binding with a second molecule that is not fluorescent but quenches the fluorescence of the labelled molecule, (d) conformational changes that induce changes in the amount of emitted fluorescence, and (e) photobleaching.

where k is the Boltzmann constant, T is the temperature and f is the friction coefficient for the particle in the fluid. In the simple case of a spherical particle f is given by [39],

$$f = 6\pi\eta r \tag{2.26}$$

where η is the viscosity of the solvent and r the hydrodynamic radius (sometimes called the Stokes radius) of the sphere. A typical diffusion time (the time taken to traverse the PSF) for a small molecule at room temperature in water is thus of the order 75 μs (given a PSF radius of ~250 nm, a solution viscosity of 1.04×10^{-3} Nsm^{-2} at 293 K and a molecular hydrodynamic radius of 10 Å). Although this represents the time taken to traverse the PSF along the shortest path it nevertheless gives an idea of the approximate timescale on which diffusion processes will be observed in the autocorrelation function. Diffusion is rarely the only source of fluctuations in FCS experiments, however, the effects of diffusion have perhaps been studied in most detail and many models have been developed [38, 40, 41].

Triplet crossing (Figure 2.11(b)) modulates the fluorescence output of the molecule causing 'blinking' on a characteristic timescale and therefore generates fluctuations that can be observed in the autocorrelation function [42]. However, the photophysics of the triplet state of common dye molecules are very poorly understood and measurements of the rates and degrees of population are inconsistent [38]. In particular, the environment of the dye molecule has been shown to greatly influence the parameters measured as does the excitation power [38] (see Chapter 4). When triplet formation is a significant contribution in FCS data it usually observed in the autocorrelation on the 1 μs timescale [38].

The timescale of additional photo-induced transient states associated with inter-molecular processes such as charge transfer reactions upon the binding of a dye to another molecule have been shown to occur in the 10–100 ns time regime [38] (see for example R6G-DNA binding [43]). Other molecular interactions (e.g. binding in a receptor–ligand complex [44], Figure 2.11(c)) may also result in slower fluctuation components that can occur anywhere in the autocorrelation function if the binding event is reversible and modulates the fluorescence signal. Intra-molecular processes (which may include global conformational changes in polymers and proteins, Figure 2.11(d) [45]) can also introduce correlated fluctuations on almost any timescale. For example, the closing–opening rates associated with a DNA hairpin loop can span the range 5–1000 μs [46]. In these examples the autocorrelation curve cannot so neatly be chopped into the different temporal processes.

A variety of other mechanisms can influence FCS measurements and are difficult to assign to a particular timescale. Photo induced isomerization has been shown to be a significant problem when working with particular dyes, for example Cy5 [38]. Dynamic photobleaching [47] of molecules to a permanent dark state is another and is of particular concern (Figure 2.11(e)). Consideration must also be given to the inhomogeneous excitation profile—all of these photo-induced effects may occur as a function of the path they take through the excitation volume, introducing another convoluted fluctuation.

Figure 2.12 summarizes the contributions of these common processes to the autocorrelation function in an FCS experiment. Other processes such as photon anti-bunching [48], saturation of the excitation emission cycle [38] and protonation of chromophores [49] will not be discussed in detail here.

2.4.4 **Physical models for the autocorrelation function**

There are a number of models that have been developed for FCS [50]. Generally, in FCS experiments the data are first processed to yield the autocorrelation function as described earlier according to equation 2.15, then a physical model which incorporates descriptions of the sources of fluctuation is used to fit this function allowing the physical parameters of interest to be determined. The simplest case is that of diffusional motion of a fluorescent particle into and out of the PSF. An analytical expression for the form of the autocorrelation function in the case of a

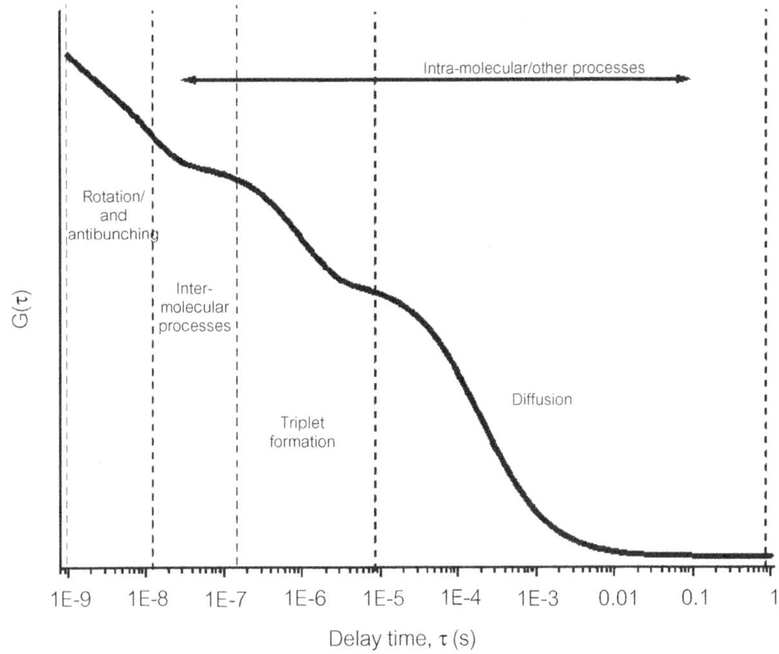

Figure 2.12 Diagram showing the temporal ranges of the processes that affect the autocorrelation of single molecule fluorescence data.

three-dimensional Gaussian PSF (see Chapter 3, Section 3.2.2) was developed by Aragon and Pecora [32].

$$G(\tau) = \frac{1}{\overline{N}}\, g_{3DG}(\tau) + DC = \frac{1}{\overline{N}}\left(1+\frac{\tau}{\tau_D}\right)^{-1}\left(1+\frac{\tau}{K^2\tau_D}\right)^{-1/2} + DC. \qquad (2.27)$$

where $\overline{N}$ is the average number of fluorescent molecules within the PSF at any instant, $K = z_0/\omega_0$ (where z_0 and ω_0 are the $1/e^2$ radii of the sample volume in the direction of, and perpendicular to, the optic axis, respectively). DC is the value of the autocorrelation as $\tau \to \infty$ (often $DC = 1$). τ_D is called the molecular diffusion time (or correlation time) and is given for one-photon excitation by [4, 31, 32],

$$\tau_D = \frac{\omega_0^2}{4D} \qquad (2.28)$$

where D is the translational diffusion coefficient. It is important to be aware of the precise definition of these quantities as they are often misinterpreted or presented in a non-standard way. Often ω_0 is incorrectly considered to be equal to the $1/e^2$ diameter of the beam and often the denominator of equation 2.28 has another pre-factor, resulting from confusion with forms of the equation for physical processes occurring in only two dimensions or in other optical configurations. In

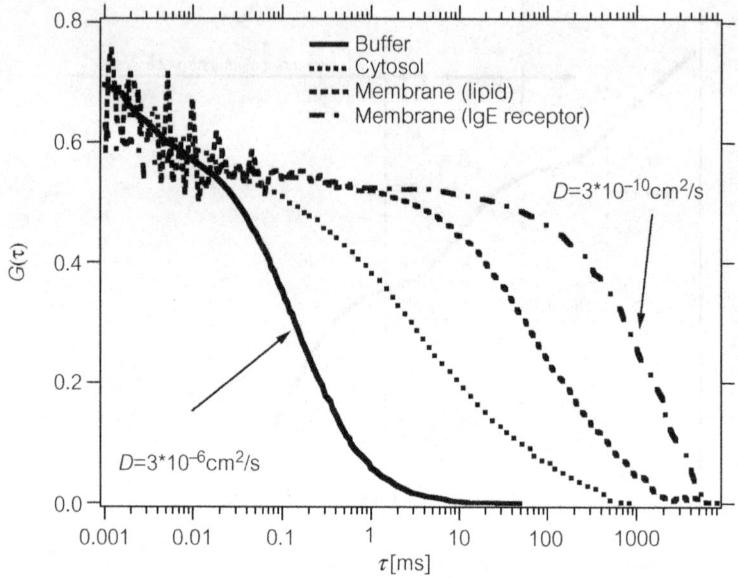

Figure 2.13 Autocorrelation functions measured for diffusion in different molecular environments; buffer, the cell cytosol, a lipid and a large protein diffusing in a membrane. Depending on the surrounding medium, the molecular mobility changes by several orders of magnitude. In aqueous buffer solutions (black) the diffusion coefficient (3×10^{-6} cm^2/s) was obtained from fits using equation 2.30. For the diffusion of a large receptor in the cell membrane, mobility was severely decreased. Here the diffusion coefficient was approximately four orders of magnitude lower. Reprinted from Haustein and Schwille, Ultrasensitive investigations of biological systems by fluorescence correlation spectroscopy, *Methods*, **29** (2003) 153–166 with permission from Elsevier.

particular, equation 2.28 refers to the one-photon excitation case only (see Chapter 3). For a two-photon excitation configuration the denominator in equation 2.28 must be doubled [51].

Figure 2.13 [52] illustrates one use of FCS to monitor simple diffusion. The figure shows the autocorrelation curves for molecular diffusion in different environments. It can be seen that the diffusion of a small dye in buffer results in an autocorrelation function at fast lag times (leftmost curve, similar to the curve presented in Figure 2.9) while the diffusion of a large dye-labelled receptor protein in the cell plasma membrane shows a much slower diffusion rate (the autocorrelation function essentially unchanged in shape but shifted to the right to longer delay times). Combined with the high spatial resolution afforded by the instrumentation typically used to collect this type of data (confocal or two-photon microscopy—see Chapter 3) a great deal of information about molecular diffusion rates in different regions of a sample, such as a cell, can be obtained.

The model of equation 2.27 can easily be extended [34] to the general case of a number of distinct species each defined by a different measured quantum yield, Q, and a different molecular diffusion time,

$$G(\tau)=\frac{\sum_{i=1}^{R}\alpha_i^2\langle \overline{\mathrm{N}}_i\rangle\left\{\left(1+\frac{\tau}{\tau_{\mathrm{D}_i}}\right)^{-1}\left(1+\frac{\tau}{\mathrm{K}^2\tau_{\mathrm{D}_i}}\right)^{-1/2}\right\}}{\left[\sum_{i=1}^{R}\alpha_i\langle \overline{\mathrm{N}}_i\rangle\right]^2}+DC, \tag{2.29}$$

where parameters are defined as for equation 2.27 and α is the detected quantum yield ratio of species i relative to species $i = 1$. In the limit of $R = 1$ equation 2.29 reduces to the case for simple one species diffusion (equation 2.27). Thus, if an autocorrelation function can be shown to be best fit, without *a priori* knowledge, by a model that requires two or more species defined by differing diffusion times, this technique can be used to identify heterogeneity in samples. While difficult to do, and relying on very careful statistical analysis of the data (see later), this application of FCS has been successfully demonstrated [53].

One of the predominant sources of fluctuations in addition to diffusion in FCS is from intersystem crossing to the lowest excited triplet state of the fluorophore [38]. Correlations due to triplet crossing tend to be dominant at very short lag times in comparison with the part of the autocorrelation curve dominated by diffusion (see Figures 2.9 and 2.12). Numerous triplet state studies have been conducted principally using the organic dyes rhodamine 6G (R6G) and fluorescein [42, 54]. A model for the autocorrelation function of a signal whose fluctuations derive from a combination of diffusion and triplet crossing has been developed [54],

$$G(\tau)=\frac{1}{N}\left(1+\frac{\tau}{\tau_{\mathrm{D}}}\right)^{-1}\left(1+\frac{\tau}{K^2\tau_{\mathrm{D}}}\right)^{-1/2}\left[F_{\mathrm{T}}\exp\left(-\frac{\tau}{\tau_{\mathrm{T}}}\right)+(1-F_{\mathrm{T}})\right]+DC, \tag{2.30}$$

where τ_{T} is the characteristic correlation time associated with triplet crossing and F_{T} is the proportion of time molecules spend in the triplet state (called the triplet fraction). Other parameters are defined as for equation 2.27. It is important to note that additional fluctuation components such as these are included in the autocorrelation in a multiplicative manner. Measurement of the triplet state using autocorrelation can be used as a sensitive probe of molecular environment, or indeed any effect that modulates either the triplet crossing rate or the triplet fraction. This is illustrated in Figure 2.14 [54]. Autocorrelation curves were measured for the fluorescent dye rhodamine 6G in varying concentrations of the quencher potassium iodide (a quencher that is well known to modify triplet crossing rates). It is shown in Figure 2.14 that as the potassium iodide concentration is increased, the relative proportion of the triplet fraction is increased.

Recently, single molecule FCS has been used to measure the conformational dynamics of biological molecules. For example, the opening and closing of a small nucleotide hairpin labelled with a fluorophore and a suitable quencher were studied using FCS and a simple two-state model [46]. The autocorrelation

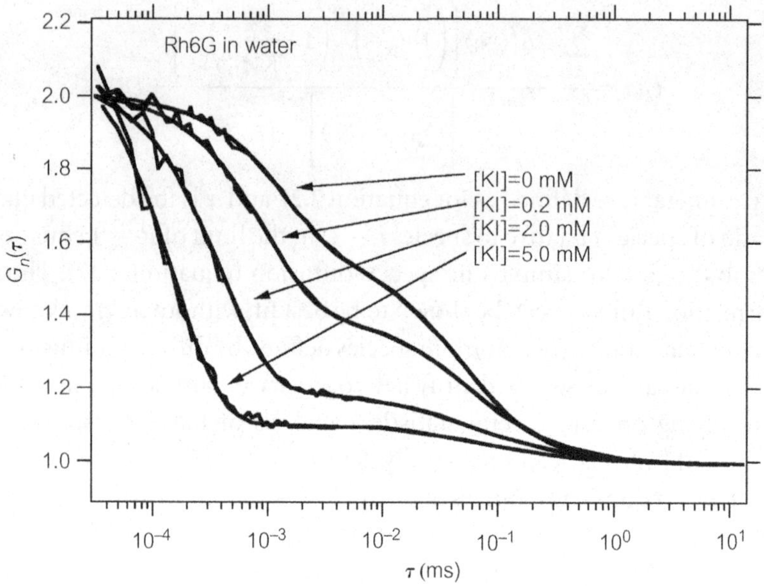

Figure 2.14 Autocorrelation curves calculated for the fluorescent dye rhodamine 6G in water at different concentrations of potassium iodide. Reprinted from Widengren, *et al.*, Fluorescence correlation spectroscopy of triplet states in solution—a theoretical and experimental study. *Journal of Physical Chemistry* **99** (1995) 13368–13379 with permission from the American Chemical Society.)

function for diffusion combined with transitions between two distinct states is of the form [28, 55, 56],

$$G(\tau)=\frac{1}{N}\left(1+\frac{\tau}{\tau_D}\right)^{-1}\left(1+\frac{\tau}{K^2\tau_D}\right)^{-1/2}\left[\frac{1-p}{p}\left(-\frac{\tau}{\tau_R}\right)+1\right]+DC \qquad (2.31)$$

where p is the fraction of particles in one state and τ_R is the chemical reaction timescale equal to the inverse of the observed kinetic rate constant k_{obs} which is itself equal to the sum of the opening and closing rate constants in a two-state process (such as in [46]). The model is thus applicable to any chemical reaction or dynamic process that can be described as two-state (and is not limited therefore to conformational dynamics of nucleotide hairpins). Unfortunately, in many cases the fluctuations in fluorescence signal caused by molecular dynamics are on a similar timescale to that of diffusion (see Figure 2.12) and it is therefore difficult to fit equation 2.31 to experimental autocorrelation functions and extract τ_R accurately. Therefore, a method was developed by Bonnet *et al.* [46] taking into account the multiplicative way in which additional components add to the autocorrelation (see equations 2.30, 2.31 and for a more recent application of the technique see [57]). Here the diffusive part of the autocorrelation function is removed by normalizing with a second measured autocorrelation function for the same molecule under conditions in which no conformational transitions occur. This normalized autocorrelation is given by,

$$G^{\text{Dynamics}}(\tau)=\frac{[G^{\text{Diffusion}}(\tau)\times G^{\text{Dynamics}}(\tau)]}{G^{\text{Diffusion}}(\tau)} \tag{2.32}$$

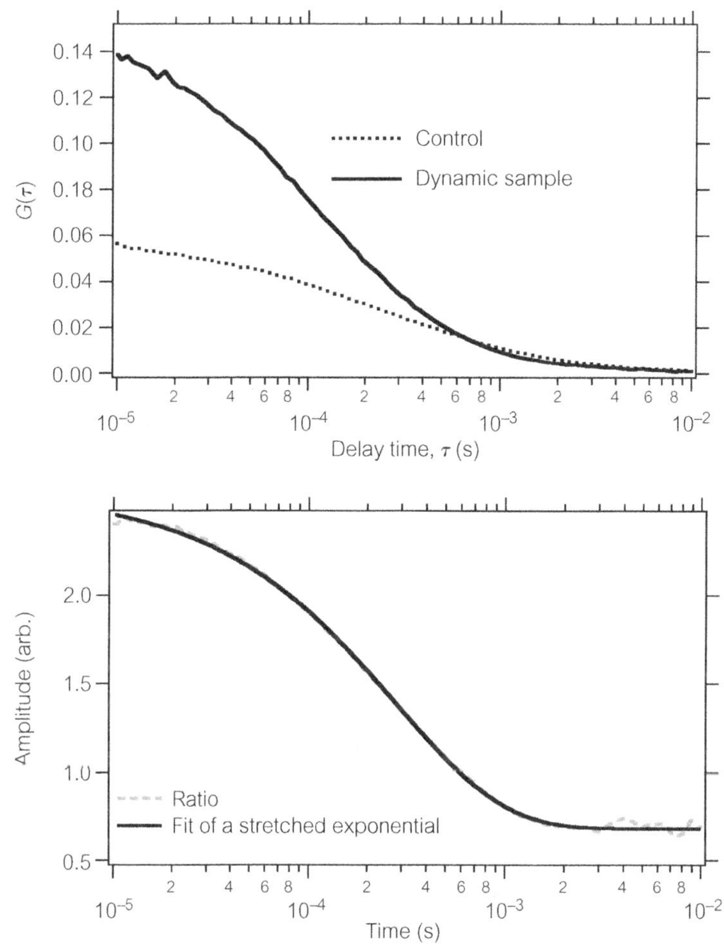

Figure 2.15 Illustration of the use of autocorrelation to probe conformational dynamics. Autocorrelation functions are measured for the kinetic sample (top, solid line) and the control sample (top, broken line), which has no dynamic conformational component. The ratio of these autocorrelation functions leaves only the conformational dynamics component (bottom, grey broken line) which can be fit with, in this case, a stretched exponential (bottom, solid black line) to extract the observed rate constant for the transition. In this case the dynamic sample was an RNA hairpin labelled with fluorescein. When the hairpin was formed the dye was brought near a quencher, thus modulating the fluorescence emission and providing fluctuations to be probed via autocorrelation. The control sample lacked the quencher.

The kinetics of the conformational transitions can therefore be fit with a simple exponential function (or in some cases better described by a stretched exponential [45]) and the observed kinetic rate constant can be determined (see Figure 2.15). Since the equilibrium constant (measured easily in a number of ensemble assays) gives the ratio of the opening and closing rates, then data can then be combined with the observed kinetic rate constant to give the opening and closing rate constants. This important technique introduced in a paper by Bonnet and co-workers [46] is reviewed more completely in Chapter 5.

2.4.5 **Statistical analysis—fitting models to measured autocorrelation functions**

Regardless of the method used to calculate the autocorrelation function or the model chosen to fit this function, a rigorous statistical analysis is essential to ensure that the complex multiparameter models used are justifiable and that the parameters produced by the fit are reliable. Essential to the accurate analysis of the autocorrelation data is a knowledge of the experimental standard deviation which allows weighted least squares fitting to be used, quality of fit assessment through normalized residual analysis and computation of a properly reduced chi-squared value. The statistical description of the autocorrelation function has been treated in an approachable manner by Wohland [37] and a detailed analysis of the standard deviation of the autocorrelation function has been presented by Koppel [33]. The importance of taking such a rigorous approach can be understood very simply by a qualitative consideration of the autocorrelation curve. Typically, the time range of an autocorrelation function covers 5 or 6 orders of magnitude (see Figure 2.9 and Figures 2.12–2.15). Furthermore, in the case of an autocorrelation function produced by a hardware correlator the lag times (and so sampling times) are also not evenly distributed in time (see Figure 2.10 and [37]). As an illustration, consider the number of measurements made in a typical hardware autocorrelation measurement carried out for 2 min. For the 1 μs delay time channel there are ~10^8 possible measurements while for the ~0.1 s delay time there are only ~10^3. In a multiple tau digital correlator (Figure 2.10 and [37]) the sampling times at different lag times are not equal. For example, in the case of the ALV5000 (ALV GmbH, Germany) the 1 μs delay channel has a sampling time of 200 ns, but at a delay time of ~0.1 s the sampling time is ~6.6 ms. At a typical average count rate of ~20 kHz, the average number of counts in a 200 ns sampling time is therefore 0.004 compared to 132 for the 6.6 ms sampling time. The signal-to-noise for any particular delay time is then proportional to the total measurement time [58], the longer the measurement the more events that are measured and the signal-to-noise should improve. However, with short delay times with necessarily short integration times and small numbers of measured photons, quantum mechanical shot noise dominates [33, 58]. The relative contribution of this effect can only be reduced through an increase of the number of photons measured in the integration time [33] which is often simply not possible since it is limited by the quantum efficiency of the fluorophore and instrument collection efficiency. At longer lag times where sampling times are longer, shot noise is small compared to the count rates involved and in this limit the signal-to-noise is given as the square root of the total measurement time. Long measurement times however sometimes introduce problems of sample degradation (photobleaching) and instrument drift and therefore a balance must generally be struck.

It is clear then that the quality of the autocorrelation data varies across the range of lag times measured, making knowledge of the standard deviation very important for accurate analysis. We focus on this point because poor (or no) statistical analysis is often presented in the literature leading to obvious errors, the significance of which often cannot be easily judged by the reader. Ideally, the standard deviation of the autocorrelation function would be computed based on a complete knowledge of the count rate history, measurement time, average count rate and number of counts per sampling time per delay time [37]. Unfortunately, many hardware correlators do not provide this information, although some modern correlators do and also calculate the standard deviation on-line [37,59]. There are two other methods that can be used to estimate the standard deviation. A theoretical estimation of the standard deviation can be used and this approach is most commonly found in the literature. The theoretical model derived by Koppel for this purpose has limitations and has been extended by others [37, 53, 60]. Despite these modifications the calculation still has a number of shortcomings: assumption of negligible background, that the signal-to-noise is independent of concentration (only true at intermediate concentrations—at low particle numbers few events are measured and at high particle numbers relative fluctuations are small) and that a uniform illumination profile is used. In addition, these theoretical approaches are only directly applicable to a single species sample. It has also been noted that the Koppel method significantly overestimates the standard deviation at long delay times [37] and therefore results in reduced chi-squared values that are very much less than 1 or, more seriously, fits that give non-minimized chi-squared values can be taken to represent good fits. The calculation of the Koppel standard deviation is given by [33],

$$\sigma^2(G(\tau)) = \frac{1}{M}\frac{1}{N^2}\left[\frac{(1+g_{3DG}^2(\tau))(1+g_{3DG}^2(\Delta\tau))}{(1-g_{3DG}^2(\Delta\tau))} + 2m g_{3DG}^2(\tau)\right]$$

$$+\frac{1}{M}\left[\frac{2(1+g_{3DG}^2(\tau))}{N<n>} + \frac{1}{<n>^2}\left(1+\frac{g_{3DG}(\tau)}{N}\right)\right], \qquad (2.33)$$

where for some delay time τ, $\Delta\tau$ is the sampling time, M is the number of counting intervals ($t/\Delta\tau$) where t is the total measurement time. N is the average number of particles in the sample volume and $<n>$ is the average count rate per channel (total count rate, $\bar{n}$, multiplied by sampling time). The function $g_{3DG}(x)$ is given in equation 2.27. Fit parameters (N, τ_D, K^2) must first be estimated from un-weighted fits to the data, thus an iterative procedure must be followed. It has been demonstrated that a single iteration step is sufficient to converge on the solution of the standard deviation [37]. Shown in Figure 2.16 (solid lines) is a typical standard deviation calculation using the Koppel method for three-dimensional diffusion with $\overline{N}=1, \tau_D = 159\ \mu s, K = 10, \bar{n} = 11$ KHz and a total measurement time of 30 s. We can first consider this result qualitatively. The standard deviation is high at short delay times, which is consistent with expectation: at short delay times few photons are counted as short integration times are used and so shot noise is significant. As the delay times increase the standard deviation reduces because longer delay times have longer integration times, so more photons are counted and the relative contribution of shot noise is reduced. However, as the delay time continues to increase the standard deviation also increase, erroneously.

The second method to estimate the standard deviation of an autocorrelation curve is by analysis of a series of consecutively measured autocorrelation curves [37] and is given by,

$$\sigma = \sqrt{\frac{1}{L-1}\sum_{l=1}^{L}\left\{\frac{G_l(\tau)-DC}{G_l(0)-DC}-\bar{g}(\tau)\right\}^2}, \tag{2.34}$$

where $\bar{g}(\tau)$ is the amplitude normalized average autocorrelation given by,

$$\bar{g}(\tau)=\frac{1}{L}\sum_{l=1}^{L}\frac{G_l(\tau)-DC}{G_l(0)-DC}. \tag{2.35}$$

σ is calculated for the autocorrelation function decaying to 0, in agreement with the Koppel definition. Further, the standard deviation is calculated only after each of the consecutive autocorrelations are normalized to have $G_l(0) = 1$. This is necessary as experimental phenomena, such as photobleaching, drift or contaminants, can cause variation in the amplitude of the measured curves that should not be reflected in the calculation of σ. As a result of this procedure once σ is calculated it must then be normalized to the amplitude of each autocorrelation fit to provide the proper weighting. DC and $G_l(0)$ can be estimated from an un-weighted fit to the data set, as for the theoretical calculation. The measured experimental standard deviation for a sample with similar fit parameters as those used for the Koppel calculation is shown in Figure 2.16 (dots). The experimental (dots) and theoretical (lines) methods are in close agreement at short lag times

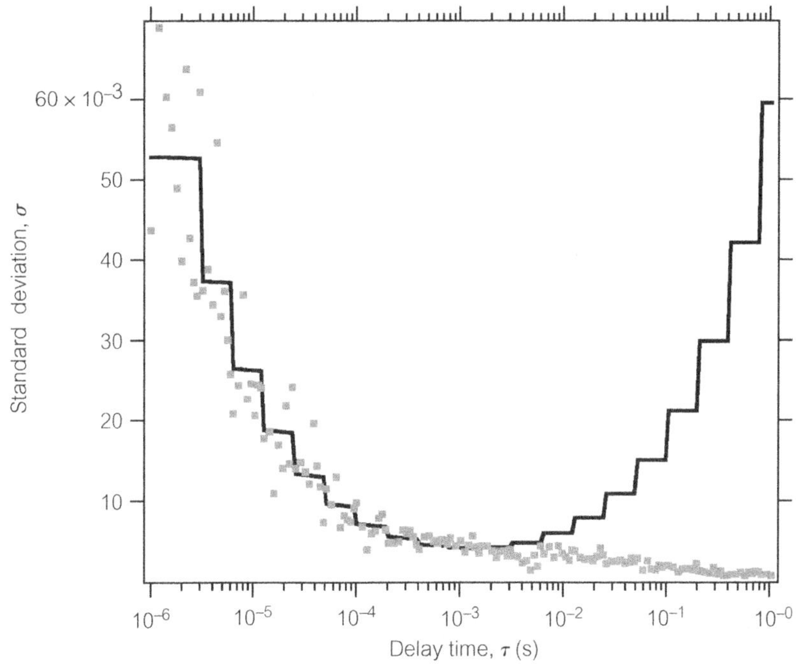

Figure 2.16 The calculated standard deviation for an autocorrelation experiment (solid line) where $\overline{N}=1$, $\tau_D = 159$ μs, $K = 10$, $\overline{n}=11$ KHz and assuming a total measurement time of 30 s using equation 2.33. Also shown is the standard deviation calculated from a repeat of ten experiments given by equation 2.34 (grey squares). Parameters from the weighted fit to the data set for each of the ten curves were approximately the same as in the calculation, for comparison.

but the Koppel method clearly overestimates the variation at long delay times. For reasons of time and other physical constraints, such as the stability of the microscope or the robustness of the sample, having to perform many experiments to calculate the standard deviation in this way may not be practical. However, in the absence of a complete knowledge of all the experimental parameters this is the method which gives the best estimate [37].

In Chapters 5 and 6, we discuss in detail a number of studies that exploit FCS, however, this is by no means an exhaustive review. A few other examples are worth mentioning briefly here: Gösch and Rigler [61] used FCS analysis with models accounting for diffusion under flow to characterize the flow profile of particles in microchip-based microchannel structures. Björling and Rigler [62] used FCS for detecting products of the polymerase chain reaction. They showed that different DNA fragment lengths can be distinguished based upon their respective diffusion times. Similarly, they demonstrated that one could identify the nature of the products by exposing them to restriction enzyme digestion. FCS has also found application in cell biology. Schwille *et al.* [63] discuss the applications of FCS in cells and membranes. Here it is necessary to employ two-photon excitation (see Chapter 3) in order to minimize autofluorescence from the cell. This work has applications in examining molecular transport, localized concentrations of metabolites, proteins or lipids, and aggregation/molecular interactions of labelled molecules within the cell [64, 65]. Another common application of FCS is in the study of DNA–RNA or DNA–protein interactions. In a typical experiment Schwille *et al.* [66] used FCS to perform a series of assays for six different tetramethylrhodamine labelled oligomeric DNA molecules binding to a 101mer target RNA. Monitoring the hybridization as a function of time showed that the six DNAs had very different association rate constants. These data were then used to infer the existence of a number of independent binding sites and to support a model of the RNA structure. Häsler *et al.* [67] used FCS to examine the binding of the δ-subunit of ATP synthase. Dimerization of the subunit was seen and from the disassociation constant measured by FCS it was shown that the binding of the subunit is of sufficient strength to remain bound during the enzymes' working cycle. Huang *et al.* [68] monitored the change (increase) in the translational diffusion time of labelled apomyoglobin during acid denaturation. They observed an increase in the diffusion time of the order 1.7 times on unfolding with a concomitant increase in fluorescence of around 40%.

2.5 Fluorescence resonance energy transfer

Fluorescence resonance energy transfer (FRET) has been established as a powerful tool in physical chemistry and biophysics for more than 30 years [69–72]. Perrin and Förster [73] first described theoretically the process of non-radiative transfer of energy from a donor (or sensitizer) chromophore to an acceptor chromophore over distances of up to 100 Å. The FRET process is described in detail elsewhere (e.g. [72, 74] and we also discuss the mechanism in more detail in Chapter 4). Here we shall attempt to outline the experimental principles and parameters used in FRET and specifically discuss its application to single molecule studies using a few examples. In terms of single molecule experiments, FRET provides a powerful methodology, not just because of the sensitive distance dependent information that can be obtained, but also due to the ratiometric nature of the measurement. Unlike PCH or FCS, complex statistical descriptions of the data are not necessary because the ratio of two instantaneous signals are compared which removes a number of complicating factors (e.g. diffusion, the different paths that can be taken through the PSF volume and the form of the PSF).

2.5.1 Principles of fluorescence resonance energy transfer

FRET occurs (using a classical electromagnetic description) through a non-radiative long-range dipole–dipole interaction of a pair of chromophores over distances in the 10–100 Å range (see Figure 2.17 and Chapter 4, Section 4.2.1 for more discussion). Upon absorption of a photon, a donor chromophore may lose this excess energy in a number of ways; fluorescence, quenching, crossing to a triplet state (intersystem crossing), vibrational relaxation (internal conversion) or non-radiative energy transfer. The fluorescence lifetime (see Section 2.7.4) of the excited donor fluorophore τ_D (in the absence of FRET) is related to the sum of rates of all the possible relaxation pathways. For example, in the presence of internal conversion (k_{IC}), inter system crossing (k_{ISC}), collisional quenching (k_Q) and fluorescence (k_F), the excited state lifetime is given by,

$$\tau_D^{-1} = k_{IC} + k_{ISC} + k_Q + k_F. \tag{2.36}$$

If a suitable acceptor molecule is present, a long-range dipole–dipole interaction results in an additional relaxation term (k_{FRET}) and is incorporated into equation 2.36,

$$\tau_{DA}^{-1} = k_{IC} + k_{ISC} + k_Q + k_F + k_{FRET}. \tag{2.37}$$

The magnitude of k_{FRET} is a function of many properties of the donor/acceptor system and the environment surrounding the two molecules.

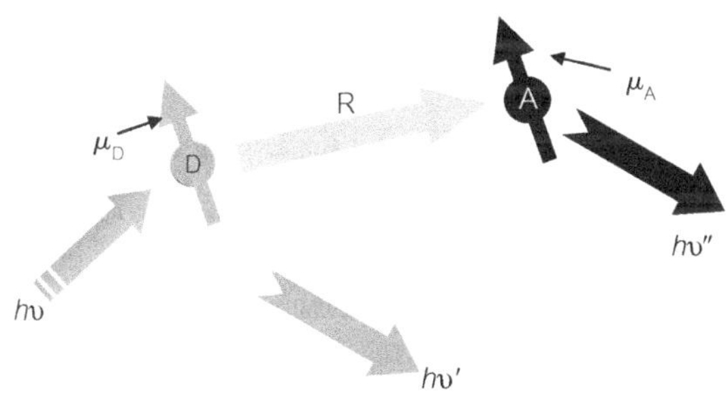

Figure 2.17 Illustration of the principle of FRET. Two dye molecules a distance *R* apart undergo non-radiative energy transfer through a dipole–dipole interaction. The dyes are generally referred to as the donor (D) and the acceptor (A) and their dipoles are illustrated, as μ_D and μ_A respectively. If the donor is excited by light of energy $h\nu$ then this excess energy may be lost via a number of mechanisms including emission of fluorescence at energy $h\nu'$ or energy transfer to the nearby acceptor molecule through a dipole–dipole interaction. The excited acceptor can now relax to its ground state via emission of a fluorescence photon of energy $h\nu''$. The relative number of acceptor fluorescence photons compared to the total number of fluorescence photons (donor and acceptor) is related to the efficiency of the energy transfer from donor to acceptor and is strongly dependant on the scalar separation, *R*.

Förster derived this relationship as [72, 74],

$$k_{\text{FRET}} = \frac{9000(\ln 10)\kappa^2 Q_D J}{128\pi^5 n^4 N_A \tau_D R^6}. \tag{2.38}$$

In this description the FRET rate is therefore a function of the refractive index of the medium between the two molecules n, the fluorescence lifetime τ_D[s] and quantum efficiency Q_D of the donor in the absence of FRET, Avogadro's number N_A, the separation of the two molecules R[cm], the normalized spectral overlap integral J [M^{-1} cm^3]and the so-called orientation factor κ. Note that different forms of equation 2.38 exist depending on the units used to express the quantities. Neglecting the detail of this expression for a moment, the key aspect for many FRET experiments is the dependence of the rate of energy transfer on the scalar separation R between the donor and acceptor molecules. Equation 2.38 can be simplified as,

$$k_{\text{FRET}} = \frac{1}{\tau_D}\left(\frac{R_0}{R}\right)^6 \quad \text{where } R_0^6 = \frac{9000(\ln 10)\kappa^2 Q_D J}{128\pi^5 n^4 N_A} \tag{2.39}$$

where R_0 is the separation at which 50% of the excitation energy is transferred to the acceptor and is known as the Förster distance. This convenient form is often used, as R_0 effectively defines the FRET relationship of a particular dye pair. Other than R the parameters in equation 2.38 change little for common experimental situations, however, care must be taken with controls, in particular refractive index changes are common and should be accounted for [75] as should changes in the quantum yield Q_D which can also occur with changes in the solvent. The additional terms in equations 2.37 and 2.38 are discussed in Section 2.5.7.

In many FRET experiments the FRET transfer efficiency E_{FRET} is the parameter that is sought. This is defined as the ratio of the energy transfer rate to the sum of all the donor de-excitation rates,

$$E_{\text{FRET}} = \frac{k_{\text{FRET}}}{k_{\text{IC}} + k_{\text{ISC}} + k_Q + k_F + k_{\text{FRET}}} \tag{2.40}$$

Using equations 2.36 and 2.39 we can see that,

$$E_{\text{FRET}} = \frac{1}{1 + \left(R/R_0\right)^6} \tag{2.41}$$

Thus the FRET energy transfer efficiency changes with the sixth power of the scalar separation between the two dyes and it therefore provides a powerful molecular length-scale structural probe. Shown in Figure 2.18 is the transfer efficiency versus scalar separation relationship for a typical FRET dye pair suitable for single molecule studies (R_0 = 54 Å) (see Chapter 4 for more information regarding dye pairs for FRET and their properties).

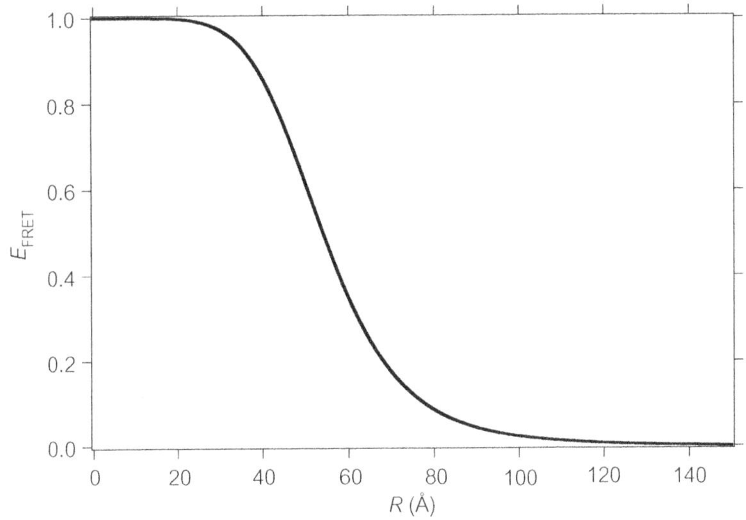

Figure 2.18 A plot of energy transfer efficiency for FRET between donor and acceptor dyes as a function of the scalar separation between them. The separation at which the efficiency is 50%, known as the Förster distance, characterizes the particular dye pair (54 Å in this example).

The transfer of energy from the donor to the acceptor results in the donor returning to the ground state and the acceptor entering an excited state. Relaxation from this state can then result in fluorescence from the acceptor. E_{FRET} can therefore be calculated experimentally in a number of ways, using the relative quantum yields (Φ), fluorescence intensities (I) or lifetimes (τ) of the donor molecule in the presence (indicated by superscript A) and absence of the acceptor;

$$E_{\mathrm{FRET}} = 1 - \frac{\Phi_{\mathrm{D}}^{\mathrm{A}}}{\Phi_{\mathrm{D}}} \quad E_{\mathrm{FRET}} = 1 - \frac{I_{\mathrm{D}}^{\mathrm{A}}}{I_{\mathrm{D}}} \quad E_{\mathrm{FRET}} = 1 - \frac{\tau_{\mathrm{D}}^{\mathrm{A}}}{\tau_{\mathrm{D}}} \tag{2.42}$$

An excellent description of the ensemble measurement and calculation of energy transfer efficiencies can be found in articles by Cheung and Clegg [72, 74].

In diffusion single molecule experiments it is rarely possible to measure the fluorescence intensity (or lifetime) from a given donor with the acceptor present and then from the same molecule in the absence of the acceptor. An exception would be to measure the donor intensity from an immobilized single molecule with energy transfer occurring, then wait for the acceptor to bleach allowing the donor signal without the acceptor to be determined. Such measurements are fairly easy in immobilized single molecule experiments (as will be shown in a later section), but not for diffusion based experiments. Thus, for measurements of diffusing single molecules the FRET efficiency is usually expressed in a relative manner by a simple ratio of the acceptor fluorescence intensity over the sum of the donor and acceptor intensities in what is termed the *ratiometric* approach [76, 77].

$$E_{\mathrm{FRET}} = \frac{I_{\mathrm{A}}}{(\Phi_{\mathrm{A}}\eta_{\mathrm{A}}/\Phi_{\mathrm{D}}\eta_{\mathrm{D}})I_{\mathrm{D}}+I_{\mathrm{A}}} = \frac{I_{\mathrm{A}}}{\gamma I_{\mathrm{D}}+I_{\mathrm{A}}} \tag{2.43}$$

Here I_{A} and I_{D} are the fluorescence intensities of the donor and acceptor respectively (for the same doubly labelled molecule), Φ_{A} and Φ_{D} are the quantum yields of the donor and acceptor molecules and η_{A} and η_{D} are the detection efficiencies of the experiment at the wavelengths of the fluorescence signals from the two molecules. For simplicity, the correction factor, γ, to account for differential detection efficiencies and quantum yields of the two fluorophores is generally assumed to be unity (see for a discussion [78]), and E_{FRET} is then sometimes referred to as the 'proximity ratio' P [77].

$$P = \frac{I_{\mathrm{A}}}{I_{\mathrm{D}}+I_{\mathrm{A}}} \tag{2.44}$$

Thus, true FRET efficiencies or inter-dye distances (R) are generally not calculated, although with careful analysis, control experiments and an awareness of the uncertainties, R can be computed (e.g. see [79, 80]. A discussion of these complexities can be found in Section 2.5.7).

2.5.2 Implementation of diffusion single molecule FRET measurements

The instrumentation for diffusion single pair FRET (spFRET) measurements is discussed in detail in Chapter 3. The experimental system is identical to that used for FCS and PCH except that fluorescence from the PSF volume is separated into donor and acceptor channels by filters and dichroic mirrors. Two separate detectors then monitor the time series of photon counts in the two channels simultaneously (see Figure 2.19). In the simplest implementation, molecules diffuse through the PSF and bursts of photons from donor and acceptor molecules, in a given integration period, are measured as a function of time. A histogram of the measured proximity ratios, calculated according to equation 2.44 for each bin, is then constructed. The stochastic nature of the single molecule fluorescence events combined with the random nature of spurious background events (and relatively low signal-to-noise for some events) means that some form of discrimination or thresholding is necessary to determine which are true FRET events. If no thresholding is applied to the raw data then a proximity ratio is calculated for every channel or bin in the data set. However, the nature of the single molecule experiment (see Figure 2.19) means that most channels contain only background noise and only occasionally is a burst of photons detected from a target molecule.

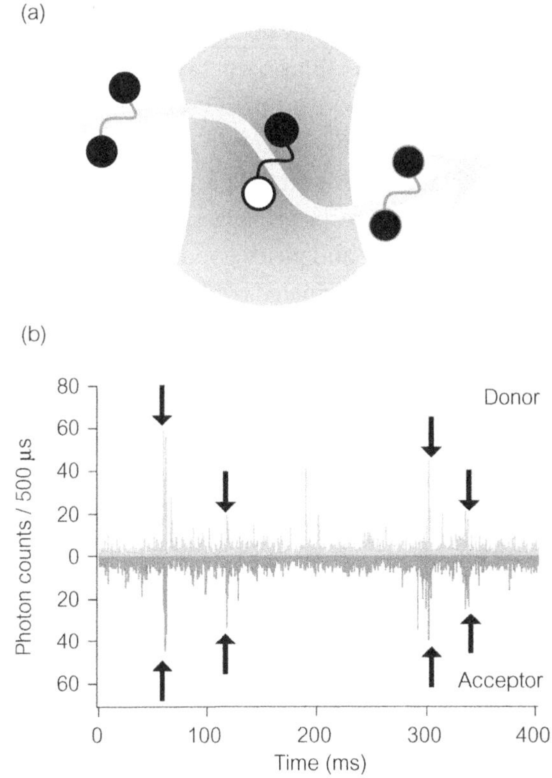

Figure 2.19 Illustration of the single pair fluorescence resonance energy transfer experiment for a configuration where the molecules are allowed to diffuse freely through the small (<0.1 fl) PSF volume. (a) Molecules labelled with donor and acceptor dyes pass through the small open volume resulting in correlated donor and acceptor fluorescence, the relative intensities of which depend on the inter-dye separation. Data is collected in two channels simultaneously over wavelength ranges that correspond to donor and acceptor emission. (b) Raw data consists of photon counts in each channel (or integration time, typically ~0.5 ms) versus time. Molecules entering the volume result in coincident, correlated bursts of fluorescence (arrows). These data were measured for a 19 nucleotide RNA hairpin in denaturing conditions, labelled with Fluorescein and TMR (see Chapter 4).

Thus, without thresholding, the histogram of proximity ratios is dominated by the small amplitude background noise signals and a symmetric histogram centred on $P_{\mathrm{FRET}} = 0.5$ is generally obtained. If thresholding is introduced, these small shot noise dominated events can be excluded and proximity ratios for only real events (although still with shot noise) are calculated. Clearly, care has to be taken not to exclude real events, which might bias the measured proximity ratio histogram.

The most common form of threshold found in single molecule diffusion FRET studies is a simple SUM threshold [75–77, 81, 82]. The sum of the signals in the two channels must exceed a certain threshold value T,

$$I_{\mathrm{A}}+I_{\mathrm{D}} \geq T. \qquad (2.45)$$

Perhaps the clearest advantage of the SUM threshold is that, from a mathematical point of view, it does not apply any bias to the measured histogram. Thus if one wishes to identify heterogeneity in a solution where two (or more) species are defined by differing FRET efficiencies (or more properly proximity ratios) then this method might seem the most appropriate. The choice of threshold level when using the SUM criterion is clearly strongly dependant on the particular data obtained. Theoretical methods to estimate a value have been suggested [83], however an empirical approach is most commonly used and yields good results. If a low threshold value is chosen the distribution will be dominated by shot noise. If too high a threshold is chosen then too few real events are included, the statistical quality of the histogram is degraded and it is biased towards high intensity burst events. One expects to find a range of threshold values where the histogram is relatively invariant. Within this range only small changes in the number of accepted events are observed and, importantly, no changes in peak positions of any species are seen [76].

Logical AND and OR have also been suggested as alternative thresholding criteria [77, 83] according to,

$$\begin{aligned} I_A \geq T_A \quad &\text{AND} \quad I_D \geq T_D \\ I_A \geq T_A \quad &\text{OR} \quad I_D \geq T_D \end{aligned} \tag{2.46}$$

Some care must be taken when applying these logical operators for thresholding. The AND criterion has the effect (mathematically) of biasing the measured distribution to intermediate FRET efficiencies. Events with a low signal in either the donor or acceptor channels are eliminated, although this does ensure that the events measured are only due to FRET. The OR operation produces more similar results to the SUM threshold; however to some extent the distribution will be biased towards high and low FRET efficiencies. Often an empirical approach testing the effect of different thresholding criteria is appropriate and a further discussion of this issue can be found elsewhere [77, 83].

When the molecule diffusing through the excitation volume is labelled with donor and acceptor dyes, the fluorescence signal in each channel is the sum of several components (see equation 2.47). The donor channel signal I_D, for example, comprises a signal arising from the donor fluorescence f_D, to which the background noise signal b_D is added, and in addition, a cross-talk term $c_D f_A$ due to acceptor fluorescence 'leaking' through the donor channel filter. In the case of the acceptor channel, there may be an additional undesirable component f_A^* due to direct excitation of the acceptor at the excitation wavelength used, which generates acceptor fluorescence unrelated to energy transfer from the donor.

$$I_D = b_D + c_D f_A + f_D$$

$$I_A = b_A + c_A f_D + f_A^{\mathrm{FRET}} + f_A^{*} \tag{2.47}$$

The magnitude and therefore the importance of these unwanted signals will be a function of the particular choice of dye pair and the experimental setup. The average background signals for the two channels can easily be approximately determined in an experiment in which the signal from the solvent is measured. This may be subtracted from the signal in the two channels. It must be noted however that the true background level is itself stochastic in nature. Consequently, subtraction of a mean value cannot correct for the occasional, but statistically significant, measurement of spurious bright background bursts. Further, a large proportion of the mean background may result from out of focus analyte molecules, which are obviously not accounted for in this method of background determination. Fortunately, the signal to background ratio is generally high for single molecule events and thus the effect of incorrectly determined mean background levels on the measured histograms is generally quite low.

Cross-talk in the acceptor channel due to donor fluorescence can be a significant (but easily quantifiable) effect because the donor fluorescence emission spectrum may have a long tail to lower energy which overlaps with the acceptor emission. The effect is greatly dependent on the experimental configuration and the dye pair chosen, but it is quite often neglected [76] despite potentially changing the histograms significantly. The importance of cross-talk is also dependent on the strength of FRET. For example, under circumstances where little energy transfer is occurring, the donor will be strongly fluorescent. Thus leakage to the acceptor channel may be significant and a higher mean FRET efficiency will be obtained. However, with the same dye pair but under conditions where strong FRET is occurring, the donor fluorescence may be very low and so the measured FRET energy will be unaffected by the leakage to the acceptor signal. The opposite case of acceptor leakage into the donor channel is generally of less importance because the acceptor fluorescence emission typically does not overlap in wavelength with the donor emission.

Direct excitation of the acceptor leading to fluorescence will result in apparently higher transfer efficiency because, like in the case of leakage of the donor photons to the acceptor channel, it appears that FRET is occurring when it is in fact not. This can adversely affect the accuracy of absolute distance measurements using FRET. This is not easily quantifiable but all the other contributions mentioned, can be corrected by modification of equation 2.44 subtracting the *mean* values of the effects [76].

Despite all of these issues careful experimentation has shown that very accurate absolute FRET efficiency measurements can be performed [78, 84, 85].

In single molecule FRET data, events above the chosen threshold should be rare. Thus like in PCH (see Section 2.3), it is necessary to measure for a long time to produce statistically reliable histograms. Clearly, the actual measurement time required is dependent on the particular experimental configuration and sample concentration and, as with PCH, many data sets may need to be combined (due to the finite number of bins available on the multi-channel scalar data acquisition cards typically used). It is important to note that, unlike fluctuation spectroscopies, we must now ensure that single molecules are measured (at least if this is the interpretation we place on the resulting histograms). Thus for spFRET low concentrations with rare single molecule events are necessary—checks should therefore be made to ensure histograms are robust at a range of concentrations.

Histograms of proximity ratios are commonly fit with single or multiple Gaussian or Lognormal curves in order to extract information on peak positions, widths and areas [76]. It is possible to describe the data with an analytic expression in some cases (see [76]), but the difference in the characteristic parameters obtained (peak position, width and area) using a Gaussian is small [76]. Care must also be taken with the construction of the histograms, with suitable numbers (and widths) of bins chosen to aid in the resolution of any components that may exist.

2.5.3 **Information in proximity ratio histograms**

Unlike FCS, diffusion spFRET is comparatively still in its infancy. In this section we will use a limited number of simple examples to demonstrate the type of information that can be obtained from FRET studies of diffusing single molecules. In particular we will illustrate the nature of the data obtained from a single molecule that undergoes conformational changes between states with different mean FRET efficiencies and so exists in solution as a heterogeneous ensemble of molecules with some dynamic equilibrium between the states—a common application of spFRET. We discuss other examples of diffusion spFRET in Section 2.5.8 and review several important experiments in detail in Chapter 5.

Shown in Figure 2.20(a) is a proximity ratio histogram for a solution of donor- and acceptor-labelled ribonucleic acid (RNA). In order to understand this data we first review the characteristics of this molecule in the solution. Under favourable conditions this short, 19 nucleotide RNA may acquire secondary structure in the form of a hairpin with 7 base-paired nucleotides, one mismatched base in the stem and a loop consisting of 4 bases, as is illustrated in the cartoon inset in Figure 2.20(a). The RNA is labelled with donor and acceptor (in this case fluorescein and tetramethylrhodamine, see Chapter 4) at each end. Thus in the folded (otherwise described as the native or closed) conformation the hairpin is formed and the stem brings the two ends with the dyes close together. Thus the separation (R in equation 2.41) is small: the dyes are attached to sites only a few angstroms apart and although they are on linkers several angstroms long (see Chapter 4) the maximum separation is still less than 20 Å. For this dye pair, with $R_0 = 50$ Å, this

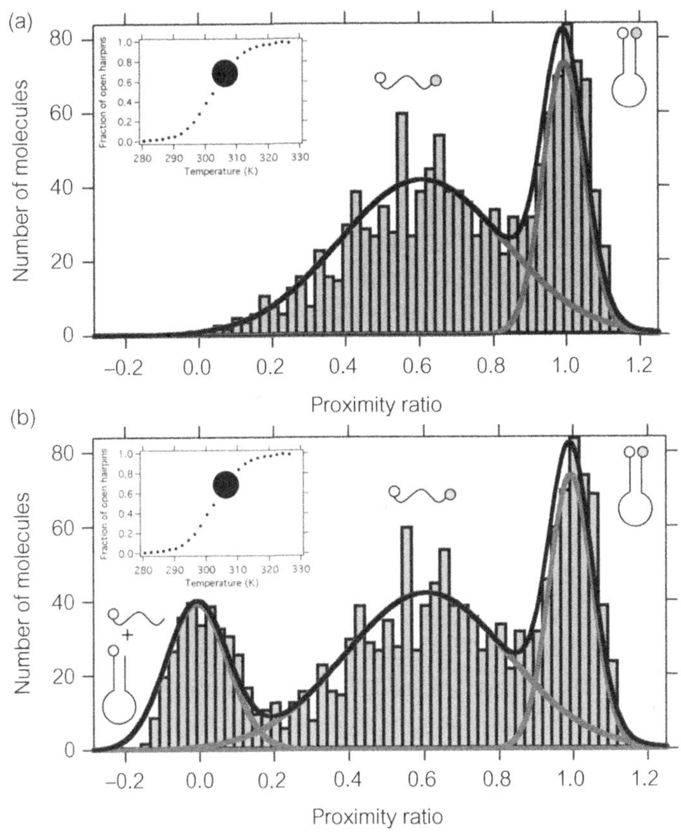

Figure 2.20 Proximity ratio histograms for a FRET labelled RNA hairpin loop diffusing in buffer. (a) Ideal histogram with the 'zero' peak removed (see Section 2.5.4). Multiple single molecules were allowed to diffuse through a sample volume and events to be included in the histogram are chosen using a SUM criterion threshold (see text). For these molecules the proximity ratio is calculated according to equation 2.44 after the mean background signal was subtracted from all channels and a correction for 5% donor–acceptor signal leakage incorporated. The distribution is fit with a double Gaussian (black line). The individual Gaussian components are shown (grey lines). Inset is the equilibrium thermal denaturation profile. The large dot shows the approximate temperature at which the spFRET measurement was conducted. The high FRET peak was assigned to formed (closed) hairpin and the lower FRET peak to the denatured (open) hairpin as illustrated in the cartoons. (b) The same data shown before subtraction of the zero peak, which is caused predominantly by donor only labelled molecules, or those in which the acceptor has photobleached.

gives near 100% energy transfer efficiency (see Figure 2.18). Thus FRET is efficient and much more acceptor fluorescence will be observed than donor fluorescence resulting in an expected proximity ratio near unity (equation 2.44). The secondary structure can be disrupted by addition of denaturant, by altering pH or by increasing temperature. Disruption of the structure will likely result in a broad ensemble of unfolded states (otherwise described as denatured or open states) and so the inter-dye separation is likely to be increased (for many of these unfolded states). The mean FRET efficiency will therefore be reduced and the relative fluorescence intensities of the donor and acceptor reverse. The inset in Figure 2.20(a) is the *ensemble* thermal denaturation profile (black circles) for this molecule measured by FRET. At low temperatures (~280 K) the ensemble consists of nearly all folded molecules and as the temperature is raised a proportion of the molecules in the ensemble become unfolded until at high temperatures (~330 K) all molecules in the ensemble are denatured or open. Thus at any intermediate temperature there exists a dynamic equilibrium between the native state and the unfolded states. At the midpoint of the thermal denaturation curve (at which point 50% of the molecules are folded and 50% unfolded at any instant) single molecules are continuously inter-converting between the two states but the average populations are at equilibrium. The shape of the ensemble curve (inset in Figure 2.20(a)) supports this view of a barrier limited two-state system with no other states *significantly* populated at equilibrium. Note, however, that the folding energy landscape of simple nucleotide hairpins has been suggested to be more complex [86–89], but for this discussion we assume a two-state system in dynamic equilibrium.

The proximity ratio histogram shown in Figure 2.20(a) was calculated from a data set of diffusing labelled RNA molecules measured at approximately 307 K. A SUM threshold (Section 2.5.2) was set at 35 counts per integration time (or bin) and an integration time of 0.5 ms was used. The sample concentration was 100 pM and approximately 9 million bins were analysed. Analysis was carried out after the *average* background signal in each channel was subtracted from all bins (so proximity ratios greater than 1 and less than 0 are possible). Additionally 5% of the donor signal in each bin was substracted from the corresponding acceptor bin to account for donor phatons that 'leaked' into the acceptor channel. The percentage was determined in a seperate, ensemble experiment. Approximately 3000 valid (above the threshold) events resulted from this analysis. Note that this does not necessarily relate to 3000 molecules because the signal from a given molecule may be spread over many integration times (see Figure 2.1) depending on the time taken to traverse the PSF. A discussion of some of the consequences of this fact is given in Section 2.5.5. The proximity ratio histogram was fitted with a convolution of two Gaussians with all the parameters left free to vary. There are clearly two peaks in this histogram: a peak centred at a high proximity ratio (~1) and a peak centred at an intermediate proximity ratio (~0.6). We assign these to the folded and unfolded species respectively, as indicated by the inset cartoons in Figure 2.20(a). There are four features of the histogram that can be discussed further: the trivial observation of the number of *resolved* peaks, and the width, position and area under the peaks.

Number of peaks

The two resolved peaks have proximity ratios that are consistent with a dynamic equilibrium between the expected folded and unfolded conformations of RNA.

The high proximity ratio species is assigned to folded molecules and the lower proximity ratio species assigned to unfolded molecules. A further discussion relating to the number of peaks that are observed, and the conditions in which one will observe distinct peaks for species in dynamic equilibrium is given in later subsections.

Peak width

The widths of the peaks in Figure 2.20(a) contain a large amount of information. The physical origins of the width of peaks in these histograms is complex but has three basic contributions: broadening due to shot noise, broadening due to the rotational freedom of the dyes attached to the analyte (see Chapter 4) and conformational fluctuations in the labelled molecule.

The primary source of broadening is often shot noise. If molecules pass, one at a time, through the PSF and each molecule has the same donor–acceptor separation with no other effects present (such as bleaching or quenching), then one might expect the exact same signal to be measured for each molecule and so a single proximity ratio to be obtained. However, as we have learned from Section 2.3.1 random shot noise will be added to the low photon count numbers that are measured in these experiments. Indeed a different amount of shot noise will be added to each of the donor and acceptor signals for each molecule and differently for different molecules. This creates an inherent distribution of proximity ratios even for a system where no other effects are present. This can be exaggerated by the fact that the absolute signals from the two dyes on different molecules may not be the same because of stochastic contributions to the fluorescence process and also due to the random path that they can take through the PSF. This effect is also threshold and signal dependant and therefore depends on experimental parameters such as laser power and integration time. If the threshold is high then only molecules with high absolute signals will be included in the histograms and these photon bursts will have a smaller percentage contribution from the shot noise. A theoretical treatment of shot noise broadening as a function of mean FRET efficiency and signal strength has been developed [75, 76].

Rotational freedom of the dyes is another possible contributor to peak width broadening. As will be discussed it is important for the dyes used to have rotational freedom such that all relative dipole orientations are sampled much faster than the timescale of the measurement (integration time). Dyes are typically attached to the host molecule via saturated carbon–carbon linkers around 10 Å in length (see Chapter 4). Thus the dyes are likely to be sampling a distribution of separations and orientations faster than the time scale of the measurement, averaging out the effect of the rotation and making this contribution effectively the same for all molecules measured. The case, however, of restricted rotation on a timescale approaching that of the measurement will lead to a distribution of measured FRET efficiencies that will depend on a number of experimental variables (integration time, linker length, site of attachment, temperature etc.). This can result in an increase in the peak width or in extreme cases even to multiple peaks being resolved. Practical considerations concerning choice of linker length, site of attachment and labelling chemistry are discussed in Chapter 4.

Finally, particularly in experiments on large macromolecules, conformational states of the labelled molecule may not be defined by a single rigid structure, rather by an ensemble of conformationally similar states. Information on this structural ensemble (on the timescale of the conformational rearrangements) is held in the width of peaks in the histograms. If the rearrangement within the ensemble is slower than the timescale of the measurement (so slower than the integration time or bin size used to measure the data OR the typical length of time taken for a molecule to diffuse through the volume, whichever is shorter) then this would strongly affect the widths of the peaks and in the extreme give rise to multiple peaks each of which would be shot noise broadened. If the rearrangement was on a timescale much faster than the measurement timescale then the effect would be averaged out. The same mean FRET efficiency would be recorded for every molecule, leading to a single shot noise broadened peak. Rather than being a disadvantage, this effect has been exploited in a number of experiments. For example, in a recent study Schuler and co-workers [75] performed diffusion spFRET measurements on cold shock protein and compared the width of the peak assigned to the unfolded state to a sample that had no conformational freedom (but had the same dyes, with the same linkers) and showed that this peak was not additionally broadened.[5] This subsequently allowed the authors to place limits on the reconfiguration time of the chain in the unfolded protein. This excellent paper is reviewed in detail in Chapter 5.

Peak area

In a simple experiment in which the peaks are properly resolved and no other biasing exists (see later sections) then the areas under the peaks are proportional to the total number of molecules observed in that particular state and so can be used to follow the conversion of one species into another. For example, Figure 2.21 shows five separate diffusion spFRET histograms calculated from experiments conducted at increasing temperatures for the RNA hairpin discussed earlier. Also shown is the ensemble melting curve, with arrows indicating the temperature at which the histograms were measured. Clearly, the ratio of the peak

[5] The reader should also consider the more recent paper by the same group [78] that addresses some complications in the interpretation of spFRET protein data with respect to the control used.

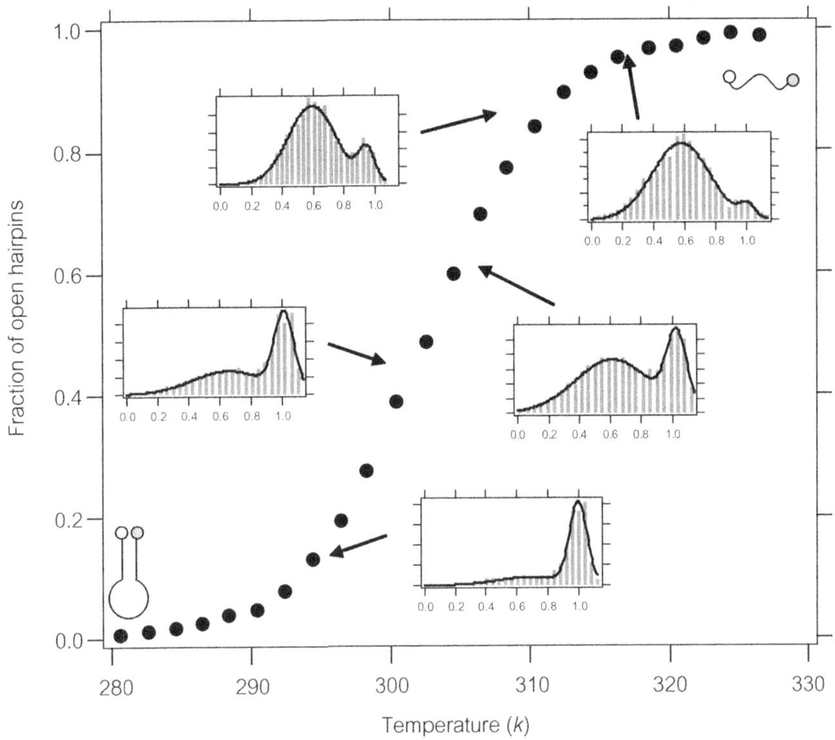

Figure 2.21 spFRET histograms, with the zero peaks removed, of a 19 nucleotide RNA hairpin (described in the text and measured as described in Figure 2.20) at various temperatures along its thermal denaturation profile (black circles, measured by ensemble FRET). As the population changes from all folded (closed hairpins) to all unfolded (open hairpins) the dominant peak in the histograms switches from high FRET efficiency to low FRET efficiency.

areas follows the equilibrium population of the two states. This type of simple analysis which recovers ensemble data has been used in a number of studies to confirm the validity of other, more interesting aspects of single molecule data contained within the histograms [82, 90].

Peak position

The position of the peaks assigned to each species gives an indication of structure, specifically the separation between the donor and acceptor dyes, and therefore on the relative position of the labelling sites. Movement of the peak position of a given species can indicate changes in the structure of that state. As an example several reports have noted movement in the position of the peak assigned to the denatured ensemble in spFRET measurements of proteins as a function of denaturant [75, 82], possibly indicating changes in the compactness of this state with denaturant, as might be expected if differing amounts of residual structure was present at different denaturant concentrations.

2.5.4 **Zero (bleaching) peaks**

Commonly in samples with n known identifiable species $n + 1$ peaks will be observed. An additional peak occurs near zero proximity ratio (the exact position depending on the analysis method), which is commonly referred to as the 'zero peak'. This extra peak was removed from Figure 2.20(a) to simplify the discussion given in the previous section but is shown in Figure 2.20(b). The peak occurs at low FRET efficiency and so is apparently indicative of a species with a large separation between the dyes and so little efficiency of energy transfer. The origin of this peak is somewhat uncertain, but it appears ubiquitously in diffusion spFRET experiments. In many studies either it is removed (by applying multiple Gaussian fits and subtraction) or low proximity ratios are simply not displayed. Certainly, it does not indicate a 'true' species but may be due to either molecules where the acceptor has photobleached or molecules for which the labelling was incomplete and no acceptor was ever present. Since the zero peak generally has an excitation intensity dependence and its magnitude is diminished either by flow or by the use of oxygen scavenging systems that extend time before photobleaching (see Chapter 4), it is likely that this peak arises from photobleaching and not incomplete labelling. The exact origin of this peak and its presence does have important consequences for the accuracy of diffusion spFRET measurements and will be discussed further in Section 2.5.7.

2.5.5 **Integration time and dynamic contributions**

Much of the discussion so far has concerned the use of spFRET in a system which has a dynamic equilibrium between two states, but without consideration of the effects of interconversion between these states. How this dynamic equilibrium manifests itself in a single molecule FRET experiment is closely linked to the integration time used [91]. It has already been discussed that the width of the measured distributions can give information about the timescale of the conformational changes occurring within the ensemble of similar structures in that state. For example, if the conformational fluctuations within the denatured ensemble are slow compared to the timescale of the measurement, then the resultant peak will be broadened because each molecule will maintain a slightly different structure as it traverses the PSF and so contribute a different measured signal to the proximity ratio histogram. If these structural fluctuations are rapid then the same average FRET value will be measured for each molecule and a narrower peak will be observed. We now extend this to consider the effect of conformational fluctuations between the states. Slow conformational transitions between two states would mean that each molecule, regardless of sample volume size, diffusion rate, integration time and burst averaging, remains either folded or unfolded (in this example) during measurement. The two individual species would be clearly resolved as two peaks in the histogram, assuming a sufficiently large difference between the mean proximity ratios of the two states exists. The widths of the individual peaks may then provide information on local conformational rearrangements *within* the state (so dynamics *within* the denatured ensemble or *within* the native ensemble). In the other limit, where the rate of conversion between the states (denatured ↔ native) is rapid, then during the transit of a molecule through the PSF, both states are observed multiple times and the same mean FRET efficiency is measured for every molecule. In a sense the resolution of the heterogeneity (folded and unfolded molecules) in the solution has been lost. However, the width of this single peak in the histogram can still contain information about this heterogeneity and in particular the timescale of the interconversion between the states; the faster the process, the greater the averaging and so the narrower the peak until the shot noise limit for that mean signal is reached. Clearly, intermediate situations arise. If the probability of a molecule folding or unfolding during an integration time is low, but non-zero, then broadening of peaks (folded or unfolded) in addition to that caused by shot noise may occur, the degree of broadening clearly reflecting the rates of interconversion between the states that exist. This discussion highlights the care that must be taken when interpreting these data. In particular, one must be careful not to simply apply concepts that are relevant for ensemble experiments to single molecule experiments. For example, consider the terms native and denatured ensemble and what these mean with respect to single molecule measurements. Some of the states in the denatured ensemble may well be indistinguishable, in terms of the distance between the dyes and therefore indistinguishable in terms of the proximity ratio, from native states! In the spFRET of the RNA molecule shown in Figure 2.20(a) the fact that the peaks are apparently resolved puts an upper limit on the rate of folding and unfolding of the hairpin loop. The integration time was 500 μs, so the rate of folding or unfolding must be significantly slower than 2000 s^{-1}. Note however that the exact manifestation of this effect in the histograms is complex and there is not a sharp threshold at which the effect does or does not manifest. One must consider the probabilities of a given molecule undergoing a transition based on the details of a given experiment (the kinetic rate constant, the integration time, the sample volume size and the rate of diffusion).

In the examples described so far the proximity ratio was calculated and a threshold was applied for *every* bin (integration time) in the data. Such an approach has advantages and disadvantages. One must remember that the data consists of bursts above the background that, even for small molecules, can persist for several integration times (see Figure 2.1). If the diffusion rates for two states of interest are significantly different then more bins with FRET proximity ratios corresponding to the slower diffusing species might be included, biasing the histogram. In this respect care must also be taken when drawing conclusions based on the widths of the spFRET peaks, it may be that the measurement time for a given molecule is actually shorter than the integration time (bin width) used. An alternative approach to this bin-wise analysis is to integrate the signal from all adjacent bins that correspond to the fluorescence burst from a particular single molecule (and to use short bins). In this case there is no risk of biasing with respect to the areas under the histograms. However, the measurement time for each molecule is no longer uniform and depends on the path through the PSF and the molecular diffusion coefficient, which may considerably complicate the discussion of dynamics and peak widths. In many studies the significance of these effects may be small and in many examples in the literature they are not taken into consideration. However, it is well worth being aware of these effects so that a judgement can be made, based on the rates of interconversion between states (and within states) and the particular experimental parameters.

2.5.6 **Studying dynamics with diffusion spFRET**

In an extension of the application of FCS to measure the dynamic behaviour of single labelled molecules (Section 2.4.4), the fluctuation in the donor and acceptor fluorescence signals due to dynamic changes in FRET efficiency can be used to measure the kinetics of the conformational changes that create these fluctuations. For slow kinetic processes it may be possible to simply measure many pseudo-equilibrium histograms as function of time (e.g. [92]). For fast processes however, such a methodology will not work. FRET, and other ratiometric approaches (e.g. [57]), have a significant advantage over FCS-based methods in that relative fluctuation amplitudes are not coupled to diffusion. Fluctuations in the *proximity ratio* are diffusion independent and so the somewhat convoluted measurement procedure described in Section 2.4.4, which removes the diffusion component of the FCS curve, is not required. Thus, in a manner directly analogous to single colour FCS, autocorrelation of the proximity ratio (equation 2.44) can potentially directly measure the observed rate constant for the conformational changes that cause the fluctuations in FRET efficiency. Klenerman and co-workers [86, 87, 93] introduced this method for studying the fast kinetics of FRET labelled molecules at equilibrium. The expression for the normalized autocorrelation of the proximity ratio is given as [87],

$$G_P(\tau) = \frac{\langle \delta P(t) \delta(P(t+\tau)) \rangle}{\langle P(t) \rangle^2} \qquad (2.48)$$

(a direct comparison can be made with equation 2.15 for FCS). The proximity ratio data $P(t)$ is formed in the same manner as the single molecule proximity ratio, however, no threshold is necessary because the signal comes from a 'few' molecules (typically the average number of molecules in the volume is of the order of 5). The calculated autocorrelation function can now be directly fit (excluding $G(0)$—see Section 2.4.1) with a single or stretched exponential (depending upon the particular kinetic scheme for the reaction/ conformational dynamics being probed) allowing determination of the observed rate constant. In addition, it can be shown that the amplitude of the autocorrelation function is related to the equilibrium constant for the reaction [86].

2.5.7 **Accuracy and other considerations**

Fundamental photophysical considerations in diffusion spFRET measurements

For FRET experiments, one must take care to choose dyes with the necessary spectral characteristics and suitable chemistry to allow conjugation to the host molecule. A detailed discussion of dyes for single molecule fluorescence is covered in Chapter 4. A review of common FRET dyes along with examples of their use and, where possible, R_0 values, is given in Table 4.2 (Chapter 4, Section 4.2.1). The R_0 value is perhaps the most important factor as this defines the distance range over which a given dye pair will be appropriate as a measure of conformational change. In fact the range of R_0 values for visible dyes suitable for single molecule spectroscopy is limited and usually close to 50 Å.

The R_0 value for a dye pair can be calculated from equation 2.39, which includes a term J for the spectral overlap between the absorption spectrum of the acceptor and the donor fluorescence emission spectrum. Assuming weak coupling (long range dipole–dipole energy transfer, as opposed to coupled electronic states) between the dye pairs it can be shown that [72]

$$J(\lambda)=\frac{\int \varepsilon_A(\lambda)f_D(\lambda)\lambda^4.d\lambda}{\int f_D(\lambda).d\lambda} \tag{2.49}$$

where $\varepsilon_A(\lambda)$ and $f_D(\lambda)$ are the normalized absorption/excitation spectrum of the acceptor (i.e. extinction coefficient) and fluorescence emission spectrum of the donor respectively (both in the absence of the partner) versus wavelength. Thus the overlap integral for any particular dye pair can easily be calculated if their absorption/excitation and emission spectra are well characterized (see Figure 2.22). If strong coupling is present energy transfer may still occur but will not described by the Förster mechanism. The result of stronger coupling is often manifest as a

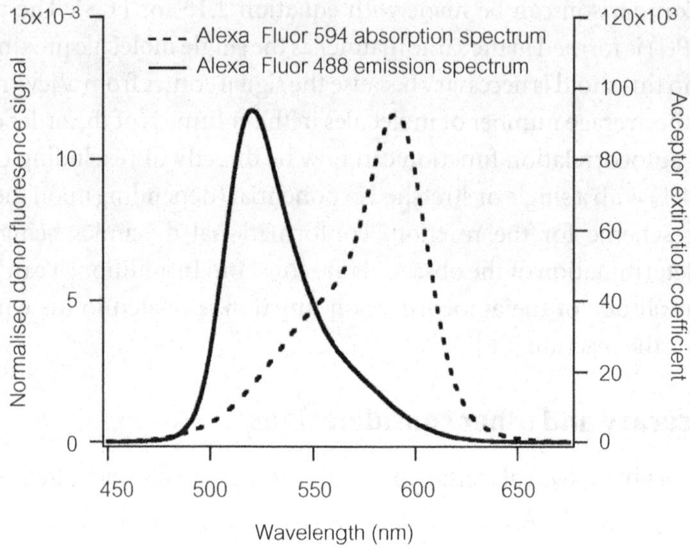

Figure 2.22 Donor emission spectrum (solid line) and normalized acceptor absorption spectrum (dashed line) showing the spectral overlap region. The normalized area of the overlap is calculated using equation 2.49.

change in the shape of the emission or absorption spectra of the dyes when they are interacting, compared with the spectra of the isolated molecules [74]. Whilst this does not necessarily preclude the use of the dye pair for single molecule measurements (one may only be concerned with using the dye pair interaction to create a binary structural marker), it does render any attempt at analysis using the Förster formulism invalid.

For an accurate determination of R_0 and hence R, the final, non-trivial, unknown parameter in equation 2.39 is the orientation factor κ^2. FRET is a dipole–dipole interaction and so the efficiency of FRET between two dyes is strongly dependant on the relative orientations of the absorption and emission dipoles of the acceptor and donor respectively (μ_A and μ_D in Figure 2.17). The orientation factor is given as [72];

$$\kappa^2 = (\cos\theta_T - 3\cos\theta_D\cos\theta_A)^2 \tag{2.50}$$

where θ_T is the angle between the emission dipole of the donor and the absorption dipole of the acceptor and θ_D and θ_A are the angles between the vector joining the two molecules and the molecules emission and absorption dipoles respectively. The minimum value of $\kappa^2 = 0$ occurs for dipoles perpendicular to each other (no energy transfer) and the maximum value $\kappa^2 = 4$ for dipoles that are parallel and aligned. In the case of freely rotating dyes, rotational averaging of the relative dipole alignments occurs (if rotation is on a timescale fast enough compared to the measurement) and in this limit, $\kappa^2 = 2/3$. It should be noted that this value is not itself indicative of rotational averaging; it is possible for fixed non-rotating dipoles to have an orientation that yields this value. In general, single molecule FRET experiments are, fortunately, not concerned with absolute distances between dye pairs, but rather with changes in FRET efficiencies. Therefore, providing κ^2 does not change significantly, relative measurements are possible without a knowledge of its actual value. Dyes are often attached via long (several angstrom) linkers that posses significant flexibility (linkers are generally saturated carbon-carbon chains). These long flexible linkers hence introduce rotational averaging (see Chapter 4). A possible drawback of this is that these linkers can introduce fluctuations in the inter-dye distance that, whilst generally on a timescale much faster than that of the measurement, could result in a degree of broadening of the measured FRET distribution for a molecule that is otherwise rigid. A common experimental approach to quantifying the rotational freedom of the dye pair is steady-state or time-resolved fluorescence anisotropy (see Sections 2.7.5 and 2.7.6 and [76]). Anisotropy values less than 0.1 suggest sufficient rotational freedom to apply $\kappa^2 = 2/3$ [78]. In fact the errors introduced by incomplete averaging are somewhat small [74]. For example, in the case of rotational averaging only occurring in one of the dyes (the other fixed rigid), then the estimated error in R_0 is only ~10% [74]. Of more concern is proving that any heterogeneity that is identified is not just caused by, for example, a proportion of molecules in which the dyes have become pinned thereby restricting or slowing rotational averaging. Only careful analysis or simultaneous single molecule polarization measurements can really exclude this possibility [79, 85].

Zero peak consequences

As discussed in Section 2.5.4, many diffusion spFRET studies result in a peak generally attributed to molecules lacking acceptor dye or to premature bleaching of the acceptor dye and this peak is then simply dismissed in the subsequent analysis. Importantly, this approach is only valid if all the species in the solution contribute in proportion to the zero peak. As discussed in Section 2.5.5, it is clear that the diffusion rate of molecules could at the very least bias the population in the zero peak. For example, if some species show much slower diffusion, or are significantly more fluorescent than another species, it is clear this population is likely to be photobleached sooner. Similarly, if experiments are carried out on dilute samples, without flow, it is possible that photobleached molecules will be re-measured and this effect will increase with time. Elegant, if complex, methods have been used to combat this effect. For example, Kapanidis *et al.* [84, 85] used alternating laser sources to first excite the donor and determine the distance dependant FRET ratio, but then switch to another light source to excite the acceptor directly, confirming its presence immediately after the FRET measurement. Molecules without viable acceptor dye are ignored and in this way the energy transfer efficiency measurements are only recorded for species with active donors and acceptors. Removal of the zero peak in this way allows species with low FRET efficiency (proximity ratios <0.4) to be observed without interference from the zero peak.

Background events

Even with the most careful protocols it may be impossible to remove all contaminants from a solution that produce a signal which may be mistaken for valid single molecule events. Of particular concern are small fluorescent molecules, free fluorescent dye and particulates that can scatter large amounts of the excitation light. Control experiments involving the solutions without the fluorescence molecules of interest should always be carried out in order to minimize and fully characterize the background contribution. Further details on sample preparation protocols that have been found to be satisfactory are presented in Chapter 4.

2.5.8 **Applications of diffusion spFRET**

Abundant examples of diffusion spFRET can be found in the literature. Indeed it is in this area that single molecule fluorescence techniques have arguably proved the most useful. In Chapter 5 we review three papers in detail but here we present a very brief review of a broader range of spFRET experiments.

Deniz *et al.* [82] applied spFRET in one of the first studies that demonstrated its use as a potential structural probe revealing heterogeneity of proteins in solution. Chymotrypsin inhibitor 2 (CI2) was FRET labelled with the dye pair TMR (tetramethylrhodamine) and Cy5 (see Chapter 4). Single molecule diffusion experiments were then performed as a function of chemical denaturant. The folded and unfolded subpopulations of the protein at various denaturant concentrations were resolved and the populations of each state (estimated by the relative area of each peak) were shown to be in broad agreement with the supporting ensemble measurements. Further, changes were seen in the mean FRET efficiency of the unfolded distribution with increasing denaturant.

One limitation of diffusion spFRET clearly seen in studies of proteins is that one requires that the states are populated for a significant time. One cannot practically study the denatured state, for example, in native buffer conditions where the population is less than a few percent. In principle, molecules in this state are measured and recorded individually but practically they would be lost in a histogram dominated by measurements of orders of magnitude more folded molecules. With this in mind, Lipman and co-workers [94] coupled single molecule FRET detection of protein molecules with a micro-fabricated fluid mixer in order to study states not populated at equilibrium. In this way, after mixing to dilute a chemical denaturant, the denatured population of the cold shock protein from the bacterium *Thermotoga Maritima* could be probed in native solution conditions. Further, they were able to follow the time course of the conversion of single unfolded molecules to the native state. One striking (but expected) observation of this study is the increased compactness of the unfolded state in the absence of denaturant, compared to the state in the presence of mild chemical denaturant.

As well as applying spFRET to structural investigations of proteins the technique has also been applied to nucleic acids. Pljevalcic *et al.* [90] measured spFRET of various mutants of single hairpin ribozymes as a function of Mg^{2+} concentration. In particular, they examined strong and differential broadening of the histogram peaks corresponding to the 'native' population. In this way the authors are able to suggest a putative folding pathway. In a paper that extends spFRET to a mutliparameter methodology (which also negates zero peak effects) Kapanidis and co-workers [84, 85] demonstrate how a FRET labelled system can be used to study stoichiometry and molecular disassociation between a protein and a nucleic acid. In this study alternating laser excitation is used, first the donor is excited in a typical FRET experiment, then this excitation source is switched off and a second source that excites the acceptor directly is turned on (see [85] for more information on this technique referred to as ALEX). The calculated FRET efficiency then gives a structural probe and a second, distance independent ratio *S*, is calculated from the acceptor signal (after direct acceptor excitation) and the donor signal corrected for FRET. This ratio gives information about the abundance of the two fluorophores and therefore provides a probe of stoichiometry. The advantage of the factor *S* is that it can provide important information about interactions, local environments and stoichiometry even for large inter-dye separation when FRET does not occur. In this study the methodology is exploited for a molecular sorting application which is termed FAMS—fluorescence aided molecular sorting. The authors first examine ideal DNA ladders (DNA constructs with donor only, acceptor only and donor and acceptors at different relative separations). Two-dimensional histograms of *S* and FRET efficiency then allow all the species to be unambiguously resolved. The authors then go on to demonstrate the power of the technique by studying the sequence specific interaction between acceptor-labelled DNA and donor-labelled catabolite activator protein (CAP) from *E. coli*, resolving all bound and unbound populations.

2.6 Measurements of immobilized single molecules

So far in this chapter we have been concerned with the statistical tools to extract information from experiments in which the molecules are freely diffusing in solution. FRET, PCH and FCS experiments all involve the measurement of many single molecule events. Clearly such experiments represent a significant step towards obtaining single molecule information and provide insights into heterogeneity and kinetics, despite a certain amount of averaging, that are not available from ensemble measurements. For example, consider a comparison of the mean proximity ratio from single molecule and ensemble measurements of N molecules,

$$P_{\text{Single Molecule}} = \frac{1}{N}\sum_{i}^{N} \frac{I_{\text{A}}^{i}}{I_{\text{A}}^{i}+I_{\text{D}}^{i}} = \left\langle \frac{I_{\text{A}}}{I_{\text{A}}+I_{\text{D}}} \right\rangle \tag{2.51}$$

$$P_{Ensemble} = \frac{\sum_{i}^{N} I_{A}^{i}}{\sum_{i}^{N} I_{A}^{i} + \sum_{i}^{N} I_{D}^{i}} = \frac{\langle I_{A} \rangle}{\langle I_{A} \rangle + \langle I_{D} \rangle}$$

and it is well known [76] that,

$$\left\langle \frac{I_{\text{A}}}{I_{\text{A}}+I_{\text{D}}} \right\rangle \neq \frac{\langle I_{\text{A}} \rangle}{\langle I_{\text{A}} \rangle + \langle I_{\text{D}} \rangle}. \tag{2.52}$$

The short observation time for each molecule in diffusion-based experiments (perhaps no longer that 1 ms) means that extracting real time kinetic information from *one molecule* is difficult. As we have seen, both FCS and FRET can provide mean kinetic rates, although the faster the kinetics with respect to the average diffusion time, the more easily and reliably the kinetics can be measured. Measurement of freely diffusing molecules does have its advantages. For example, this approach enables diffusion coefficients to be determined and provides the ability to conduct experiments under near-physiological conditions without the influence of surfaces or immobilization protocols. However, an exciting prospect for single molecule experiments is the potential for studying an individual molecule for an extended period of time by virtue of it being immobilized on or near a surface.

Immobilization of molecules in polymer films [95], directly to solid surfaces [96–100], in high water content gels [101], in cells [102–104] or liposomes [105,106] has enabled experiments to be carried out that monitor the fluorescence from a single molecule for several seconds and longer (see Chapter 4, Section 4.6 for a detailed review of immobilization protocols). Shown in Figure 2.23 is an example of the simplest data type from immobilized single molecule experiments. Using a scanning confocal or total internal reflection

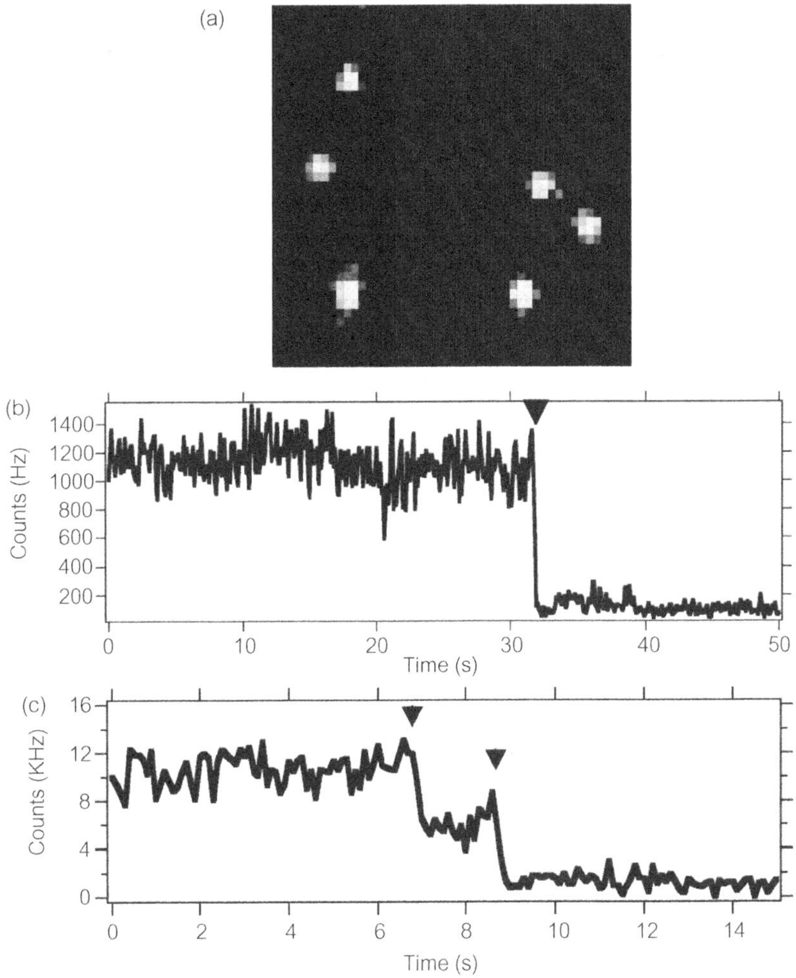

Figure 2.23 Three examples of immobilized single molecule fluorescence data. (a) Total internal reflection fluorescence microscope (TIRFM—see Chapter 3) image of a glass surface with adsorbed fluorescently labelled DNA molecules. (b) A typical trajectory following the time course of fluorescence from singly labelled monomeric immobilized molecules. The fluorescence persists for approximately 30 s until the single dye molecule photobleaches. This measurement was conducted using a scanning confocal instrument on the dye tetramethylrhodamine in a polymer film. (c) A trajectory for a potentially multimeric nucleotide system adsorbed onto glass. Each monomer was labelled with a single dye (Alexa Fluor 488). The two-step bleaching (arrows) is indicative of a dimeric complex. This trajectory (c) was calculated from a dataset obtained using TIRFM.

fluorescence microscope (SCM and TIRFM respectively; see Chapter 3) images of a sample immobilized onto a glass surface are straightforwardly obtained. Part of one such image of immobilized molecules is shown in Figure 2.23(a). This image was recorded using TIRFM and shows singly labelled DNA duplexes immobilized onto a glass surface using bitoin—avidin chemistry (see Chapter 4).

2.6.1 **Information contained in fluorescence intensity trajectories from single molecules**

If TIRFM is used then sequential image recordings are performed and the fluorescence intensity time trajectories of individual molecules can be plotted. In SCM raster scanning of the surface can be used to create images such as that shown in Figure 2.23(a), however, the scanning rate is generally slow so sequential scanning would result in very poor time resolution. Therefore, in the SCM configuration, images are not usually recorded, rather the sample is raster scanned until a molecule is detected by the fluorescence signal exceeding a threshold level. A fast algorithm then quickly searches the local region and centres the confocal sample volume on the molecule and the fluorescence versus time trajectory can be recorded with very high time resolution. Indeed the time resolution of scanning confocal measurements performed in this way generally surpasses that obtained in TIRFM (due to current CCD readout speeds—see Chapter 3), although TIRFM can generate large amounts of data more quickly as several single molecules can be monitored at once. An algorithm for fast localization of single molecules using a scanning confocal system is described in [107]. An example of a trajectory from an immobilized experiment is shown in Figure 2.23(b). The measurement is started and a continuous fluorescence signal is recorded (with 100 ms integration time) until a single-step, irreversible, photobleaching event is encountered (arrow). It can be seen that the fluorescence emission persists for a considerable time (>30 s) and the single-step bleaching event provides confirmation that a single molecule is being observed (as does the signal level or the size of the feature in an image). Simple measurements such as this can reveal the photobleaching lifetime of molecules in different environments or, in more complex systems the disappearance of fluorescence might be attributed to disassociation of the molecule from the surface [102, 103] (see a review of one of these papers in Chapter 6). These simple, single colour datasets can also be used to measure the oligomeric state of multimeric complexes by simply counting photobleaching events to determine the oligometric state [108–110]. As an example see Figure 2.23(c), where two bleaching steps are clearly identifiable (arrows). This data was generated using a TIRFM image series of singly labelled RNA molecules that are able to form multimeric complexes which were non-specifically absorbed to a glass surface under water. Single colour trajectories from immobilized molecules are limited in the information that they can provide. In Figure 2.23(b) and (c), fluctuations are clearly seen in the trajectory before the bleaching steps. Such fluctuations can be attributed to a number of sources including triplet population, conformational fluctuations of the host molecule (which modulates the fluorescence through differential quenching) and shot noise. Conformational fluctuations might be of particular interest, but in single colour data it is difficult to isolate conformational contributions from the other sources of fluctuation. In addition, due to the compromise between signal-to-noise, dynamic range of the detector (see Chapter 3) and time before photobleaching, even the identification of bleaching steps is not always clear (although various low-pass filtering schemes can be applied, with care, [111] to help identify such features).

Studies of immobilized and FRET labelled systems provide one way to decouple these effects. In these experiments two colour images, corresponding to donor and acceptor dye labels, are recorded simultaneously from the same area of the sample surface (see Chapter 3 for instrumentation details). Figure 2.24(a) shows a protein labelled with the dyes Alexa Fluor 488 and Alexa Fluor 594 absorbed

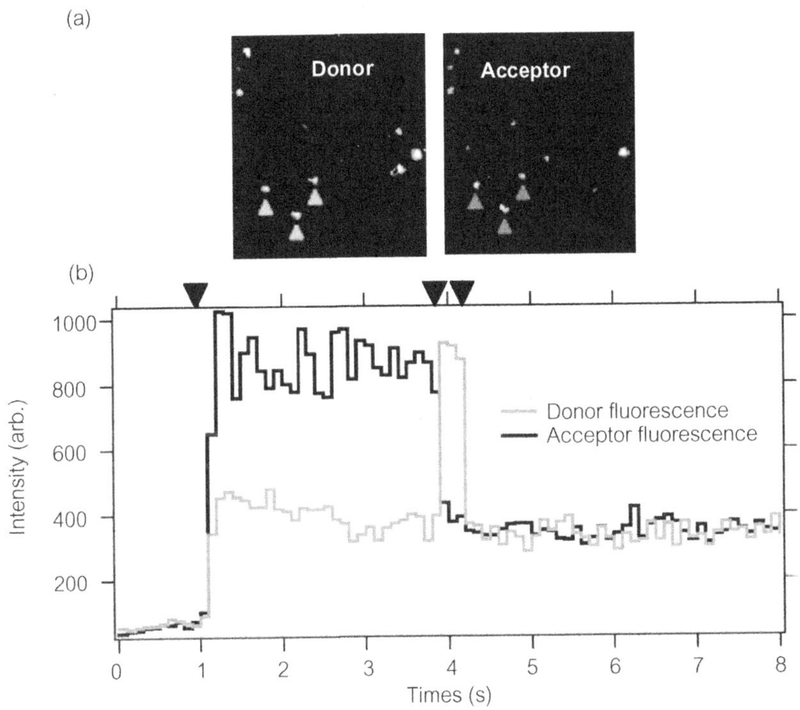

Figure 2.24 (a) Total internal reflection fluorescence microscopy (TIRFM) images of a protein doubly labelled for FRET and non-specifically adsorbed onto a glass slide under water. The two images show the same area of a sample recorded simultaneously for green (donor) and red (acceptor) fluorescence. Spatially correlated features can be observed (some highlighted by arrowheads). (b) Intensity trajectory measured from a TIRFM recorded image sequence. The anticorrelation of donor and acceptor fluorescence is clearly observed; the first arrowhead indicates the point at which the laser was turned on. At this point significant red (acceptor) fluorescence is measured indicating near 100% FRET, after approximately 4 s the acceptor dye molecule photo bleaches (arrowhead) and the green (donor) fluorescence increases. Shortly afterwards the donor fluorescence ceases as this molecule bleaches.

onto glass. Fluorescence intensity trajectories can be calculated for the donor and acceptor dyes from the same molecule from these images. Fluctuations due to conformational changes which cause changes in the dye separation produce anti-correlated features in the donor and acceptor fluorescence trajectories. Importantly, this allows one to decouple fluctuations due to structural changes from noise or quenching of one of the dyes, which is not anti-correlated. An extreme example of this anti-correlated behaviour is shown in the fluorescence intensity trajectory shown in Figure 2.24(b). In this example the excitation light is switched on after approximately 1 s (first arrow), at which point strong acceptor fluorescence and no donor fluorescence is seen (in this case the FRET efficiency is near 100%). After some time the acceptor dye bleaches (second arrow) and, instantaneously, the donor fluorescence recovers (there is now no energy transfer partner for this molecule). Shortly afterwards, the donor also bleaches (third arrow) and the signals for this location, in both channels, stay at the background level. Despite the limited use of this particular example, it does indicate the potential of this simple data: following conformational changes or binding in a FRET labelled system. In reality one would discard this particular data in such a study; in order to assign a transition unambiguously to a conformational change one would have to either see multiple transitions or engineer the sample so that the high and low FRET states were not 100% and 0%, thus confirming that the effect seen is not simply acceptor photo bleaching (as in this case). Such an example as this is shown in Figure 2.25 [80]. Here two trajectories for the donor and acceptor signals measured for a protein (adenylate kinase) encapsulated in a tethered liposome are shown (a and c) along with the calculated resulting FRET efficiency traces (b and d). In the first trace (a and b) the molecule returns to the starting state, indicating a structural change rather than a simple photobleaching event. This experiment is reviewed fully in Chapter 6.

The information that can be obtained from immobilized single molecule data is varied and depends very strongly on the timescale of the processes being probed. If we focus on conformational changes between ensembles in a two-state system, then we might expect to see transitions between just two FRET efficiency levels, but only if chain reconfiguration is much faster than the integration time of the measurement.[6] Shown in Figure 2.26(a) are data from such an experiment [112]. Here the FRET efficiency trajectories for single ribozyme molecules are plotted. Multiple transitions in a single molecule are seen between two states with the same FRET efficiency and are thus assigned to structurally similar states which in this case are a catalytically active docked and a non-active open conformation of

[6] We must acknowledge that 'two-state' is an ensemble concept. If we accept that a given single molecule does not 'tunnel' from one conformation to another then the number of states we observe a single molecule in depends, ideally, only on the integration time used.

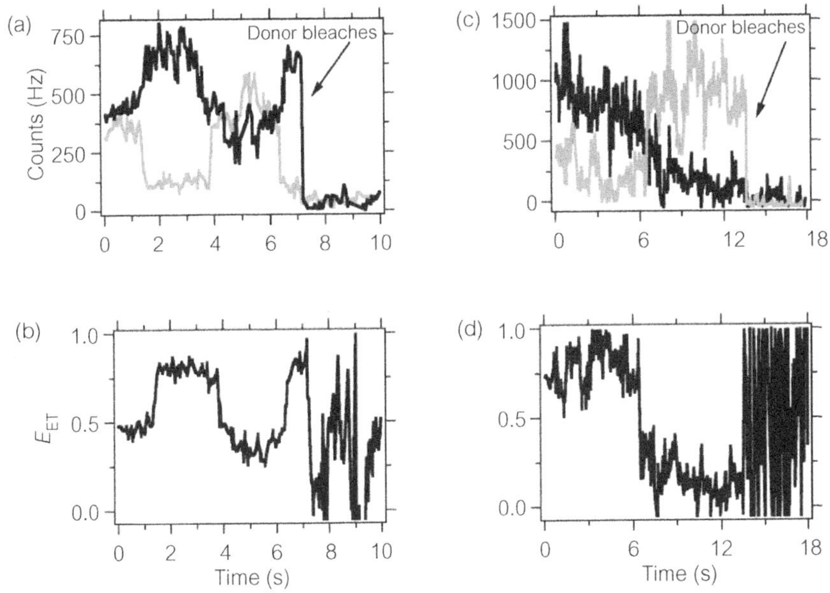

Figure 2.25 (a and c) Time traces of individual FRET labelled, vesicle-trapped, adenylate kinase protein molecules under mid-transition conditions with the acceptor signal in grey and the donor in black. The traces were collected with 20 ms time bins. (b and d) FRET efficiency (E_{ET}) trajectories calculated from the signals in (a) and (c), respectively. In (a) and (b) several transitions occur between states that are essentially within the 'folded' ensemble, whereas in (c) and (d) a single transition takes the molecule from the folded to the 'denatured' ensemble. Note that transitions can be strictly recognized by an anticorrelated change in the donor and acceptor fluorescence intensities as opposed to uncorrelated fluctuations sometimes appearing in one of the signals. (Reproduced from Rhoades *et al.*, Watching proteins fold one molecule at a time. *Proceedings of the National Academy of Sciences of the United States of America*, 100 (2003) 3197–3202 with permission from National Academy of Sciences, USA (Copyright 2003)).

the ribozyme. The catalytically closed conformation was assigned to the high FRET signal and the open, unfolded conformation to the lower FRET efficiency. One powerful feature of the data is that the docking (folding) and opening (unfolding) rate constants can be obtained directly from these traces by construction of histograms of the occupancy times in the two states (see Chapter 6 for a thorough discussion of this data). Figure 2.26(c) shows such a calculation where the histogram of the dwell times (length of occupancy) from many tens of molecules in the docked state was fitted with a multi-exponential curve, revealing kinetic pathways that were previously obscured in the equivalent ensemble experiments. Another striking aspect of these trajectories that can be seen in Figure 2.26(a) is that individual molecules tend to undergo transitions to and from docked states with similar dwell times (from which similar stabilities and structures for each docked state can be inferred) but different molecules seem to sample docked states with different stability. This is an elegant demonstration of

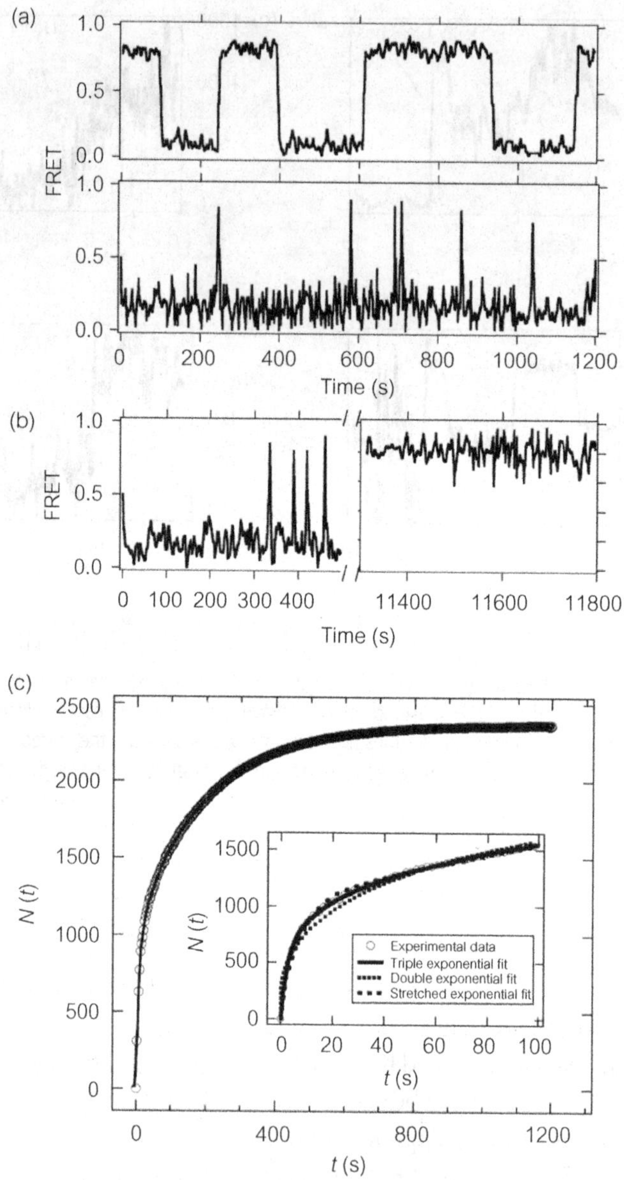

Figure 2.26 (a) FRET—time trajectories of single ribozyme-substrate complexes. The traces illustrate the change in FRET efficiency on going from docked to undocked states. Of particular interest is the strong 'memory' effect observed in the un-docking kinetics. (b) An example of memory loss during a long measurement. The excitation laser was shut off for 3 h around 500 s into the measurement to allow this very slow turnover to be probed despite photobleaching. (c) Histogram of dwell times in the docked state thus following un-docking kinetics, measured from trajectories such as those in (a) and (b). The kinetics are complex and are best fit with a triple exponential–inset shows a comparison of the earlier part of the histogram fit with other analytic forms. Reprinted with permission from Zhuang *et al.*, Correlating structural dynamics and function in single ribozyme molecules. *Science* 296 (2002) 1473–1476. Copyright 2002 AAAS.

static heterogeneity which Zhuang and co-workers described as a 'memory' effect. Even more striking is that the molecules also display slow dynamic heterogeneity. Figure 2.26(b) shows a molecule that samples one docked state for a considerable length of time until it 'forgets' its preferred state and then occupies an extremely stable docked conformation (note the time lapse nature of this experiment, after 400 s the laser was turned off for 3 h to prevent photobleaching, before being turned on again to probe the kinetics).

As well as observing transitions between states assigned at the macroscopic level to two ensembles, measuring the associated rate constants and identifying heterogeneity, single molecule trajectories also provide the possibility to directly study the reaction pathway between states (these transitions appear as near vertical lines in Figure 2.26(a)). Studies of states that are stable and long lived at equilibrium (such as those in Figures 2.25 and 2.26) provides information about energy barriers between states. Ideally we would also like to follow the reconfiguration of the chain through the transitions. The time resolution of current instrumentation makes this difficult, although attempts have been made to reduce the integration times to as little as 100 μs which allows an upper limit to be placed on the temporal width of the transition between two structurally distinct ensembles of states [113]. This highlights a strength of immobilized single molecule methods: the ability to follow a single molecule over a long period of time and identify rare states that are not significantly populated and cannot therefore be seen in ensemble equilibrium or kinetic measurements. Whilst ensemble techniques such as temperature jump [114], rapid mixing [115], stopped flow kinetics and other spectroscopic methods [116, 117] have been able to identify or infer intermediate states along the reaction coordinate, or indirectly probe the transition states, single molecule methods present an exciting opportunity to directly probe these structures. Unfortunately, states difficult to see by ensemble methods can also be hidden in single molecule studies: the fact that they are not significantly populated generally means that they are populated transiently. Consequently, diffusion-based single molecule studies may be inappropriate and in immobilized single molecule trajectories the comparatively long integration times (generally >10 ms) may mean these states are missed, or transitions to these states will be rare and so not be identified as statistically significant in analysis. However, in ensemble measurements the fact that a state is stable, even for long periods of time, does not necessarily imply that it can be observed—all the molecules in the ensemble may take different paths in conformational space from one 'ensemble' to another, such that any individual stable state is never significantly populated.

For complex molecules there are often intermediate states along the pathway from one equilibrium state to another that are populated for long enough such that they may be revealed in immobilized single molecule studies. Such an example is shown in Figure 2.27 (a–c) [80] for a FRET labelled protein in which many small, stepped transitions between many different FRET efficiency levels (inferred conformational states) are seen as well as some rapid transitions between distal states (the native and denatured ensembles). These are some of the

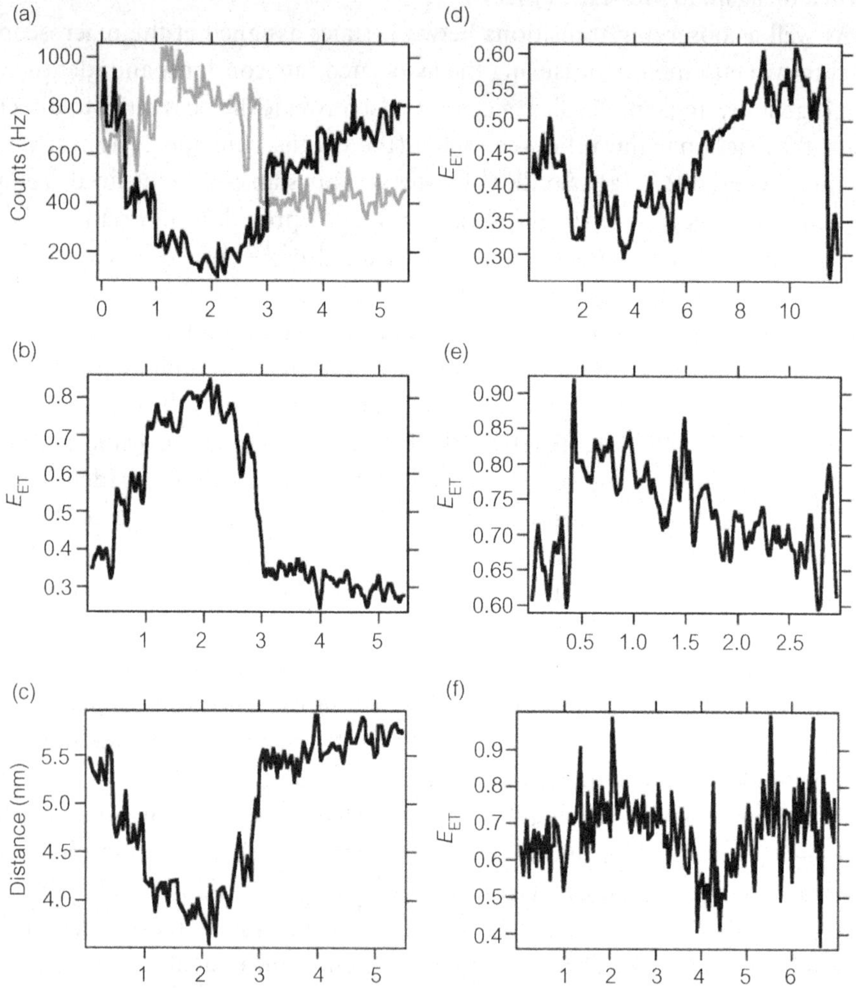

Figure 2.27 Time-dependent signals from single FRET labelled protein molecules showing slow (resolved) folding or unfolding transitions. (a) Signals showing a slow folding transition starting at around 0.5 seconds and finishing at around 2 seconds. The donor signal is shown in black and the acceptor signal in grey. (b) FRET efficiency trajectory calculated from the signals in a. (c) The inter-probe distance trajectory calculated from b. (d–f) Additional FRET efficiency trajectories demonstrating slow transitions. (Reproduced from Rhoades *et al.*, Watching protein fold one molecule at a time *Proceedings of the National Academy of Sciences of the United States of America* 100 (2003) 3197–3202 with permission from National Academy of Sciences, USA (Copyright 2003)).

most impressive single molecule observations to date and suggest many interesting questions, particularly if this diversity of behaviour can be linked to function for physiologically relevant molecules.

Finally, it is worth considering if fluorescent probes will ever allow detailed observation of the complete trajectory of chain reconfiguration between states in biological macromolecules. A typical fluorescence lifetime of a dye molecule is around 5 ns and when the absorption cross section of such molecules is taken into account then even with a 100% efficient detection system one could never exceed photon count rates of more than one photon every few tens of nanoseconds. This simple calculation also ignores shot noise and its effect on the number of photons necessary for a statistically meaningful analysis. It also ignores the fact that spFRET experiments only probe the structural changes in distance between two points on the chain and are insensitive to long (>100 Å) and short (<20 Å) changes. This time and distance resolution (among other problems) is therefore unlikely to allow us to follow the chain reconfiguration completely with current methodologies [75].

2.6.2 **Practical considerations when studying immobilized single molecules**

Obtaining the data

In contrast to the complex analysis algorithms and very large data sets required for PCH, FCS and spFRET experiments on freely diffusing molecules, only relatively small numbers of molecules are typically analysed in experiments in which the molecule is immobilized. Extracting the lifetimes of molecular states and the construction and fitting of histograms from the data is therefore relatively trivial if somewhat tedious unless well automated. The analysis can be split into a number of stages:

(1) identification of molecules to analyse,
(2) recording of information (position, trajectory etc.),
(3) processing of the trajectory (noise reduction),
(4) extraction of parameters (dwell times in states, step sizes, transition widths etc.).

Mashanov *et al.* [102] propose a method to discriminate single molecules in a series of images before accepting the fluorescence data for further analysis (this method is appropriate to post-collection analysis of images measured by SCM or TIRFM). They apply criteria referred to by the acronym DISH. If a fluorescent feature in the image is a single molecule then it must be Diffraction limited in size and have an Intensity that corresponds to that expected for a single molecule. Additionally the feature must demonstrate Single-step photobleaching and have a Half-life that is proportional to the laser power used.

In an SCM configuration it may be desirable to 'find' molecules, based on various criteria which must be done as quickly as possible to maximise the data that can be collected before photobleaching occurs, at which point the next molecule is sought. Such a configuration can simply be realized by scanning the sample and applying a threshold level on the detected signal that suggests a molecule has been found, followed by a fast iterative algorithm to locate the centre of the molecule and proceed with the measurement of the trajectory [107]. Further discussion of this aspect of instrumentation along with some details of the type of detectors and other elements used to record the data can be found in Chapter 3.

Once molecules have been identified, recording of the intensity trajectory is a fairly straightforward process. The integration time must be carefully chosen in the context of the noise levels and the timescale of the fluctuations that are of interest, which are highly system dependant. Indeed it may be necessary to record data on a variety of different timescales in order to be sure to identify all states (e.g. data in [112] were measured with integration times of 2 s or 0.1 s, probing rates of 0.001 or 0.02 s^{-1}, respectively). The number of observations that should be made (transitions per molecule, or molecules) varies greatly (and may be restricted by the system). One of the motivations for single molecule experiments is to make possible the study of rare events lost in ensemble measurements. The number of measurements required will be dictated by the statistical certainty with which a rare event can be identified. For example, if one is interested in rate constants and kinetic mechanisms, then for a system with single exponential kinetics (e.g. transitions between two states) typically 30–100 transitions are necessary to describe the single exponential and return rate constants with acceptable errors (~10%) [118]. The situation is more complex for measuring multiple transitions between heterogeneous states defined by multi-exponential kinetics. While little data may be needed to demonstrate kinetics are not single exponential, the task of obtaining enough transitions to fit data and extract multiple rate constants is challenging, especially if the population of some of the states is low (see [112] where up to three rates were demonstrated).

Measurement of the dwell times and the subsequent calculation of kinetic rate constants is perhaps the simplest form of analysis on single molecule trajectories. Indeed, if the data is suitable with clear transitions between two well-defined ensembles of states, then this type of analysis might even be considered routine; this simple data extraction is used powerfully in [80, 112, 119]. However, arguably more interesting data might originate from trajectories where multiple transitions or heterogeneous step sizes occur (and so transitions are not so distinct), or where the integration time or signal level was low (in order to increase time resolution or time before photobleaching). In these cases the trajectories can be extremely noisy so transitions are masked, at least to the naked eye, and some kind of noise reduction algorithm (post-processing) is necessary. An excellent discussion of this can be found elsewhere [111] and we briefly review it here. The task of identifying transitions and steps is one that has already been approached in a number of other fields, including single ion-channel activity measurements, in which several methods for identifying steps in noisy data have been successfully applied [120]. Gilad Haran [111] has applied and extended these methods to filter data from single molecule FRET experiments of protein folding. A filter in this sense should be designed to remove noise without averaging out features in the trajectories that are real. For example, in the case of data of the type shown in Figure 2.27 we do not want to average out the small step transitions. The simplest type of filter available for this purpose is perhaps the running average filter (RAF). In this filter (equation 2.53) the intensity at some point in the averaged signal is calculated from the average of the N intensities that precede or follow it (backward and forward predictors respectively).

$$\begin{aligned} \langle I_{\text{forward}}(t)\rangle &= \frac{1}{N}\sum_{j=i-N}^{i-1} I(j) \\ \langle I_{\text{backward}}(t)\rangle &= \frac{1}{N}\sum_{j=i+1}^{i+N} I(j) \end{aligned} \tag{2.53}$$

More sophisticated averaging algorithms make use of combinations of both backward and forward predictors with various lengths (number of points behind or in front that are averaged, N). Weighted sums of the various predictors are then calculated so that averages are not taken when transitions occur in that window [111]. Haran employs this methodology for spFRET measurements and uses the *a priori* knowledge that 'real' transitions will be anticorrelated in the donor and acceptor trajectories. This is incorporated into the weights used on the predictors, increasing the efficiency and reliability of the filter [111]. The efficacy of this analysis method is illustrated in Figure 2.28 where the filter is applied to simulated data for a system in which the FRET signal changes abruptly between three signal levels. The improvement in resolution obtained by the application of the Haran filter can be seen by comparing the unfiltered trajectory on the left to the filtered on the right (a–d). The histograms of the FRET efficiency values from the traces clearly show that the discrimination of three FRET levels (Figure 2.28(f)) is not possible from a histogram of the unfiltered data (Figure 2.28(e)). Once data has been processed without biasing it then remains a fairly trivial exercise to extract useful parameters from the data sets such as the number of transitions between states, the dwell time in a particular state and the number of states.

An alternative analysis method involves autocorrelation of the trajectories in order to identify and determine the timescale of fluctuations within a given

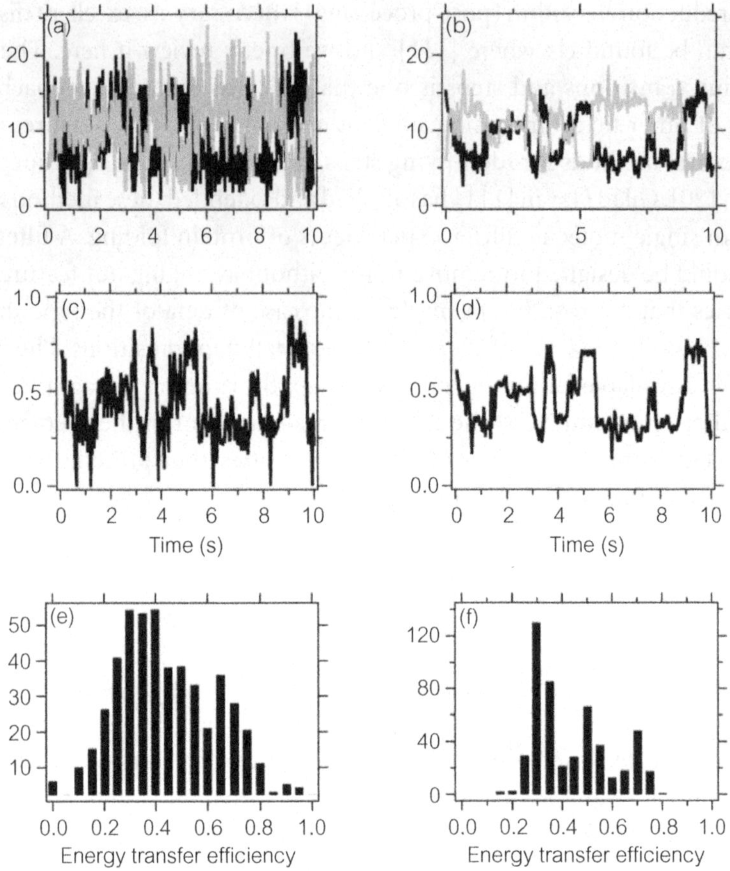

Figure 2.28 Simulation of a three-state single-molecule protein folding experiment in which the FRET value changes abruptly between 0.3, 0.5 and 0.7. The overall count rate is 1000 Hz. (a) Simulated data, acceptor in black and donor in gray. (b) Simulated data after filtration with the filter. (c) FRET efficiency calculated from a. (d) FRET efficiency calculated from b. (e) Histogram of the FRET efficiency values of c. (f) Histogram of the FRET efficiency values of (d).(Reprinted from Haran, G, Noise reduction in single-molecule Fluorescence trajectories of folding proteins. Chemical Physics 307 (2004) 137–145. (Copyright (2004) with permission from Elsevier.))

ensemble, such as the denatured state of a protein [91, 111]. If sufficient transitions could be measured then autocorrelation of the entire trajectory could be used in order to quantify the kinetics both within and between ensembles of similar conformations (see Section 2.4). Such analysis is not straightforward because the correlation functions can be dominated by other fluctuations, such as shot noise, as has been discussed. Methods to isolate the useful fluctuations *via* filtering of the power spectrum of the trajectory have also been proposed [91].

Photophysical considerations

Experiments with immobilized molecules suffer photophysics related problems too, many of which are analogous to the difficulties seen in diffusion single molecule measurements that require careful and detailed statistical analysis (see Chapter 4 for a further discussion of dye photophysics). The fluorescence trajectories are dominated by the photophysical properties of the dyes and careful analysis is required so that the changes seen in the trajectories can be assigned reliably to conformational changes (for example) and not the photophysics of the dyes. Generally, integration times greater than 1 ms are employed and thus triplet crossing and rotational (dye-linker) effects are averaged out. Noise reduction in FRET data can help considerably (see the previous section), however, photobleaching of the dye label can be a significant problem since it can limit the total observation time biasing the experiment towards more rapid kinetic phases that are observed more often before bleaching. In order to overcome this and allow observation of slow changes a number of studies have used periodic illumination [102, 103, 121] where the laser illumination is only provided for a fraction of the total experimental time. Such methodologies must be used with care otherwise rate constants, for example, may not be calculated correctly as some events may be missed.

The use of spFRET removes many of the noise sources from single molecule fluorescence trajectories: requiring that the donor and acceptor signals be anticorrelated as has been discussed. Unfortunately, this does not magically remove photophysical effects or shot noise that cause uncorrelated changes and these should be minimized if possible. The use of filters in post-processing can reduce these contributions and shot noise can in addition be minimized by increasing the measured signal by increasing laser power or integration time, although this has disadvantages regarding the total observation time, temporal resolution and triplet population of single molecules. Of particular concern, especially in studies that involve naturally fluorescing proteins (such as GFP (green fluorescent protein)), are so-called blinking events—the transition of the fluorophore into non-fluorescent states caused by, for example, structural changes (photo-induced isomerization) [49, 122, 123].

2.6.3 Analysis and application of immobilized single molecule experiments

In Chapter 6, we review a number of studies that illustrate the many ways in which single molecule fluorescence trajectories may be manipulated to give insight into many types of behaviour. Immobilized single molecule fluorescence experiments are now receiving considerable attention in the literature and new and exciting ways of applying the techniques are constantly emerging. Unlike diffusion single molecule methods, studies of immobilized molecules require much less complex (although no less rigorous) statistical analysis, as we have illustrated in this chapter. Common tasks such as generating trajectories from image sequences can often be accomplished in commercial image analysis software, versions of which are often supplied with CCD cameras (which is the detector type used most commonly for these types of experiments). Construction of histograms of dwell times, application of noise reduction algorithms and determination of step sizes in these trajectories can be achieved most efficiently with custom analysis environments created in data analysis software packages such as Igor Pro (Wavemetrics Inc., USA). Indeed, such software can also often interface with the instrumentation used for the measurement and so provide an integrated solution to a particular experiment. Additional discussion of some of these aspects can be found in Chapter 3, but the details of the procedures necessary are entirely system specific and so somewhat beyond the scope of this text.

2.7 Other related techniques

2.7.1 Moment analysis

Moment analysis [11] refers to the reduction of data to statistical properties such as the mean and variance of the distribution of measured values. If a quantity, x, has a mean value (or expectation value) $<x>$ given by,

$$\langle x\rangle=\frac{1}{N}\sum_{j=1}^{j=N}x_j \tag{2.54}$$

then the mth moment of x, μ_m, is given by;

$$\mu_m=\langle x^m\rangle. \tag{2.55}$$

The most commonly utilized moment is the first moment ($m = 1$), which is equivalent to the mean. In many cases the mean is a poor description of a data set and a more complete description can be obtained by considering the higher order moments. The variance or standard deviation is used to characterize the width of a distribution once the mean has been determined and is given by,

$$\sigma^2=\frac{1}{N-1}\sum_{j=1}^{j=N}(x_j-\langle x\rangle)^2. \tag{2.56}$$

Thus the variance involves analysis of the second moment of the data set (the second power of x is involved). Higher moments can be used to obtain more information; the third and fourth moments are less commonly used in general but well known to statisticians as the skewness and kurtosis respectively. As the moment increases, the accuracy in the parameter involved decreases; the most reliable quantities within a distribution are the numbers themselves (moment zero).

In a similar way to which PCH analysis is applied to single molecule fluorescence data sets (see Section 2.3), analysis of the first three moments of the photon count distribution can be used to determine certain properties of a sample, for example the concentrations of fluorescent species [11]. However, moment analysis is rather rare in single molecule spectroscopy and has effectively been superseded by methods that use a complete description of the photon count distribution such as PCH and FIDA.

2.7.2 **Cross-correlation**

The use of autocorrelation in FCS has been discussed in detail in Section 2.4. In this type of analysis a fluctuating signal is compared with a time-delayed version of itself in order to reveal any temporal correlations in the data. Cross-correlation [51, 52] is similar in principle but seeks correlations between two different fluctuating signals. The cross-correlation function is given by,

$$G(\tau)=\frac{\langle\delta A(t)\delta B(t+\tau)\rangle}{\langle A(t)\rangle\langle B(t)\rangle} \tag{2.57}$$

which is of the form of equation 2.15, but now fluctuations in one signal, A, are compared with fluctuations in another signal, B, at some later delay time τ.

Common in single molecule fluorescence spectroscopy is the cross-correlation analysis of two colour experiments. In these experiments two light sources are used simultaneously to illuminate the sample volume. Two dyes with distinct absorption maxima are used and two detectors monitor the dyes fluorescence separately.[7] The principle advantage of cross-correlation is specificity and the ability to detect binding of complexes which involve only small changes in mass, where small concomitant changes in diffusion coefficient may not lead to measurable changes in the autocorrelated signal of either one of the dyes [52]. For example, in the case of two complimentary DNA strands each labelled with a different dye, a cross-correlation signal will only be recorded for the hybridized product, regardless of the concentrations of the single strands. Indeed, the concentration of the hybridized product can be calculated easily as, under the proper conditions, the amplitude of the cross-correlation function is directly proportional to the concentration of the product [52].

[7] See Chapter 3 for a description of the instrumentation necessary for dual colour cross-correlation.

Dual colour cross-correlation analysis has many of the disadvantages of conventional autocorrelation and significant artefacts can be observed if experimental arrangements are not ideal [124]. Particular to cross-correlation are problems associated with dual colour excitation and detection. It can be difficult to eliminate cross-talk between the two detectors used (or rather to prevent fluorescence from one channel leaking into, and being detected by, the other) leading to false cross-correlations. Furthermore, when employing dual excitation to excite two dyes simultaneously it can be difficult to achieve a perfect overlap (i.e. three-dimensional alignment) of the spectrally distinct excitation volumes (two photon excitation can in some cases circumvent this problem—see Chapter 3).

2.7.3 **Higher order fluorescence correlation spectroscopy**

Higher order autocorrelation analysis is analogous to the analysis of the higher order moments of the photon count distribution. Such a correlation function is given by,

$$G_{ij}(\tau)=\frac{\langle \delta F^{i}(t)\delta F^{j}(t+\tau)\rangle - \langle \delta F^{i}(t)\delta F^{j}(t)\rangle}{\langle F(t)\rangle^{i+j}} \tag{2.58}$$

where in the limit of analysing only the first moment, $i = j = 1$, equation 2.15 is recovered.

As in cross-correlation the motivation for the analysis of higher order autocorrelations is the identification of sample heterogeneity despite only modest variations in diffusion times between the species, which makes analysis by first order autocorrelation difficult. Furthermore, the amplitude of the autocorrelation function is not only dependent on the concentrations of the two species, but also the squares of their relative quantum yields. By simultaneous analysis of multiple higher order autocorrelation functions, sample heterogeneity, and in particular the concentrations of the particular species, can be obtained with a greater degree of accuracy [125, 126].

2.7.4 **Time resolved fluorescence measurements**

The fluorescence lifetime is essentially the average amount of time that a fluorophore spends in the excited state, after absorption of a photon, before returning to the ground state. At both the ensemble and single molecule level the fluorescence lifetime can reveal a wealth of information including details of quenching processes (and so environment), rates of energy transfer in FRET (see Section 2.5) and determination of time-resolved anisotropies (see later). Furthermore, in an ensemble experiment multiple decay constants can indicate heterogeneity in molecular environments of the members of the ensemble.

The challenge for measurement of the fluorescence lifetime is in the fast rates of the processes involved (excited state lifetimes for common high quantum yield dyes are generally <10 ns). Thus fairly sophisticated electronics are necessary (reviewed briefly in Chapter 3) although the components are becoming much more affordable. Two common schemes are available for lifetime determination, the time-domain pulsed method and the phase-modulation (frequency space) method [1]. The time-domain method is common in ensemble methods and the only method employed currently in single molecule studies. Pulsed fluorescence lifetime measurements involve the excitation of the fluorophore with, ideally, an infinitesimally short pulse of light, this results in an initial population of molecules in an excited state, N_0, which then relax to the ground state following an exponential probability distribution,

$$N(t)=N_0 exp\left(\frac{-t}{\tau}\right) \tag{2.59}$$

where τ is the lifetime of the excited state given by,

$$\tau = \frac{1}{\gamma + k} \tag{2.60}$$

where γ and k are the radiative and non-radiative decay rates respectively (one can immediately see how the lifetime can thus be used to monitor quenching or energy transfer). The ensemble fluorescence signal thus decays exponentially as members of the ensemble return to the ground state. The fluorescence lifetime is generally defined as the time at which the initial fluorescence signal has fallen by a factor $1/e$ [1]. In this definition it thus follows from equation 2.59 that the excited state lifetime, τ, is the fluorescence lifetime.

The measurement of the ensemble fluorescence lifetime seems straightforward, however, two factors in particular make the process somewhat more complicated. First, the pulse produced by typical light sources (see Chapter 3) is finite in width and so the measured decay is convolved with the temporal profile of the excitation pulse, and second, the decay cannot be measured from just one excitation pulse (detectors with the necessary sub-nanosecond time resolution are not available). The first of these problems are solved by separately recording the instrument response function of the measurement system (e.g. by measuring scattered light) and then iterative reconvolution with the measured fluorescence decay is used to determine the real decay [1]. The second problem is overcome by the time correlated single photon counting method [1]. Briefly, rather than using only a single pulse, the sample is excited with a train of pulses. When fluorescence photons are detected, the time delay between the last excitation pulse and the detection event is recorded. This is then repeated for many pulses. The emission rate is kept purposefully low so only a single photon is *detected* for each excitation pulse on average. The histogram of time intervals for hundreds to thousands of such excitation–detection events builds up the fluorescence decay (equation 2.59). The details of this and that of iterative convolution are somewhat beyond the scope of this text and the reader is referred elsewhere [1].

Single molecule techniques to measure the fluorescence lifetime [79, 127] are directly analogous to the ensemble methods. In diffusion experiments pulsed excitation is used, with pulse widths of the order of 100 ps and repetition rates of the order of 70 MHz. Histograms of the arrival times of each photon in each burst relative to the excitation pulse are constructed using the time correlated single photon counting method. These histograms are then fitted to extract the single molecule, burst integrated, lifetimes. If sufficient photons are available bursts can be split into smaller 'windows' to monitor the intra-burst lifetime trajectory [79]. It should be noted that unlike an ensemble lifetime measurement, where any number of excitation–emission time intervals can be measured to build up a low noise, fully described exponential (by simply extending measurement time) this is not possible for single molecule experiments where the number of excitation–detection events available to define the exponential is limited by photobleoching. In this way careful thought has to be applied to the accuracy of the fitting of the decays [127].

2.7.5 **Steady-state polarization anisotropy measurements**

The intensity of fluorescence emission from single molecules can be measured as a function of emission or excitation polarization to provide information about the rotational freedom (rates of rotation) of the fluorophores or the molecules to which they are attached [1, 128, 129]. Measurements are generally made by exciting with linearly polarized light and monitoring emission at orthogonal polarizations. The technique is commonly applied in ensemble experiments and has particular relevance with respect to testing for rotational freedom of attached dyes for single molecule FRET experiments (see Section 2.5.7).

When a fluorescent solution is excited with polarized light the emission is also polarized. Initially photo-selection occurs in the sample whereby only molecules with electric dipoles aligned with respect to the polarization of the excitation light are excited efficiently (molecules not aligned with the excitation light have a low, but non-zero, probability of excitation). If no rotational motion were to then occur in the molecule *while it is in the excited state*, the fluorescence emission would have the same polarization as the excitation light (assuming that the absorption and emission dipoles of the molecules are collinear, see the following paragraph). If angular movement occurs in the molecule *while it is in the excited state* the emitted fluorescence attains a new polarization (the emission dipole is now differentially oriented to when the molecule was excited and the emitted light has this new polarization). It is worth emphasizing that any information about slow rotation that is not complete during the excited state is lost. The angular changes that can occur in the excited state (which typically is only a few nanoseconds long) can be induced by a number of things. Primarily thermal energy (Brownian motion) in the solution causes rotational diffusion in the fluorophore (and the host molecule). Thus the rate of the diffusion is dependant on things like the environment of the dye on the host molecule and the viscosity of the solvent (other effects such as energy transfer or re-absorption can cause additional depolarization, but are not discussed here [1]). The anisotropy, r, therefore is a measure of the degree of depolarization of the emission with respect to the excitation [1] and is defined as,

$$r=\frac{I_{\parallel}-I_{\perp}}{I_{\parallel}+2I_{\perp}} \tag{2.61}$$

where $I_{\parallel}$ and $I_{\perp}$ are the intensities of the emission parallel and perpendicular to the excitation polarization, respectively.

In the case of a molecule that is free to explore a wide range of orientations rapidly, compared with the lifetime of the excited state[8], the emission will be essentially depolarized and $r = 0$ since $I_{\parallel} = I_{\perp}$. However, if the molecule experiences some degree of hindered rotation or alignment, r will be non-zero. For completely polarized light, for example scattered polarized laser light, $r = 1$. Note that the absorption and emission dipoles for typical fluorophores are never perfectly parallel and so some degree of depolarization is intrinsic even for a rotationally 'frozen' molecule (hence the use of scattered light in this example).

Considering the case of rotational diffusion reducing anisotropy (i.e. causing depolarization) one expression that can be derived for the steady state (time averaged) anisotropy [1] is,

$$r=\frac{r_0}{1+\tau/\phi} \tag{2.62}$$

where r_0 is the intrinsic anisotropy of the fluorophore due to misalignment of the absorption and emission dipoles and photoselection (so can be thought of as the intrinsic anisotropy as would be measured without additional rotational motion), τ is the fluorescence lifetime of the excited state and ϕ is the rotational correlation time of the fluorophore given by,

$$\phi=\frac{\eta V}{K_B T} \tag{2.63}$$

[8] Note that the dependence involving the excited state lifetime means that anisotropy cannot be accurately determined for the donor in doubly labelled molecules that display FRET, as the energy transfer is an additional relaxation pathway and so reduces the excited state lifetime resulting in apparently larger anisotropies.

where η is the solvent viscosity, K_B the Boltzmann constant and T the temperature of the solution and V is the volume of the rotating fluorophore.

It is worth noting that this formulization relies on a single decay component to the anisotropy. In the case of a large macromolecule labelled with an extrinsic fluorophore, one might be concerned with two main sources of depolarization, first, local motion of the dye (chain reconfiguration of the host molecule at the site of attachment or bond rotation/flexibility of the often saturated dye attachment linker), and second, rotation of the entire macromolecular assembly. However, depolarization only occurs due to motion during the excited state lifetime. Generally, local motions of the dyes are on a timescale much faster than this; small dye molecule rotational correlation times are generally subnanosecond compared to excited state lifetimes of several nanoseconds [127] and so contribute strongly to the depolarization. Conversely, rotational correlation times of the host molecules such as moderately sized proteins are much longer (e.g $\phi = 33$ ns for a 50,000 Da fragment of IgG [1]) and so do not contribute strongly to the measurement of steady state anisotropy for dyes with excited state lifetimes of only a few nanoseconds. Thus dyes used for single molecule determination do not always allow accurate determination of anisotropy due to host molecule rotation.

In terms of using the degree of anisotropy as a ruler for rotational freedom it is difficult to simply state what level of anisotropy represents sufficient rotational averaging to justify, in particular, single molecule FRET distance calculations. However, the rotational correlation times can be calculated or measured for small dye molecules in water, the fluorescence lifetime approximated and an ideal anisotropy calculated (where no rotational hindrance of the host molecule is present). This can then be compared to the experimental value for the complex (which might be hindered). For example, the dye rhodamine 123 has $r_0 = 0.37$, a rotational correlation time of $\phi = 0.2$ ns and an excited state lifetime of the order of $\tau = 4$ ns in aqueous solutions at room temperature [127]. Using equation 2.62 we can see that the anisotropy for this molecule totally unhindered is approximately $r = 0.02$. Thus, using this scale and the supporting theory [1] one can make judgements as to the lack of depolarization of macromolecule tethered dyes. Steady state anisotropy experiments to determine dye rotational freedom when labelled to small (<80 amino acids) proteins reveal that tethered dyes can possess remarkably low anisotropies, although somewhat larger than the minimum value possible, but indicating little hindrance to rotational freedom from the host molecule. For example, Schuler and co-workers found a range of anisotropies for Alexa Flour 488 and 594 of $r = 0.06$–0.09 in all conditions [75]. Deniz and co-workers [82] found similar values for the dyes Cy5 and TMR in some conditions but in other conditions they found higher anisotropy ($r = 0.2$) for one of the dyes at one of the conjugation sites. This clearly illustrates the system specific nature of the phenomenon and emphasises that checks should always be made. It is worth noting however that high anisotropies do not necessarily preclude the use of the system for FRET studies, however they do possibly compromise the accuracy if absolute distances are calculated. Clearly anisotropy is useful, for example, in eliminating incorrect assignment of 'interesting' populations simply because this species causes differential hindrance of one of the dyes. Ensemble measurements of anisotropy might of course not be sufficient to prove this, indeed some single molecule studies have used single molecule anisotropy measurements in parallel with other probes to simply eliminate this uncertainty [79]. In other studies however the investigators have used differential single molecule anisotropies to directly detect different single molecules in solution [127].

The instrumentation for single molecule fluorescence polarization measurements is essentially the same as that used for diffusion FCS and FRET using a confocal configuration, but with the addition of polarizing optics in the detection path to record the fluorescence intensities parallel and perpendicular to the excitation polarization (see Chapter 3). One complication however arises in that high numerical aperture optics (see Chapter 3) used in single molecule optics will cause some depolarization of the excitation light [127] and this, combined with potentially differential detection efficiencies for the orthogonal measured intensities can skew the measured polarization (due to the fundamental polarization sensitivity of any number of materials used in these microscopes, filters, detectors, etc.). Furthermore the influence of scattered light (which has inherently large anisotropy) can make measurements difficult. With careful experimentation and modifications to the simple equation 2.61 all of these effects can be effectively accounted for or eliminated [127].

2.7.6 **Time resolved anisotropy**

Steady state anisotropy measurements described in Section 2.7.5 can sometimes be misleading if a number of sources of depolarization are present. For example, as discussed, it can be difficult to differentiate global molecular rotation from local rotational freedom of the fluorophore when it is attached to a larger molecule [1, 127]. In these cases time resolved single molecule fluorescence anisotropy measurements can be of use since the contributions to the depolarization may well act on different timescales. For example, for a labelled protein segmental motion of the protein backbone near the label and motion in the linker by which the dye is attached is fast, but the rotational diffusion of the protein as a whole may be slow [127]. A time resolved fluorescence anisotropy can be expressed as [79, 127, 130],

$$r = \frac{I_{\parallel}(t) - I_{\perp}(t)}{I_{\parallel}(t) + 2I_{\perp}(t)} \tag{2.64}$$

Time resolved anisotropies can simply be measured by adding polarization sensitivity to time resolved fluorescence measurements described in Section 2.7.4. The qualitative interpretation of the data follows intuitively from the discussion of steady state anisotropy in Section 2.7.5, so we do not describe the method further. For more information we refer the reader elsewhere [1, 127].

2.7.7 **Single molecule emission spectroscopy**

All of the techniques discussed so far in this chapter have involved measuring the fluorescence signal of single molecules integrated over a range of wavelengths

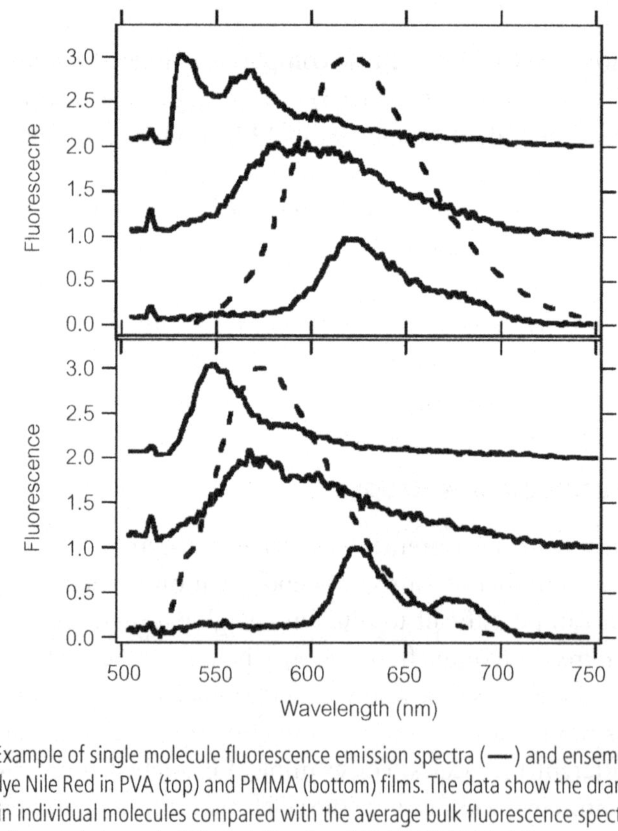

Figure 2.29 Example of single molecule fluorescence emission spectra (—) and ensemble emission spectra (---) for the dye Nile Red in PVA (top) and PMMA (bottom) films. The data show the dramatic spectral variation observed in individual molecules compared with the average bulk fluorescence spectra. Reprinted with permission from Hou *et al.*, *Journal of Physical Chemistry B* 104 (2000) 212–219.

usually selected by a filter (see Chapter 3). It is also possible to measure the fluorescence from single molecules at room temperature as a function of wavelength, and so measure their emission spectra, by the incorporation of a spectrograph in the detection arm of the experiment (see Chapter 3). Spectral information can be used to reveal heterogeneity across many copies of otherwise identical molecules and can be a sensitive probe of environment [1]. For example, shown in Figure 2.29 are three emission spectra for the fluorescent molecule Nile Red in PVA and PMMA films [95]. The figure shows the single molecule spectra for three molecules (solid lines), displaying very different spectral properties due to different local environments. The panels also show the bulk ensemble averaged spectra (broken lines), which are clearly the approximate sum of the three single molecule spectra. The recording of single molecule spectra is challenging as the throughput of most spectrographs is poor and generally a compromise must be reached between spectral resolution and signal-to-noise. Consequently, this type of analysis is not commonly encountered in single molecule fluorescence studies. Further details of the instrumentation can be found in Chapter 3.

References

[1] **Lakowicz, JR**, *Principles of Fluorescence Spectroscopy*, Plenum Press, New York, 1983.

[2] **Hecht, E and Zajac, A**, *Optics*, Addison-Wesley, Tokyo, 1982.

[3] **Wilson, J and Hawkes, JFB**, *Optoelectronics: An Introduction*, Prentice Hall Int., Cambridge, 1989.

[4] **Rigler, R and Mets, U**, *Diffusion* of single molecules through a Gaussian laser beam. *Laser Spectroscopy of Biomolecules* **1921** (1992) 239.

[5] **Ferris, MM, McCabe, MO, Doan, LG, and Rowlen, KL**, *Rapid enumeration of respiratory viruses. Analytical Chemistry* **74** (2002) 1849–1856.

[6] **Osborne, MA, Balasubramanian, S, Furey, WS, and Klenerman, D**, *Optically biased diffusion of single molecules studied by confocal fluorescence microscopy. Journal of Physical Chemistry B* **102** (1998) 3160–3167.

[7] **Li, HT, Ying, LM, Green, JJ, Balasubramanian, S, and Klenerman, D**, Ultrasensitive coincidence fluorescence detection of single DNA molecules. *Analytical Chemistry* **75** (2003) 1664–1670.

[8] **Chen, Y, Muller, JD, Berland, KM, and Gratton, E**, Fluorescence fluctuation spectroscopy. *Methods—A Companion to Methods in Enzymology* **19** (1999) 234–252.

[9] **Chen, Y, Muller, JD, So, PTC, and Gratton, E**, The photon counting histogram in fluorescence fluctuation spectroscopy. *Biophysical Journal* **77** (1999) 553–567.

[10] **Kask, P, Palo, K, Ullmann, D, and Gall, K**, Fluorescence-intensity distribution analysis and its application in biomolecular detection technology. *Proceedings of the National Academy of Science of the United States of America* **96** (1999) 13756–13761.

[11] **Qian, H and Elson, EL**, Distribution of molecular aggregation by analysis of fluctuation moments. *Proceedings of the National Academy of Sciences of the United States of America* **87** (1990) 5479–5483.

[12] **Kask, P, Palo, K, Fay, N, Brand, L, Mets, U, Ullmann, D,** *et al.*, Two-dimensional fluorescence intensity distribution analysis: Theory and applications. *Biophysical Journal* **78** (2000) 1703–1713.

[13] **Perroud, TD, Bo, HA, Wallace, MI, and Zare, RN,** Photon counting histogram for one-photon excitation. *Chemphyschem* **4** (2003) 1121–1123.

[14] **Perroud, TD, Huang, B, Wallace, MI, and Zare, RN,** Photon counting histogram for one-photon excitation (vol 4, pg 1121, 2003). *Chemphyschem* **4** (2003) 1280–1280.

[15] **Perroud, TD, Huang, B, and Zare, RN,** Effect of bin time on the photon counting histogram for one-photon excitation. *ChemPhysChem* **6** (2005) 905–912.

[16] **Muller, JD, Chen, Y, Gratton, E, in R Rigler and ES Elson (Eds),** Photon counting Histogram Statistics. *Fluorescence Correlation Spectroscopy: Theory and Applications*. Springer, Berlin, 2001, pp. 410–437.

[17] **Mandel, L,** Fluctuations of photon beams and their correlations. *Proceedings of the Physical Society of London* **72** (1958) 1037–1048.

[18] **Van Kampen, NG,** *Stochastic Processes in Physics and Chemistry*, Elsevier, New York, 1981.

[19] **Chandrasekhar, S,** Stochastic problems in physics and astronomy. *Reviews of Modern Physics* **15** (1943) 1–89.

[20] http://fms.physics.uiuc.edu/Lfd/Globals/lead.html

[21] **Bevington, PR,** *Data Reduction and Error Analysis for the Physical Sciences*, McGraw-Hill, Boston, London, 2003.

[22] **Osborne, MJ, Breeze, AL, Lian, LY, Reilly, A, James, R, Kleanthous, C,** *et al.*, Three-dimensional solution structure and 13C nuclear magnetic resonance assignments of the colicin E9 immunity protein Im9. *Biochemistry* **35** (1996) 9505–9512.

[23] **Muller, JD, Chen, Y, and Gratton, E,** Resolving heterogeneity on the single molecular level with the photon- counting histogram. *Biophysical Journal* **78** (2000) 474–486.

[24] **Chirico, G, Bettati, S, Mozzarelli, A, Chen, Y, Muller, JD, and Gratton,** E, Molecular heterogeneity of o-acetylserine sulfhydrylase by two-photon excited fluorescence fluctuation spectroscopy. *Biophysical Journal* **80** (2001) 1973–1985.

[25] **Palo, K, Metz, U, Jager, S, Kask, P, and Gall, K,** Fluorescence intensity multiple distributions analysis: Concurrent determination of diffusion times and molecular brightness. *Biophysical Journal* **79** (2000) 2858–2866.

[26] **Palo, K, Brand, L, Eggeling, C, Jager, S, Kask, P, and Gall, K,** Fluorescence intensity and lifetime distribution analysis: Toward higher accuracy in fluorescence fluctuation spectroscopy. *Biophysical Journal* **83** (2002) 605–618.

[27] **Schaertl, S, Meyer-Almes, FJ, Lopez-Calle, E, Siemers, A, and Kramer, J,** A novel and robust homogeneous fluorescence-based assay using nanoparticles for pharmaceutical screening and diagnostics. *Journal of Biomolecular Screening* **5** (2000) 227–237.

[28] **Magde, D, Elson, EL, and Webb, WW,** Thermodynamic fluctuations in a reacting system—measurement by fluorescence correlation spectroscopy. *Biopolymers* **29** (1972) 705–708.

[29] **Webb, WW, in R Rigler and ES Elson (Eds),** Fluorescence Correlation Spectroscopy: Genesis, Evolution, Maturation and Prognosis. *Fluorescence Correlation Spectroscopy: Theory and Applications*. Springer, Berlin, 2001, pp. 305–330.

[30] **Magde, D and Elson, EL,** Fluorescence correlation spectroscopy. II. An experimental realization. *Biopolymers* **13** (1974) 29–61.

[31] **Aragon, SR and Pecora, R,** Fluorescence correlation spectroscopy and Brownian rotational diffusion. *Biopolymers* **14** (1975) 119–138.

[32] **Aragon, SR and Pecora, R**, Fluorescence correlation spectroscopy as a probe of molecular dynamics. *The Journal of Chemical Physics* **64** (1976) 1791–1803.

[33] **Koppel, DE**, Statistical accuracy in fluorescence correlation spectroscopy. *Physical Review A* **10** (1974) 1938–1945.

[34] **Thompson, NL, in JR Lakowicz (Ed.)**, Fluorescence Correlation Spectroscopy. *Topics in Fluorescence Spectroscopy*. Plenum Press, New York, 1991, pp. 337–374.

[35] **Chen, Y, Muller, JD, Tetin, SY, Tyner, JD, and Gratton, E**, Probing ligand protein binding equilibria with fluorescence fluctuation spectroscopy. *Biophysical Journal* **79** (2000) 1074–1084.

[36] **Press, WH, Teukolsky, SA, Vetterling, WT, and Flannery, BP**, *Numerical Recipes in C: The Art of Scientific Computing*, CUP, New York, USA, 1992.

[37] **Wohland, T, Rigler, R, and Vogel, H**, The standard deviation in fluorescence correlation spectroscopy. *Biophysical Journal* **80** (2001) 2987–2999.

[38] **Widengren, J, in R Rigler, and ES Elson (Eds)**, Photophysical Aspects of FCS Measurements. *Fluorescence Correlation Spectroscopy: Theory and Applications*. Springer, Berlin, 2001, pp. 277–301.

[39] **Wohland, T, Friedrich, K, Hovius, R, and Vogel, H**, Study of ligand-receptor interactions by fluorescence correlation spectroscopy with different fluorophores: evidence that the homopentameric 5-hydroxytryptamine type 3(As) receptor binds only one ligand. *Biochemistry* **38** (1999) 8671–8681.

[40] **Rigler, R, Mets, U, Widengren, J, and Kask, P**, Fluorescence correlation spectroscopy with high count rate and low background—analysis of translational diffusion. *European Biophysics Journal* **22** (1993) 169–175.

[41] **Rigler, R, Widengren, J, and Mets, U, in O Wolfbeis (Ed.)**, *Fluorescence Spectroscopy: New Methods and Applications*. Springer-Verlag, Berlin, 1993, pp. 13–24.

[42] **Widengren, J, Rigler, R, and Mets, U**, Triplet-state monitoring by fluorescence correlation spectroscopy. *Journal of Fluorescence* **4** (1994) 255–258.

[43] **Widengren, J, Dapprich, J, and Rigler, R**, Fast interactions between Rh6G and Dgtp in water studied by fluorescence correlation spectroscopy. *Chemical Physics* **216** (1997) 417–426.

[44] **Berland, KM, So, PTC, Chen, Y, Mantulin, WW, and Gratton, E**, Scanning two-photon fluctuation correlation spectroscopy: Particle counting measurements for detection of molecular aggregation. *Biophysical Journal* **71** (1996) 410–420.

[45] **Edman, L, Mets, U, and Rigler, R**, Conformational transitions monitored for single molecules in solution. *Proceedings of the National Academy of Sciences of the United States of America* **93** (1996) 6710–6715.

[46] **Bonnet, G, Krichevsky, O, and Libchaber, A**, Kinetics of conformational fluctuations in DNA hairpin-loops. *Proceedings of the National Academy of Sciences of the United States of America* **95** (1998) 8602–8606.

[47] **Widengren, J and Rigler, R**, Photobleaching investigations of dyes using fluorescence correlation spectroscopy (FCS). *Progress in Biophysics and Molecular Biology* **65** (1996) PH109.

[48] **Mets, U, in R Rigler and ES Elson (Eds)**, Antibunching and Rotational Diffusion in FCS. *Fluorescence Correlation Spectroscopy: Theory and Applications*. Springer, Berlin, 2001, pp. 346–359.

[49] **Haupts, U, Maiti, S, Schwille, P, and Webb, WW**, Dynamics of fluorescence fluctuations in green fluorescent protein observed by fluorescence correlation spectroscopy. *Proceedings of the National Academy of Sciences of the United States of America* **95** (1998) 13573–13578.

[50] **Visser, A and Hink, MA**, New perspectives of fluorescence correlation spectroscopy. *Journal of Fluorescence* **9** (1999) 81–87.

[51] **Hess, ST, Huang, SH, Heikal, AA, and Webb, WW**, Biological and chemical applications of fluorescence correlation spectroscopy: A review. *Biochemistry* **41** (2002) 697–705.

[52] **Haustein, E and Schwille, P**, Ultrasensitive investigations of biological systems by fluorescence correlation spectroscopy. *Methods* **29** (2003) 153–166.

[53] **Meseth, U, Wohland, T, Rigler, R, and Vogel, H**, Resolution of fluorescence correlation measurements. *Biophysical Journal* **76** (1999) 1619–1631.

[54] **Widengren, J, Mets, U, and Rigler, R**, Fluorescence correlation spectroscopy of triplet states in solution—a theoretical and experimental study. *Journal of Physical Chemistry* **99** (1995) 13368–13379.

[55] **Berne, BJ, Pecora, R**, Dynamic Light Scattering. John Wiley and Sons, New York (1976).

[56] **Maiti, S, Haupts, U, and Webb, WW**, Fluorescence correlation spectroscopy: Diagnostics for sparse molecules. *Proceedings of the National Academy of Sciences of the United States of America* **94** (1997) 11753–11757.

[57] **Li, HT, Ren, XJ, Ying, LM, Balasubramanian, S, and Klenerman, D**, Measuring single-molecule nucleic acid dynamics in solution by two-color filtered ratiometric fluorescence correlation spectroscopy. *Proceedings of the National Academy of Sciences of the United States of America* **101** (2004) 14425–14430.

[58] **Schwille, P, Haupts, U, Maiti, S, and Webb, WW**, Molecular dynamics in living cells observed by fluorescence correlation spectroscopy with one- and two- photon excitation. *Biophysical Journal* **77** (1999) 2251–2265.

[59] **Eid, JS, Muller, JD, and Gratton, E**, Data acquisition card for fluctuation correlation spectroscopy allowing full access to the detected photon sequence. *Review of Scientific Instruments* **71** (2000) 361–368.

[60] **Kask, P, Gunther, R, and Axhausen, P**, Statistical accuracy in fluorescence fluctuation experiments. *European Biophysics Journal with Biophysics Letters* **25** (1997) 163–169.

[61] **Gosch, M, Blom, H, Holm, J, Heino, T, and Rigler, R**, Hydrodynamic flow profiling in microchannel structures by single molecule fluorescence correlation spectroscopy. *Analytical Chemistry* **72** (2000) 3260–3265.

[62] **Bjorling, S, Kinjo, M, Foldes-Papp, Z, Hagman, E, Thyberg, P, and Rigler, R**, Fluorescence correlation spectroscopy of enzymatic DNA polymerization. *Biochemistry* **37** (1998) 12971–12978.

[63] **Schwille, P, Korlach, J, and Webb, WW**, Fluorescence correlation spectroscopy with single-molecule sensitivity on cell and model membranes. *Cytometry* **36** (1999) 176–182.

[64] **Boxer, SG**, Molecular transport and organization in supported lipid membranes. *Current Opinion in Chemical Biology* **4** (2000) 704–709.

[65] **Elson, EL**, Quick tour of fluorescence correlation spectroscopy from its inception. *Journal of Biomedical Optics* **9** (2004) 857–864.

[66] **Schwille, P, Oehlenschlager, F, and Walter, NG**, Quantitative hybridization kinetics of DNA probes to RNA in solution followed by diffusional fluorescence correlation analysis. *Biochemistry* **35** (1996) 10182–10193.

[67] **Hasler, K, Panke, O, and Junge, W**, On the stator of rotary atp synthase: The binding strength of subunit delta to (alpha beta)(3) as determined by fluorescence correlation spectroscopy. *Biochemistry* **38** (1999) 13759–13765.

[68] **Huang, SH, Butler, JS, Loh, SN, and Webb, WW**, Fluorescence correlation spectroscopy analyzes the equilibrium folding pathway of apomyoglobin. *Biophysical Journal* **80** (2001) 2519 Part 2512.

[69] **Steinberg, IZ**, Long-Range Nonradiative Transfer of Electronic Excitation Energy in Proteins and Polypeptides. *Annual Reviews of Biochemistry*. **40** (1971) 83–114.

[70] **Fairclough, RH and Cantor, CR,** The use of singlet-singlet energy transfer to study macromolecular assemblies. *Methods in Enzymology* **48** (1978) 347–379.

[71] **Stryer, L,** Fluorescence energy transfer as a spectroscopic ruler. *Annual Review of Biochemistry* **47** (1978) 819–846.

[72] **Cheung, HC, in JR Lakowicz (Ed.).** Resonance Energy Transfer in Topics in Flourescence Spectroscopy, Volume 2: Principles. Plenum Press, New York, 2 1991.

[73] **Förster, T,** Transfer Mechanisms of electronic excitation. *Discussions of the Faraday Society* **27** (1959) 7–17.

[74] **Clegg, RM,** Fluorescence resonance energy transfer and nucleic acids. *Methods in Enzymology* **211** (1992) 353.

[75] **Schuler, B, Lipman, EA, and Eaton, WA,** Probing the free-energy surface for protein folding with single-molecule fluorescence spectroscopy. *Nature* **419** (2002) 743–747.

[76] **Dahan, M, Deniz, AA, Ha, T, Chemla, DS, Schultz, PG, and Weiss, S,** Ratiometric measurement and identification of single diffusing molecules. *Chemical Physics* **247** (1999) 85–106.

[77] **Deniz, AA, Laurence, TA, Dahan, M, Chemla, DS, Schultz, PG, and Weiss, S,** Ratiometric single-molecule studies of freely diffusing biomolecules. *Annual Reviews of Physical Chemistry* **52** (2001) 233–253.

[78] **Schuler, B, Lipman, EA, Steinbach, PJ, Kumke, M, and and Eaton, WA,** Polyproline and the spectroscopic ruler revisited with single-molecule fluorescence. *Proceedings of the National Academy of Sciences of the United States of America* **102** (2005) 2754–2759.

[79] **Margittai, M, Widengren, J, Schweinberger, E, Schroder, GF, Felekyan, S, Haustein, E,** *et al.*, Single-molecule fluorescence resonance energy transfer reveals a dynamic equilibrium between closed and open conformations of syntaxin 1. *Proceedings of the National Academy of Sciences of the United States of America* **100** (2003) 15516–15521.

[80] **Rhoades, E, Gussakovsky, E, and Haran, G,** Watching proteins fold one molecule at a time. *Proceedings of the National Academy of Sciences of the United States of America* **100** (2003) 3197–3202.

[81] **Deniz, AA, Dahan, M, Grunwell, JR, Ha, T, Faulhaber, AE, Chemla, DS,** *et al.*, Single-pair fluorescence resonance energy transfer on freely diffusing molecules: observation of Förster distance dependence and subpopulations. *Proceedings of the National Academy of Sciences of the United States of America* **96** (1999) 3670–3675.

[82] **Deniz, AA, Laurence, TA, Beligere, GS, Dahan, M, Martin, AB, Chemla, DS,** *et al.*, Single-molecule protein folding: Diffusion fluorescence resonance energy transfer studies of the denaturation of chymotrypsin inhibitor 2. *Proceedings of the National Academy of Sciences of the United States of America* **97** (2000) 5179–5184.

[83] **Ying, LM, Wallace, MI, Balasubramanian, S, and Klenerman, D,** Ratiometric analysis of single-molecule fluorescence resonance energy transfer using logical combinations of threshold criteria: A study of 12-Mer DNA. *Journal of Physical Chemistry B* **104** (2000) 5171–5178.

[84] **Kapanidis, AN, Lee, NK, Laurence, TA, Doose, S, Margeat, E, and Weiss, S,** Fluorescence-aided molecule sorting: Analysis of structure and interactions by alternating-laser excitation of single molecules. *Proceedings of the National Academy of Sciences of the United States of America* **101** (2004) 8936–8941.

[85] **Lee, NK, Kapanidis, AN, Wang, Y, Michalet, X, Mukhopadhyay, J, Ebright, RH,** *et al.*, Accurate FRET measurements within single diffusing biomolecules using alternating-laser excitation. *Biophysical Journal* **88** (2005) 2939–2953.

[86] **Wallace, MI, Ying, L, Balasubramanian, S, and Klenerman, D**, Non-Arrhenius kinetics for the loop closure of a DNA hairpin. *Proceedings of the National Academy of Sciences of the United States of America* **98** (2001) 5584–5589.

[87] **Wallace, MI, Ying, LM, Balasubramanian, S, and Klenerman, D**, Fret fluctuation spectroscopy: Exploring the conformational dynamics of a DNA hairpin loop. *Journal of Physical Chemistry B* **104** (2000) 11551–11555.

[88] **Gell, C, Sabir, T, Smith, DAM, and Stockley, PGS**, *Unpublished results.*

[89] **Zhang, WB and Chen, SJ**, RNA hairpin-folding kinetics. *Proceedings of the National Academy of Sciences of the United States of America* **99** (2002) 1931–1936.

[90] **Pljevaljcic, G, Millar, DP, and Deniz, AA**, Freely diffusing single hairpin ribozymes provide insights into the role of secondary structure and partially folded states in RNA folding. *Biophysical Journal* **87** (2004) 457–467.

[91] **Talaga, DS, Lau, WL, Roder, H, Tang, JY, Jia, YW, Degrado, WF,** *et al.*, Dynamics and folding of single two-stranded coiled-coil peptides studied by fluorescent energy transfer confocal microscopy. *Proceedings of the National Academy of Sciences of the United States of America* **97** (2000) 13021–13026.

[92] **Ying, LM, Green, JJ, Li, HT, Klenerman, D, and Balasubramanian, S**, Studies on the structure and dynamics of the human telomeric G quadruplex by single-molecule fluorescence resonance energy transfer. *Proceedings of the National Academy of Sciences of the United States of America* **100** (2003) 14629–14634.

[93] **Ying, LM, Wallace, MI, and Klenerman, D**, Two-state model of conformational fluctuation in a DNA hairpin- loop. *Chemical Physics Letters* **334** (2001) 145–150.

[94] **Lipman, EA, Schuler, B, Bakajin, O, and Eaton, WA**, Single-molecule measurement of protein folding kinetics. *Science* **301** (2003) 1233–1235.

[95] **Hou, YW, Bardo, AM, Martinez, C, and Higgins, DA**, Characterization of molecular scale environments in polymer films by single molecule spectroscopy. *Journal of Physical Chemistry B* **104** (2000) 212–219.

[96] **Heyes, CD, Kobitski, AY, Amirgoulova, EV, and Nienhaus, GU**, Biocompatible surfaces for specific tethering of individual protein molecules. *Journal of Physical Chemistry B* **108** (2004) 13387–13394.

[97] **McKinney, SA, Declais, AC, Lilley, DMJ, and Ha, T**, Structural dynamics of individual holliday junctions. *Nature Structural Biology* **10** (2003) 93–97.

[98] **Osborne, MA, Furey, WS, Klenerman, D, and Balasubramanian, S**, Single molecule analysis of DNA immobilized on microspheres. *Analytical Chemistry* **72** (2000) 3678–3681.

[99] **Jia, YW, Talaga, DS, Lau, WL, Lu, HSM, DeGrado, WF, and Hochstrasser, RM**, Folding dynamics of single GCN4 peptides by fluorescence resonant energy transfer confocal microscopy. *Chemical Physics* **247** (1999) 69–83.

[100] **Weston, KD, Carson, PJ, Metiu, H, and Buratto, SK**, Room-temperature fluorescence characteristics of single dye molecules adsorbed on a glass surface. *Journal of Chemical Physics* **109** (1998) 7474.

[101] **Kummer, S, Dickson, RM, and Moerner, WE**, Probing single molecules in polyacrylamide gels. *Proceedings of the SPIE* **3273** (1998) 165–173.

[102] **Mashanov, GI, Tacon, D, Knight, AE, Peckham, M, and Molloy, JE**, Visualizing single molecules inside living cells using total internal reflection fluorescence microscopy. *Methods* **29** (2003) 142–152.

[103] **Mashanov, GI, Tacon, D, Peckham, M, and Molloy, JE**, The spatial and temporal dynamics of pleckstrin homology domain binding at the plasma membrane measured by Imaging single molecules in live mouse myoblasts. *Journal of Biological Chemistry* **279** (2004) 15274–15280.

[104] **Byassee, TA, Chan, WCW, and Nie, SM**, Probing single molecules in single living cells. *Analytical Chemistry* **72** (2000) 5606–5611.

[105] **Boukobza, E, Sonnenfeld, A, and Haran, G**, Immobilization in surface-tethered lipid vesicles as a new tool for single biomolecule spectroscopy. *Journal of Physical Chemistry B* **105** (2001) 12165–12170.

[106] **Chiu, D, Wilson, CF, Karlsson, A, Danielsson, A, Lundqvist, A, and Stromberg, A**, Manipulating the biochemical nanoenvironment around single molecules contained within vesicles. *Chemical Physics* **247** (1999) 133–139.

[107] **Ha, T, Chemla, DS, Enderle, T, and Weiss, S**, Single molecule spectroscopy with automated positioning. *Applied Physics Letters* **70** (1997) 782–784.

[108] **Abe, K, Kaya, S, Hayashi, Y, Imagawa, T, Kikumoto, M, Oiwa, K**, *et al.*, Correlation between the activities and the oligomeric forms of pig gastric H/K-ATPase. *Biochemistry* **42** (2003) 15132–15138.

[109] **Kaya, S, Abe, K, Taniguchi, K, Yazawa, M, Katoh, T, Kikumoto, M**, *et al.*, Oligomeric structure of P-type ATPases observed by single molecule detection technique. *Annals of the New York Academy of Sciences* **986** (2003) 278–280.

[110] **Ying, LM, and Xie, XS**, Fluorescence spectroscopy, exciton dynamics, and photochemistry of single allophycocyanin trimers. *Journal of Physical Chemistry B* **102** (1998) 10399–10409.

[111] **Haran, G**, Noise reduction in single-molecule fluorescence trajectories of folding proteins. *Chemical Physics* **307** (2004) 137–145.

[112] **Zhuang, XW, Kim, H, Pereira, MJB, Babcock, HP, Walter, NG, and Chu, S**, Correlating structural dynamics and function in single ribozyme molecules. *Science* **296** (2002) 1473–1476.

[113] **Rhoades, E, Cohen, M, Schuler, B, and Haran, G**, Two-state folding observed in individual protein molecules. *Journal of the American Chemical Society* **126** (2004) 14686–14687.

[114] **Dimitriadis, G, Drysdale, A, Myers, JK, Arora, P, Radford, SE, Oas, TG**, *et al.*, Microsecond folding dynamics of the F13W G29A mutant of the B domain of staphylococcal protein A by laser-induced temperature jump. *Proceedings of the National Academy of Sciences of the United States of America* **101** (2004) 3809–3814.

[115] **Capaldi, AP, Kleanthous, C, and Radford, SE**, Im7 folding Mechanism: Misfolding on a path to the native state. *Nature Structural Biology* **9** (2002) 209–216.

[116] **Friel, CT, Beddard, GS, and Radford, SE**, Switching two-state to three-state kinetics in the helical protein Im9 via the optimisation of stabilising non-native interactions by design. *Journal of Molecular Biology* **342** (2004) 261–273.

[117] **Spence, GR, Capaldi, AP, and Radford, SE**, Trapping the on-pathway folding intermediate of Im7 at equilibrium. *Journal of Molecular Biology* **341** (2004) 215–226.

[118] **Bagshaw, CR and Conibear, PB**, Single-molecule enzymology: Critical aspects exemplified by myosin ATPase activity. *Single Molecules* **1** (2000) 271–277.

[119] **Zhuang, XW, Bartley, LE, Babcock, HP, Russell, R, Ha, TJ, Herschlag, D**, *et al.*, A single-molecule study of RNA catalysis and folding. *Science* **288** (2000) 2048–2051.

[120] **Sakmann, B and Neher, E**, *Single Channel Recording*, Plenum Press, New York, 1995.

[121] **Zhuang, X, Kim, H, Pereira, MJ, Babcock, HP, Walter, NG, and Chu, S**, Correlating structural dynamics and function in single ribozyme molecules. *Science* **296** (2002) 1473–1476.

[122] **Widengren, J, Mets, U, and Rigler, R,** Photodynamic properties of green fluorescent proteins investigated by fluorescence correlation spectroscopy. *Chemical Physics* **250** (1999) 171–186.

[123] **Jung, G, Wiehler, J, Gohde, W, Tittel, J, Basche, T, Steipe, B,** *et al.*, Confocal microscopy of single molecules of the green fluorescent protein. *Bioimaging* 6 (1998) 54–61.

[124] **Hess, ST and Webb, WW,** Focal volume optics and experimental artifacts in confocal fluorescence correlation spectroscopy. *Biophysical Journal* **83** (2002) 2300–2317.

[125] **Palmer, AG and Thompson, NL,** Molecular aggregation characterized by high-Order autocorrelation in fluorescence correlation spectroscopy. *Biophysical Journal* **52** (1987) 257–270.

[126] **Thompson, NL and Mitchell, JL, in R Rigler ES and Elson (Eds),** High order Autocorrelation in Fluorescence Correlation Spectroscopy" *Fluorescence Correlation Spectroscopy: Theory and Applications*. Springer, Berlin, 2001, pp. 438–458.

[127] **Schaffer, J, Volkmer, A, Eggeling, C, Subramaniam, V, Striker, G, and Seidel, CAM,** Identification of single molecules in aqueous solution by time-resolved fluorescence anisotropy. *The Journal of Physical Cemistry* A **103** (1999) 331–336.

[128] **Forkey, JN, Quinlan, ME, and Goldman, YE,** Protein structural dynamics by single-molecule fluorescence polarization. *Progress in Biophysics and Molecular Biology* **74** (2000) 1–35.

[129] **Ha, T, Laurence, TA, Chemla, DS, and Weiss, S,** Polarization spectroscopy of single fluorescent molecules. *Journal of Physical Chemistry B* **103** (1999) 6839–6850.

[130] **Ha, T, Enderle, T, Chemla, DS, Selvin, PR, and Weiss, S,** Single molecule dynamics studied by polarization modulation. *Physical Review Letters* 77 (1996) 3979–3982.

THREE
Single molecule fluorescence instrumentation

3.1 Introduction

Perhaps the greatest obstacle that prevents non-specialists utilizing a relatively new technique such as single molecule fluorescence spectroscopy is the apparently complex instrumentation. Whilst in recent years commercial systems capable of single molecule fluorescence detection have become available (see Section 3.9), it is often still necessary to make significant modifications to these systems. The aim of this chapter is to provide an overview of the common experimental arrangements for single molecule fluorescence spectroscopy and to describe how they are implemented in a manner that makes them accessible to a broad range of potential users.

The instrumentation necessary to achieve single molecule fluorescence detection is relatively straightforward (see Figure 3.1). The main requirements are high efficiency optical collection and a good signal-to-noise ratio. These are usually achieved by using high numerical aperture microscope objectives (see Section 3.6) and sensitive detectors such as silicon avalanche photodiodes or charge-coupled devices, all of which are commercially available (see Section 3.7). The subtleties of single molecule fluorescence instruments are related to the specific application but are, of course, vital to the success of these measurements.

Optical detection of a single molecule requires that its optical signal (usually fluorescence) can be distinguished from the scattered light and fluorescence arising from the other molecules within the detection volume. This in turn implies that the optical system must have a high throughput and detection efficiency and that background noise is efficiently rejected. Generally, the detection of a single molecule is achieved either by using very low sample concentrations (<100 pM) or by immobilizing single molecules on a surface with sub-monolayer coverage and combining either of these approaches with a very small observation volume (<0.1 fl). Even when such small observation volumes are used, one generally still faces the problem of a relatively large background signal from the many solvent molecules, in comparison with the few fluorescence photons from the single

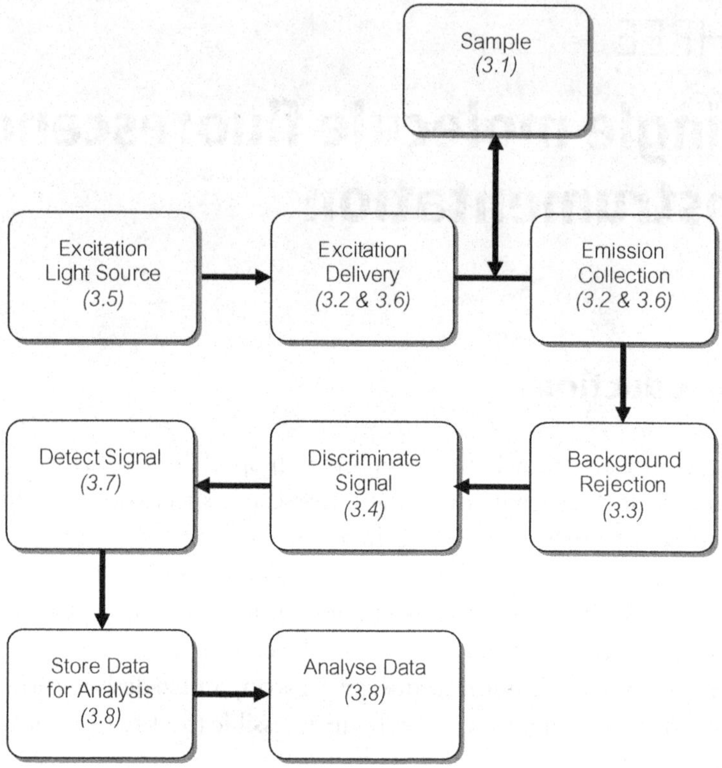

Figure 3.1 Schematic overview of the elements of a single molecule fluorescence experiment. The italicized numbers indicate the section in this chapter where a discussion of each aspect can be found. The sample, on top of a glass coverslip, is presented to the excitation light delivery/emission light collection optics. The emitted light then has the background rejected from the desired fluorescence signal. The fluorescence signal is then discriminated (for example, split into wavelength or polarization components) and then detected and the observable stored for analysis.

fluorescent molecule of interest. For example, in 0.1 fl water there are ~10^9 water molecules. If even small amounts of unwanted signal is emitted from these water molecules which overlaps the spectral region of the fluorescence of the single molecule of interest, the signal to background will make measurements impossible. There are three primary sources of background noise:

1. Rayleigh scattering by the solvent molecules. This process results in a background signal at the excitation wavelength that may leak through the optical detection filters (see later), which cannot provide 100% efficient rejection.
2. Raman scattering by the solvent molecules. This process results in photons at both higher and lower energy compared to the excitation light. However, since fluorescence of the single molecule of interest will be at longer wavelengths than the excitation (see Chapter 4), it is the Raman scattered light at lower

energy (Stoke's radiation) that is of concern. Some of these Stoke's scattered photons may overlap the detection filter pass band and therefore contribute to the background signal.

3. Finally, a combination of fluorescence, Rayleigh and Raman scattering from impurities in the solvent introduced by impurities in buffer components or by careless sample preparation.

In addition, the signal-to-noise may also be affected by problems with the instrumentation such as the stability of the light source, mechanical drift, detector noise, and non-linearity. Many of these points are discussed later.

For a typical fluorescent dye molecule with a quantum yield of 0.8, we might expect of the order 10^5–10^6 fluorescence photons per second to be detected if an excitation flux of ~100 kWcm^{-2} is used (i.e. ~100 μW excitation power into an observation volume of 250 nm diameter, an overall detection efficiency of 1%, visible excitation and a fluorescence cross section of 4×10^{-16} cm^2 for Rhodamine 6G [1]). Even without any impurities present, the background signal due to Raman and Rayleigh scattering from the large number of solvent (water) molecules present would be many orders of magnitude larger. Efficient methods for rejecting this background signal and for detection of the few fluorescence photons are clearly essential. The most trivial method is that by reducing the size of the detection volume, we minimize the number of solvent molecules and therefore minimize the scattered light. Rayleigh scattered light removal is relatively easy as this is generally spectrally distinct from the fluorescence emission. Similarly, good fortune might also allow the fluorescence signal from the molecule of interest to be separated from Raman scattered light and the fluorescence from impurities by a suitable band pass filter (see Section 3.3—a filter that allows only certain wavelengths of light to be transmitted with any efficiency). Whilst the intensity of the Raman scattered light is low the large number of solvent molecule in the volume makes this effect significant. For water (typically the highest concentration solvent component) the inelastic Raman scattering causes a shift in the scattered light that leads to a number of bands (due to, for example, the vibrations along O–H or H–O–H). The bands due to Raman scattering are typically expressed as the shift in wavenumbers (cm^{-1}) of scattered light with respect to the excitation light. The relationship between wavenumber and wavelength is given by *wavenumber* [cm^{-1}] = 10^7/*wavelength* [nm], so, for example, 488 nm ~20492 cm^{-1}. For water 8 distinct Raman bands may be observed resulting from 6 vibrational bands [2]. The shift of the principle (most intense) band is ~3439 cm^{-1} (at the principle maximum) and is relatively broad (a half-width of around 400 cm^{-1}). The other bands are generally too weak to be observed (or are not shifted sufficiently from the excitation wavelength). Thus for

488 nm excitation (20492 cm^{-1}) the scattered light due the Raman shift will occur, principally, at 23931 cm^{-1} (20492 + 3439) or at a wavelength of ~586 nm. In this example then the Raman scattering is not only spectrally distinct from the excitation source but also the typical emission wavelength of a directly excited dye (where the Stokes shift of the emitted fluorescence is typically less than 30 nm—see Chapter 4). Although we note that, were 488 nm light to be used to excite the FRET pair Fluorescein–TMR then the Raman band due to scattering from water is centred on the acceptor dyes fluorescence emission maximum (at 586 nm). This does not preclude its use but will mean a lower signal-to-noise in the acceptor channel.

Since fluorescent dyes usually have quite broad emission spectra, narrow band pass filters that reject the majority of the unwanted background signals can also reduce the number of fluorescence photons reaching the detector. Minimizing the sample volume and the use of appropriate filters are the two most important approaches to improving the signal-to-noise ratio in single molecule fluorescence experiments.

Before discussing the details of the instrumentation let us quickly review the two common experimental geometries. Generally, there are two ways in which the sample is presented in single molecule fluorescence experiments. The analyte is either freely diffusing in a solvent or immobilized at an interface by chemical attachment or physical adsorption (see Figure 3.2(a) and (d)). In diffusion experiments fluorescence photons emitted by the analyte are detected in transient bursts, generated as the molecule transits the small excitation/observation volume. In Figure 3.2(b) we present typical single molecule fluorescence data for a diffusing analyte. In this case the analyte was an RNA molecule doubly labelled with two fluorescent dyes (designed for fluorescence resonance energy transfer, FRET studies of folding, see Chapter 2). As the molecules diffuse through the excitation/observation volume, fluorescence photons emitted by the two dyes at two wavelengths (red and green) are measured separately by two detectors. These bursts of photons are recorded as a function of time to provide intensity versus time traces for two detector channels monitoring the two wavelength ranges. Single molecule fluorescence events appear as transient bursts of photons above the background level (arising from the Rayleigh and Raman scattering from the solvent that is mentioned earlier). In experiments in which the analyte is labelled with two dyes for FRET, the ratio of the red and green signals is related to the spatial separation of the dyes (see Chapter 2). Since these RNA molecules exist in a dynamic equilibrium between folded and unfolded conformations a histogram of these ratios (Figure 3.2(c)) reveals conformational heterogeneity. This ability to dissect heterogeneity of structure or dynamics in a molecular system is one of the great strengths of single molecule experiments.

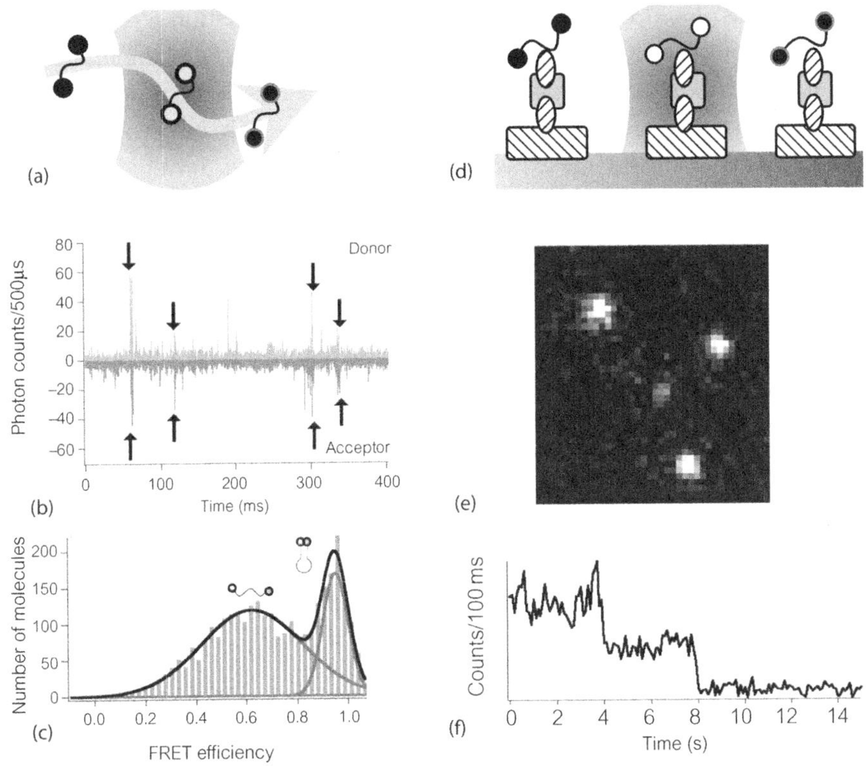

Figure 3.2 The principles of the common single molecule fluorescence techniques. (a) Illustration of the principle of diffusion based single molecule measurements. A doubly dye-labelled macromolecule diffuses along a random path through a small observation volume defined, in part, by a focused excitation beam. (b) As the molecule diffuses through the volume, a burst of fluorescence photons from each of the red (acceptor) and green (donor) dye molecules is detected and recorded in an intensity versus time trace. Single molecule events show up as transient bursts above a background level which may be discriminated from spurious background events by their characteristic coincidence. (c) FRET analysis of these data may, for example, reveal whether the molecule is folded or extended based on the ratio of the red and green signals (see Chapter 2). A histogram of the number of events observed at each FRET efficiency reveals heterogeneity in the sample at equilibrium. (d) Illustration of the principle of immobilized single molecule fluorescence measurements. For example, biotinylated molecules of interest may be immobilized through a biotin–avidin interaction onto a glass surface (biotin represented as hatched ovals, avidin as solid rounded square, attached to a biotinylated BSA shown as hatched rounded rectangle). Fluorescence is generated when the surface is illuminated either by a scanned focused beam (as depicted) or through total internal reflection (TIRF) excitation (see Section 3.2.3). (e) An example of a TIRF image of individual dimers of single dye-labelled RNA immobilized by biotinylation. (f) The fluorescence signal as a function of time from one of the single RNA complexes shows two photobleaching steps confirming the presence of two labelled RNA molecules in each complex.

In experiments using immobilized molecules, the fluorescence emitted can be detected over an extended period of time from the same molecule until the molecule photobleaches, rather than in transient bursts as different molecules diffuse through the excitation/observation volume. Figure 3.2(e) shows an image of a surface at which green fluorescent dye-labelled RNA complexes have been immobilized.

This image was obtained using a total internal reflection excitation geometry and an electron multiplying CCD detector (see Section 3.7). In this case, the fluorescence intensity versus time trace for one spot shows discrete two-step irreversible photobleaching behaviour because there are two RNA molecules in each complex under these conditions. Whether the experiments are best (or most conveniently) carried out using an immobilized or freely diffusing system depends on the properties that are to be investigated. This choice will be discussed further in Chapter 4 and examples of both types of experiment will be described in detail in Chapters 5 and 6.

In the following sections we will discuss the variety of instrumental methods available for delivering the excitation light to the sample, collecting the resulting fluorescence photons, methods for reducing the background signal and the choice of excitation source and detector.

3.2 Optical arrangements for single molecule detection

A variety of optical arrangements have been chosen for single molecule fluorescence experiments ranging from novel fibre optics [3] to unmodified commercial microscopes [4]. However, by far the most common approaches are the confocal epifluorescence far-field [5], multi-photon epifluorescence [6,7], and total internal reflection (TIR) geometries [8,9]. The optical arrangement is mainly determined by the experimental design, that is whether the molecules of interest are freely diffusing or are fixed in space. In diffusion experiments, which have the benefit of being relatively simple to set up, confocal or two-photon far-field illumination is typically used [6,7]. In immobilization experiments the TIR geometry is often employed as it provides inherent surface specificity. This can reduce the background signal from free molecules in solution and therefore can increase the overall signal-to-noise [10] in experiments where binding events are to be studied. Confocal or two-photon scanning microscopes can also be used to study surface immobilized molecules and we will discuss the relative merits of this technique compared to TIR later in the chapter.

Next we shall describe these two common experimental geometries, but before we proceed we re-emphasize the key considerations of the experiment to maximize the signal-to-noise:

(1) Reduction of the size of the sample volume;

(2) Efficiency of delivery of the excitation light and collection of fluorescence;

(3) Efficiency of rejection of unwanted components in the collected light.

3.2.1 **Epi-fluorescence far-field microscopy**

The epi-fluorescence (episcopic-fluorescence) configuration (Figure 3.3) is very commonly encountered in microscopy. A single optical element is used to deliver the excitation light to the sample and to collect the fluorescence emission. Generally a high magnification, high numerical aperture microscope objective is used (see Section 3.6). 'One-photon' excitation (i.e. the use of excitation photons with energy matching the absorption transition in the fluorescent molecule of interest) is the most commonly encountered excitation protocol. A collimated laser beam is reflected off a dichroic mirror (see Section 3.3) into the back-aperture of the microscope objective. The light is focused to a spot at the focal plane, which is placed at the region of interest in

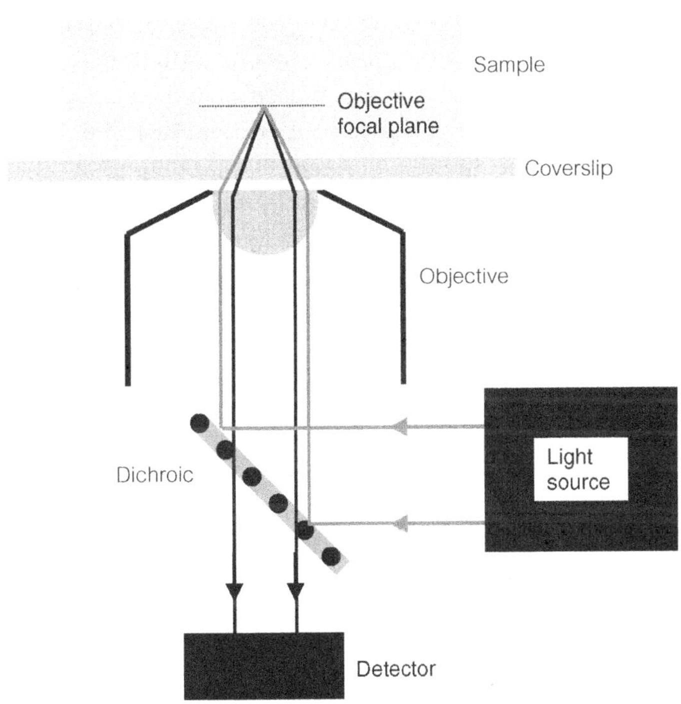

Figure 3.3 Illustration of the inverted epi-fluorescence configuration. The excitation beam (grey, collimated or parallel rays) is reflected towards the sample by a dichroic mirror (grey, black dots, see Section 3.3—essentially a semi-transparent mirror) and focused at a point within the sample at the front focal plane of a microscope objective. In this example the sample is represented as a fluid (light grey) sitting on a thin glass coverslip (grey), as is typical in these inverted configurations. A portion of the fluorescence (and scattered excitation light) is collected by the same microscope objective (black rays). The fluorescence is transmitted through the dichroic mirror towards the detector while scattered excitation light is not transmitted but reflected (not shown) back towards the light source.

the transparent sample (see Figure 3.4 for a detailed view of the excitation volume). The radius of the focused spot perpendicular to the direction of propagation is given by [28],

$$w = \frac{1.27\lambda f}{nD} \tag{3.1}$$

where λ is the wavelength of light used, f is the focal length of the objective, n is the refractive index of the medium in which the light is focused and D is the diameter of the incident beam. Using $f = 1.14$ mm, $n = 1.515$ (for a typical arrangement), $D = 3$ mm and $\lambda = 488$ nm gives an approximate focal spot diameter of ~300 nm. It should be noted that this figure relates to a theoretical minimum (the diffraction limit) and that rarely will this level of performance be reached due to a variety of optical aberrations present in the optical system.[1] It also relates to the l/e diameter of the spot (the distance at which the intensity has decayed to l/e of the maximum). Focal spot diameters of around 500 nm–1 μm are more typical. This focusing limits the extent of the region in the sample that is excited and therefore limits the region in which fluorescence or scattering is generated. Some of the fluorescence photons emitted by molecules in the excitation volume are collected by the microscope objective (see Section 3.6) and directed by the dichroic mirror to the detection arrangement.

Clearly, single-photon far-field excitation in this manner provides no spatial reduction of the excitation/collection volume in the direction of propagation of the light and so additional optics are required in the detection path to minimize

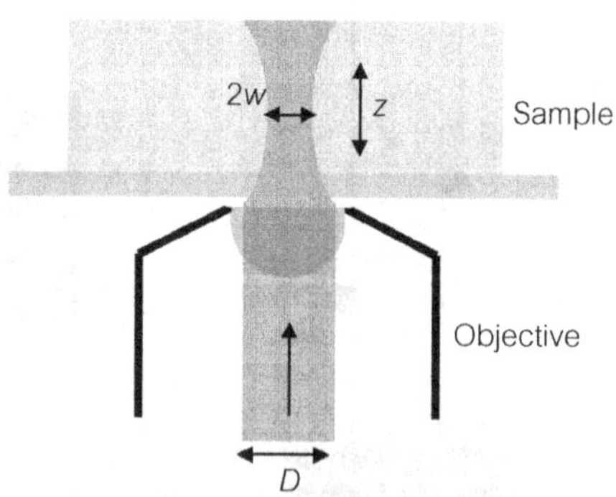

Figure 3.4 Close up of the excitation volume (not to scale) created by the epi-fluorescence configuration. A collimated laser beam (dark grey) of width D is focused through a glass coverslip (grey) by a microscope objective and brought to a focus some distance above the glass/water interface. The depth of focus z inside the sample is shown (defined, in this case, as twice the distance from the focal plane to the point at which the intensity has dropped by 1/e). The configuration results in spatial restriction of the beam to diameter 2w in the direction perpendicular to the direction of propagation but does not restrict the beam in the direction of propagation.

[1] For a discussion of aberrations, their causes and solutions see [12].

this volume. To achieve this, confocal detection is often employed (Figure 3.5). Confocal detection uses a small aperture (typically 25–50 μm in diameter) in the optical detection path [13]. The light collected by the microscope objective is focused onto this pinhole such that only light collected from very close to the focal plane (~0.5 μm) of the objective will be transmitted through the pinhole. Light originating from regions away from the focal plane of the objective will be out of focus at the pinhole and will be rejected to a large extent and not reach the detector. The use of this confocal arrangement therefore does not restrict the volume of excitation but does efficiently reduce the collection volume. The extent to which the confocal approach is effective is a function of the pinhole size, microscope objective, and the lens that is used to focus the light onto the pinhole. Details of optical arrangements that result in a depth of focus of the order of a few μm are presented in Section 3.9. A common modification to this principle is to use the point nature of some detectors to provide confocality rather than adding a pinhole in the detection path. For example, the active area of an avalanche

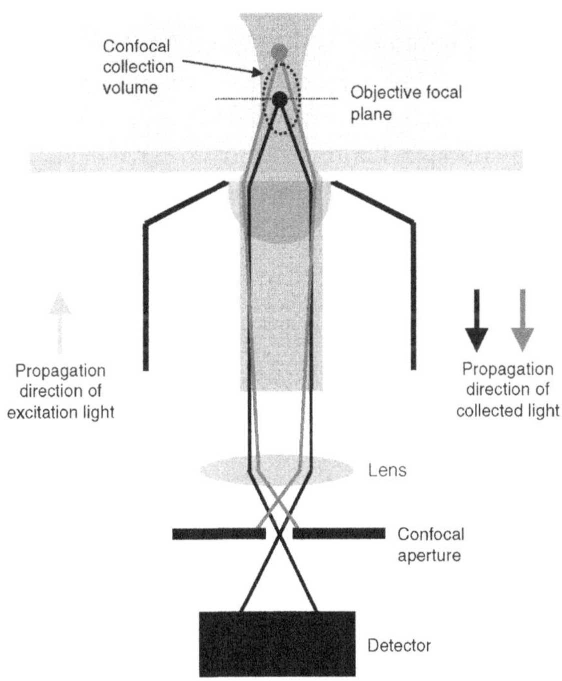

Figure 3.5 Illustration of the principle of confocal detection to limit the collection volume in the direction of the propagation of the excitation beam. Light emerging from near the focal plane (black spot) is collected and collimated by the microscope objective and then focused by a second lens to pass through an aperture and onto the detector. Light that originates from in front or behind the focal plane (grey spot) is out of focus at the aperture and only a small proportion continues to the detector. The aperture is said to be *confocal* with the objective. (Optics delivering the excitation light have been omitted for clarity).

photodiode detector is typically circular and ~100 μm in diameter (see Section 3.7) and so its use provides inherent confocality when used with suitable focusing optics.

The transmission efficiency of modern microscope objectives at visible wavelengths is high, approaching 90% (see Section 3.6), but the overall detection efficiency of the objective is a more complex issue and is discussed in detail in Section 3.6. The fluorescence from a single molecule is emitted in all directions, however the microscope objective collects the fluorescence from a solid angle defined by the numerical aperture (see Section 3.6). Furthermore, when the dependence of the collection efficiency on the position of the molecule within the focal volume is taken into account, the overall objective detection efficiency at visible wavelengths can be as low as 20% [11]. Although this may seem very low it represents a practical limitation of using a single microscope objective and this limitation exists for all forms of microscopy. It is thus essential to optimize the efficiency of all other elements in the instrument.

Single-photon excitation in an epi-fluroescence configuration combined with confocal detection provides the necessary spatial reduction of the collection volume that is required to minimize the background noise, and has the added benefit of simplicity. However, this approach has two distinct disadvantages. First, the excitation light is only spatially restricted in one direction (perpendicular to the light path) and although fluorescence from far outside the focal plane is rejected by the confocal detection scheme, molecules in the larger excitation volume are continuously irradiated by this simple arrangement (Figure 3.6, left panel). This causes unnecessary photobleaching of molecules that can reduce the useful lifetime of the sample, particularly in solid samples. Second, the introduction of a pinhole (and associated optics, focusing lens) in the detection path reduces the overall amount of fluorescence from the single molecule of interest reaching the detector (potentially overcome by using a detector with a small active area as discussed earlier).

Adopting a multi-photon excitation approach can ameliorate some of these disadvantages. Two-photon excitation requires the simultaneous absorption by the fluorescent molecule of two photons, each of approximately half the energy of the optical absorption transition, resulting in promotion of an electron to the first excited singlet state as with one-photon excitation (Figure 3.7). A formal description of two-photon excitation is beyond the scope of this text, however, the reader may wish to consult [7]. A number of useful properties of two-photon excitation can be summarized straightforwardly: the use of two-photon excitation, unlike single-photon confocal excitation/detection, results in an inherent reduction of the excitation volume in the direction of propagation of the laser beam. This is because only in a region near the focal plane is the photon flux high enough to result in a significant probability of simultaneous absorption of two photons. Because the individual photon energy is much less than the absorption transition energy, the amount of excitation along the direction of propagation away from the focal plane is inherently reduced (see Figure 3.6, right panel), which reduces

Figure 3.6 Photographs of an illuminated fluorescent material showing the laser focus for one-photon (left) and two-photon (right) excitation. In the first case a larger volume of the sample is illuminated (which is prone to bleaching) due to the large spatial extent in the direction of propagation. Two-photon excitation inherently restricts the volume in the direction of propagation since the photon fluxes are only sufficiently high at the focus to allow quasi-simultaneous absorption of two photons. Reprinted from Haustein, E and Schwille, P, Ultrasensitive investigations of biological systems by fluorescence correlation spectroscopy, Methods **29** (2003) 153–166, with permission from Elsevier.

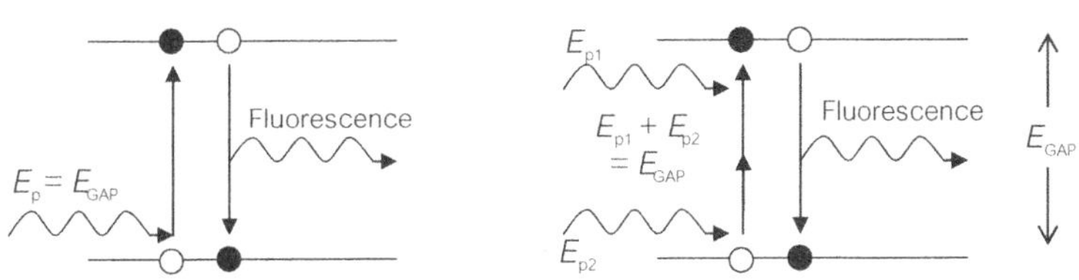

Figure 3.7 The principles of one- and two-photon excitation. In one-photon excitation (left) a single photon, whose energy matches the electronic transition between the ground state and first singlet state causes an electron to be excited. Subsequent relaxation of the electron results in fluorescence emission. In two-photon excitation, two photons, each with approximately half the energy of the electronic transition, are absorbed simultaneously resulting in excitation of an electron and subsequent fluorescence.

photobleaching. Finally, because the multi-photon excited electronic transitions are symmetry forbidden, the absorption spectra obey different selection rules and vibronic couplings [7]. This tends to result in a broadening of the fluorescence excitation spectrum of many commonly available dyes [14] providing the opportunity to excite fluorescence in multiple dyes whose fluorescence emission peaks are spectrally well separated (making their discrete detection more straightforward). Although this means that it is necessary to acquire a knowledge of the two-photon excitation spectra of dyes [14], this added complexity is counterbalanced by the potential to conduct and simplify multi-dye experiments (see [15–18]). In one-photon excitation a particular excitation wavelength is required for efficient fluorescence emission, and generally the difference between this wavelength and the peak emission wavelength (the Stokes shift) is small. Thus all dyes efficiently excited at the same range of wavelengths tend to have spectrally similar emission wavelengths.

One-and two-photon epi-fluorescence permit similar experiments to be carried out, although one may have practical advantages over the other [19].

It is clear how these configurations can be used to perform diffusion single molecule fluorescence measurements of the type described in Figure 3.2. The point spread function (PSF, the convolution of the excitation and collection volumes) for either one-photon (confocal) or two-photon geometries defines a volume in solution that is *approximately* cylindrical, around 500 nm in diameter and 1 μm long (so a volume of ~0.2 fl) through which the molecules diffuse and are excited. The configuration can also be used to take images of a surface if the detection volume is placed at the coverslip/water interface (see Figures 3.2 (d), 3.3 and 3.4) and scanned over the sample surface (see Section 3.9).

3.2.2 **The PSF in single-photon confocal and two-photon epi-fluorescence illumination systems**

The instrument point spread function (PSF, also known as the spatial detectivity function, among other synonyms) is the mathematical function that describes the way in which light is transformed as it passes through an optical system. For example, in imaging systems the image obtained by the detector is convoluted with the transmission, detection, and illumination properties of the particular instrument used. In a confocal system one might first consider the shape of the volume created in solution from the laser beam focused by the microscope objective, and determine the PSF that describes this volume. The volume from which light is detected however is further restricted by the confocal pinhole, so this modifies the PSF and a PSF for the combined pinhole—objective system is determined. Further, any lens, filter or detector may alter the apparent spatial profile of the volume from which light is detected. The convolution of all of these gives us the instrument point spread function and so a description of the detection volume through which molecules may pass. The form of the PSF volume defined either by a confocal or two-photon geometry is therefore of crucial importance for a range of fluctuation spectroscopy methods, in particular fluorescence correlation spectroscopy (FCS) and the photon counting histogram (PCH, see Chapter 2). Ratiometric techniques, such as FRET, compare the instantaneous signals of two fluorophores attached to the same molecule and thus the shape, size, and intensity profile of the detection volume are unimportant. However, fluctuation techniques generally rely on measuring the photon count distribution for a signal detected from a single diffusing dye molecule. This data is then largely stochastic, different signals being caused by diffusion along random paths through the volume. Thus to enable a prediction of the resultant data (for example, to calculate the expected photon count distribution or to determine a diffusion rate) the sample volume must be well defined.

In confocal microscopy the PSF is a function of a number of physical components, namely the laser beam, the microscope objective, the confocal pinhole, and the detector. The PSF is a convolution of the effect of all of these components. Despite this apparent complexity it is perhaps surprising that a simple description of the PSF for these microscopes has been widely applied in which the sample volume is described by a three-dimensional Gaussian with a $1/e^2$ beam waist diameter $2\omega_0$ and a length $2z_0$ along the optic axis [6,20,21] and is given by,

$$\overline{\mathrm{PSF}}(\vec{r}) = \overline{\mathrm{PSF}}(x,y,z) = \exp\left(-\frac{2(x^2+y^2)}{\omega_0^2} - \frac{2z^2}{z_0^2}\right) \tag{3.2}$$

This simple model has however some weaknesses; in particular, in many cases it apparently does not describe the shape of volume accurately and can introduce artefacts into PCH and FCS measurements that manifest as, for example, apparent additional species in the solution [6,21–25].

Alternative, both empirically determined and theoretical, descriptions have been suggested for confocal PCH [23–25] (and already used in the similar analysis technique FIDA [22]). For confocal FCS the discussion of this issue has centred more around the optimization of experimental conditions to achieve as near a three-dimensional Gaussian PSF (as defined by equation 3.2) as possible [6,21]. A detailed study by Hess and co-workers [6] suggests that the near three-dimensional Gaussian PSF can be obtained by careful illumination of the sample with a Gaussian laser beam *underfilling* (i.e. smaller than the back aperture) of the microscope objective and by using a small confocal aperture. This is somewhat contrary to convention; usually, filling the back aperture with a Gaussian intensity profile laser beam would be considered close to optimum. Certainly, our own experience is that by underfilling the back aperture of the objective we have never observed significant artefacts in either FCS or PCH in control experiments using a three-dimensional Gaussian description of the PSF (see Chapter 2 for many examples). However, we emphasize that with all experimental setups it is essential to perform extensive controls, in all solution conditions (in case of refractive index changes for example) for all techniques. This is particularly important when one hopes to probe sample heterogeneity. Thus any new instrument (or alignment) must be checked using FCS and PCH. In the case of two- or multi-photon excitation the three-dimensional Gaussian description has been shown to provide an accurate description of the volume without modification or effort [6].

While ratiometric techniques such as FRET are in principle unconcerned with the PSF there are some aspects that are still of importance. Achieving a three-dimensional Gaussian sample volume is also likely to indicate best-optimization of the instrument and therefore is likely to result in the best signal-to-noise. Further, as was discussed in Chapter 2, ratiometric techniques such as FRET are not immune to other effects. For example, changes in solution refractive index or mis-alignment could in principle deform the PSF volume affecting the occupancy time, and therefore the measurement time, of each molecule. Thus, consistent alignment procedures and control measurements to characterize the PSF are prudent.

3.2.3 **Near-field or evanescent excitation**

Near-field or evanescent excitation is particularly suited to measurements of surface immobilized molecules [10]. By using an evanescent wave (formed at the surface between the glass coverslip and the solvent) to excite the sample it is possible to limit the extent of the excitation volume in the direction normal to the surface. An evanescent wave is generated by total internal reflection [9,26,27]. Light incident at the interface between two materials of differing refractive index (n_1 and n_2) can be transmitted, refracted, and reflected in proportions described by the Fresnel formulation[2] (see Figure 3.8). If the angle of incidence (θ_i) is increased then a 'critical' angle is reached at which all the light is reflected, that is, total internal reflection occurs (note that above the critical angle the Fresnel equations no longer hold). Snell's law gives the critical angle θ_C in terms of the refractive indices,

$$\theta_C = \sin^{-1}\left(\frac{n_2}{n_1}\right), \tag{3.3}$$

for the case where $n_1 > n_2$. Thus, as the angle of incidence is increased the relative intensity of the reflected ray increases until the critical angle is reached. Figure 3.9

[2] For further reading see [28,29].

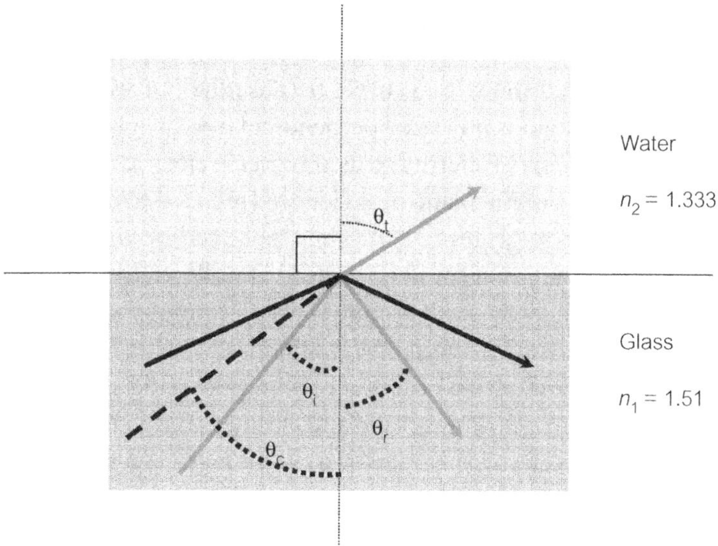

Figure 3.8 Ray diagram illustrating the phenomena of transmission, reflection, and refraction at the interface between two media of differing refractive index. In this example rays pass from a high (glass) to low (water) refractive index medium. At angles of incidence (θ_i) below the critical angle (θ_c) transmitted (refracted) and reflected rays are produced (grey). Above the critical angle total internal reflection occurs (black) and no propagating transmitted ray exists.

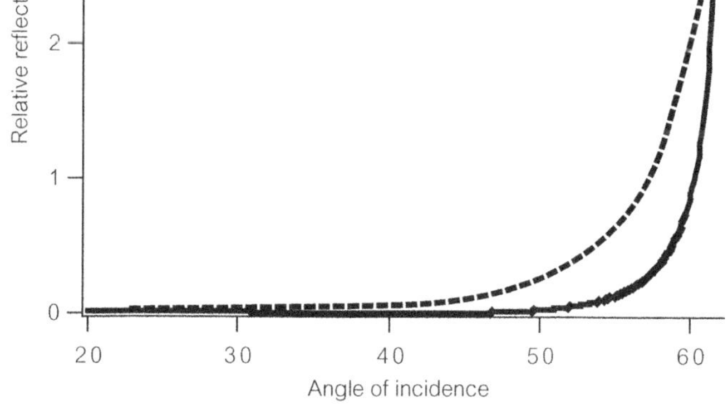

Figure 3.9 Graphs illustrating the relative intensity of the reflected ray from the boundary between glass ($n = 1.51$) and water ($n = 1.33$) as a function of angle of incidence up to the critical angle ($\theta = 61.7°$). The reflected light intensity is calculated using the Fresnel equations [29] for both p- (solid line) and s- (broken line) polarized light.

shows the relative intensity of the reflected ray from the boundary between glass ($n = 1.51$) and water ($n = 1.33$) for angles of incidence up to the critical angle ($\theta_C = 61.7°$). Figure 3.9 shows two curves because the process is dependent on whether the incident light is polarized perpendicular to the plane containing the incident and reflected rays (referred to as transverse electric or s-polarized), or whether it is polarized in the plane of incidence (referred to as transverse magnetic or p-polarized) (see Figure 3.10). Thus, polarization must be considered when using this excitation geometry. Indeed it follows that if circularly or elliptically polarized light is used, the interface will induce a degree of further polarization in the transmitted and reflected light.

When total internal reflection occurs at or above the critical angle (Figure 3.8) the Fresnel formulation is no longer valid and although the incident light is totally internally reflected, an electric field (an evanescent field) is generated on the low refractive index side of the interface that does not propagate into the lower refractive index medium [8]. The evanescent field extends into the lower refractive index material with intensity dependence in the z direction, normal to the interface, given by,

$$I(z) = I(0) \exp\left(-\frac{z}{d}\right) \tag{3.4}$$

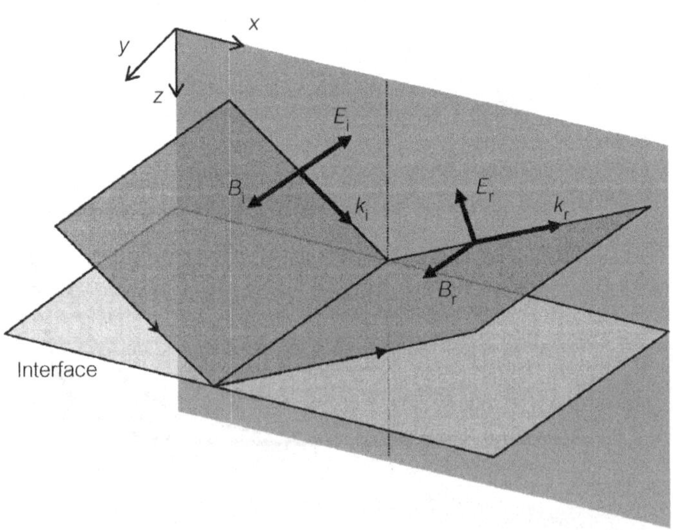

Figure 3.10 Definition of the coordinate system, a light ray that is totally internally reflected at a boundary between glass (above) and water (below). Electric (E), magnetic (B), and propagation (k) vectors are shown for the case of p-polarized light. Subscripts i and r refer to incident and reflected light rays, respectively.

where d is the 'lifetime' of the decay, or the penetration depth and is related to the angle of incidence, the refractive indices of the two materials and the wavelength of light and is given by,

$$d = \frac{\lambda}{4\pi\sqrt{(n_1^2 \sin^2\theta - n_2^2)}}, \tag{3.5}$$

where λ is the wavelength of light and n_1 is the higher of the two refractive indices. Figure 3.11 provides a schematic of the arrangement showing how the excitation volume is limited to a region close to the interface. Clearly, it is useful to have an appreciation of the penetration depth, d, of the evanescent wave into the lower refractive index medium. At the critical angle $d \rightarrow \infty$, however above this angle d is reduced to a value equal to the wavelength of light used, or smaller, as the angle of the incident light is increased. Presented in Table 3.1 are the critical angles and penetration depths for a range of materials at a wavelength of 488 nm. The polarization of the incident light and angle of incidence play a role in determining the intensity of the evanescent wave at the interface [8]. Figure 3.12 shows this dependence for the boundary between glass ($n = 1.51$) and water ($n = 1.33$) for angles of incidence greater than the critical angle ($\theta = 61.7$). Note that whilst the intensity of the evanescent wave is dependent on the polarization of the incident light it can be seen that the penetration depth is not (from equation 3.5).

So far we have not discussed the state of polarization of the evanescent wave, which is an important factor since the orientation of the transition dipole

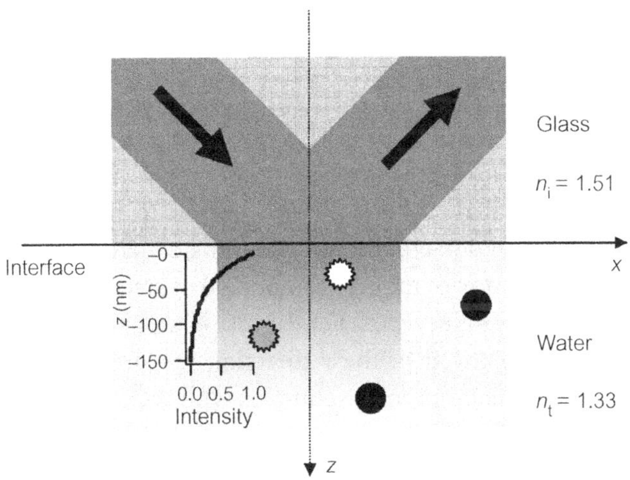

Figure 3.11 Illustration of the principle of evanescent wave excitation for single molecule fluorescence experiments. A laser beam is totally internally reflected at the interface between a higher refractive index medium (often glass) and one with a lower refractive index (often water). An evanescent wave is generated on the lower refractive index side of the interface that decays exponentially in the z direction. A typical exponential decay length for the evanescent electric field is 100 nm (see inset graph). Only molecules near the interface (within ~100 nm) experience an optical field strength large enough to have a high probability of being excited and emitting fluorescence.

Table 3.1 Critical angles of total internal reflection and evanescent wave penetration depths

	Glass ($n = 1.51$)	Quartz ($n = 1.55$)
Air ($n = 1$)	41.3°, 193 nm	40.2°, 189 nm
Water ($n = 1.33$)	61.9°, 215 nm	59.6°, 206 nm
Cell cytoplasm ($n = 1.36$)	63.8°, 220 nm	61.3°, 211 nm
8M urea ($n = 1.4$)	67.5°, 229 nm	64.6°, 215 nm

As the penetration depth tends to infinity at the critical angle, the depths have been calculated 1° above the critical angle.

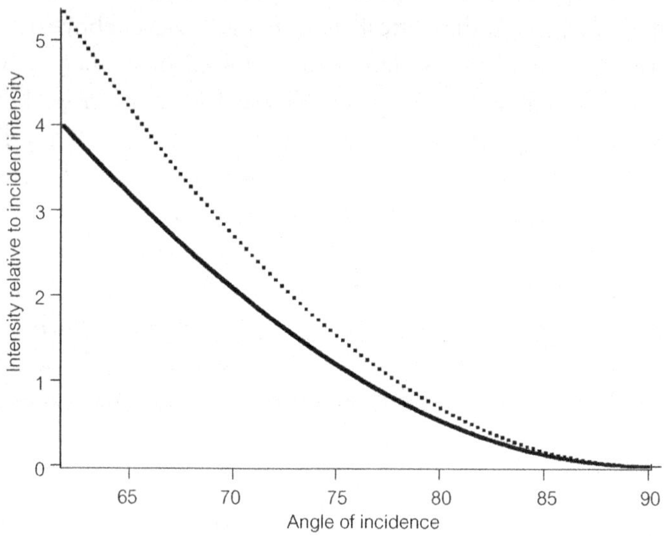

Figure 3.12 Graph illustrating the relative intensity of the evanescent field created from total internal reflection of 488 nm light at the boundary between glass ($n = 1.51$) and water ($n = 1.33$) for angles of incidence above the critical angle ($\theta = 61.7°$) [8]. The curves for both p- (solid line) and s- (broken line) polarized light are shown.

moment of molecules at the interface, with respect to the polarization, has a strong effect on the excitation efficiency [30] of single molecules. It can be shown that the state of polarization of the evanescent field depends on the state of polarization of the incident light [8,31,32]. For the case of s-polarized light the evanescent electric field is polarized in the y direction (Figure 3.10). In the case of p-polarized light the situation is more complex with the evanescent electric field polarized principally in z (along the optical axis, see Figure 3.10) but also with a very small component in x [31]. To alleviate this in some real experiments it is common to use circularly polarized excitation light; this results in an evanescent field in the z direction (from the p-polarized component) and in the y direction (from the s-polarized component). However, for efficient excitation, a dye's transition dipole must be aligned in the direction of one of these components, thus molecules with dipoles fixed along x are still very inefficiently excited (around 10%

compared to the z component [32]). The dominant polarization component in the z direction has been exploited in some experiments in order to probe fluorophore orientation [31,33]. For single molecule studies however, where we wish to use techniques such as FRET, the lack of an evanescent field component polarized along x could be problematical, unless one has been able to ensure that the dye labels are free to rotate on a timescale much faster than the measurement time (see Chapter 2). The only way to entirely remove this effect is either to use two orthogonal excitation beams so that dipoles in x and y are excited equally [32], or alternatively to use more inefficient annular excitation schemes [34].

3.2.4 Realization of near-field excitation for single molecule detection

There are a number of common methods for achieving TIR for single molecule detection. In this section we review two of the most common methods: prism coupling and through-the-objective configurations. In particular, we focus on objective based configurations as these are perhaps the most straightforward to implement, provide flexibility, and also require only modest modification to standard epi-fluorescence configurations described in previous sections.

A convenient method of creating TIR illumination is achieved by focusing a collimated (parallel light rays) light beam onto the periphery of the back focal plane (BFP) of a high numerical aperture microscope objective [27] as illustrated in Figure 3.13. This results in a narrow collimated beam emerging from the objective at an angle determined by the distance of the point of focus in the BFP from the optical axis of the objective (so the distance from the centre of the BFP) and also the numerical aperture of the objective. The sample, say a glass coverslip/aqueous solution, is placed on the objective, using index matching fluid, thus the beam emerging from the objective is now incident on the glass/water interface at some angle. Translation of the incident light beam across the BFP results in a variation of the angle of incidence at the sample interface. Moving the focus in the BFP nearer to the centre of the back aperture reduces the angle of incidence at the interface, while moving away from the centre towards the edge of the back aperture increases the angle of incidence. In this manner the beam can be scanned across the back aperture of the objective (see black double arrow in Figure 3.13) to vary the angle of incidence to near the critical angle for TIR. To achieve TIR the numerical aperture of the objective must be at least equal to the lower of the two refractive indices in the arrangement; very often this is water or an aqueous solution for which the refractive index is ~1.33. Therefore, in general, objectives with numerical apertures of 1.45 (maximum theoretical angle of incidence of 73.8°) or greater are used for TIR applications.

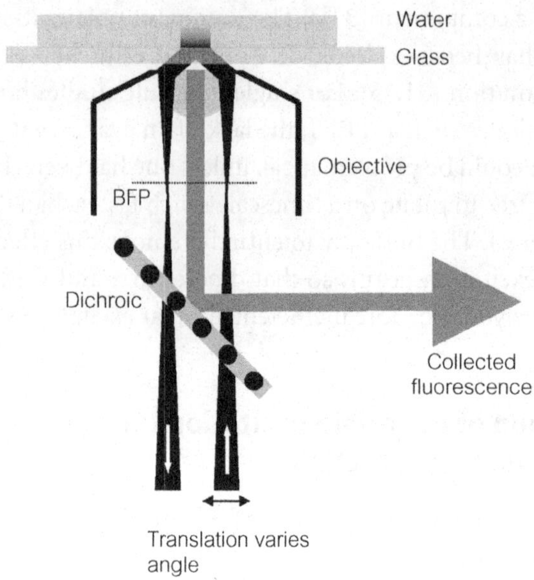

Figure 3.13 Illustration of the through-the-objective configuration for evanescent field generation at an interface between a high and a low refractive index medium. Collimated excitation light (black) is focused onto the periphery of the back focal plane (BFP) of a high numerical aperture microscope objective. The resulting narrow collimated beam emerging from the objective is incident at an angle on the glass–water interface. The precise angle is a function of the distance between the optical axis of the objective and the position of the focus on the BFP and also of the numerical aperture of the objective. Translating the incident light beam across the BFP can be used to vary this angle. Emitted fluorescence is collected by the objective in the normal manner (grey).

Recently, manufacturers have designed objectives with even higher numerical apertures that would be ideal for TIR experiments with high refractive index samples (see Section 3.6). However, these objectives require special immersion fluid and coverslip materials, which may have implications for other aspects of the experiments such as surface immobilization chemistry and cost. In essence the TIR configuration is identical to the epi-fluorescence configuration discussed earlier, with the exception of the excitation beam convergence and positioning with respect to the optical axis of the objective.

Another common, although perhaps less convenient, method for achieving TIR excitation is to use a prism. In this case the evanescent wave is generated either at the interface between a prism and a lower refractive index medium, in which case the prism often serves as the sample substrate itself, or the prism is optically coupled to a disposable glass slide that forms the sample substrate. In this geometry the sample is generally sandwiched between a second glass coverslip and fluorescence is collected using an oil immersion microscope objective (see Figure 3.14). In cases where water immersion objectives can be used (see Section 3.6) a second coverslip may not be necessary (depending on the objective design) and the objective can be placed directly in the solution. However, care must be

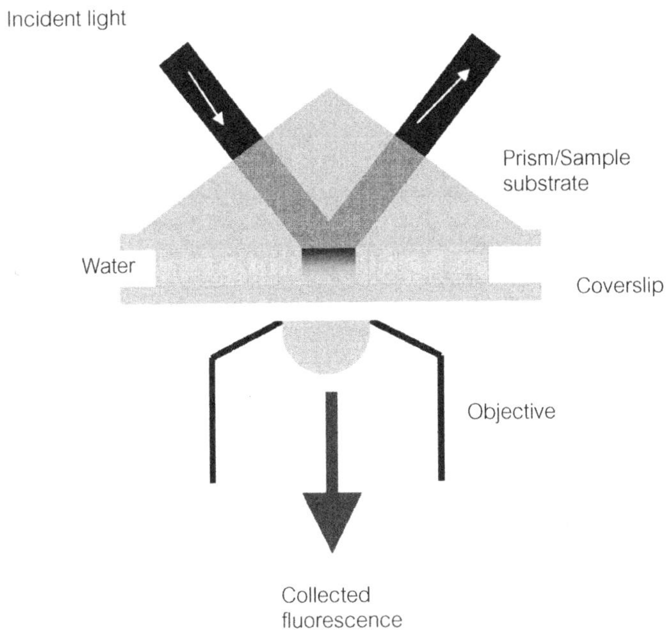

Figure 3.14 Illustration of the arrangement for prism-based evanescent wave excitation. A collimated laser beam (black) is totally internally reflected off the interface formed by the prism/substrate and water. A microscope objective is used to collect the fluorescence generated by the evanescent field at the surface of the substrate in 'wide field' mode.

taken that the objective itself is not contaminated with fluorescent material when it is exposed to the sample solution in this way. The clear advantage of the prism arrangement is simplicity and low cost, requiring only an inexpensive glass or quartz prism to couple the light into the sample and a relatively inexpensive lower numerical aperture microscope objective for the collection (see Section 3.6). As a result, alignment is also more straightforward when using the prism coupling method. An additional advantage is the elimination of fluorescence from the optical components in microscope objectives, which can be a significant source of unwanted background signal in single molecule experiments; in a prism configuration excitation light does not propagate through the objective and further the prism and slides used can be made of silica, which has a lower background fluorescence signal than the glass typically used in objectives. The inclusion of multi-wavelength excitation into the instrument is also greatly simplified with the prism configuration, as the excitation light does not need to be directed through an expensive dual-transmission-band dichroic (see Figures 3.13 and 3.15) and into the objective[3]. However, the drawback of the prism coupling approach is the need for optical realignment each time the sample is exchanged and the need for

[3] For an alternative arrangement see ref. [9].

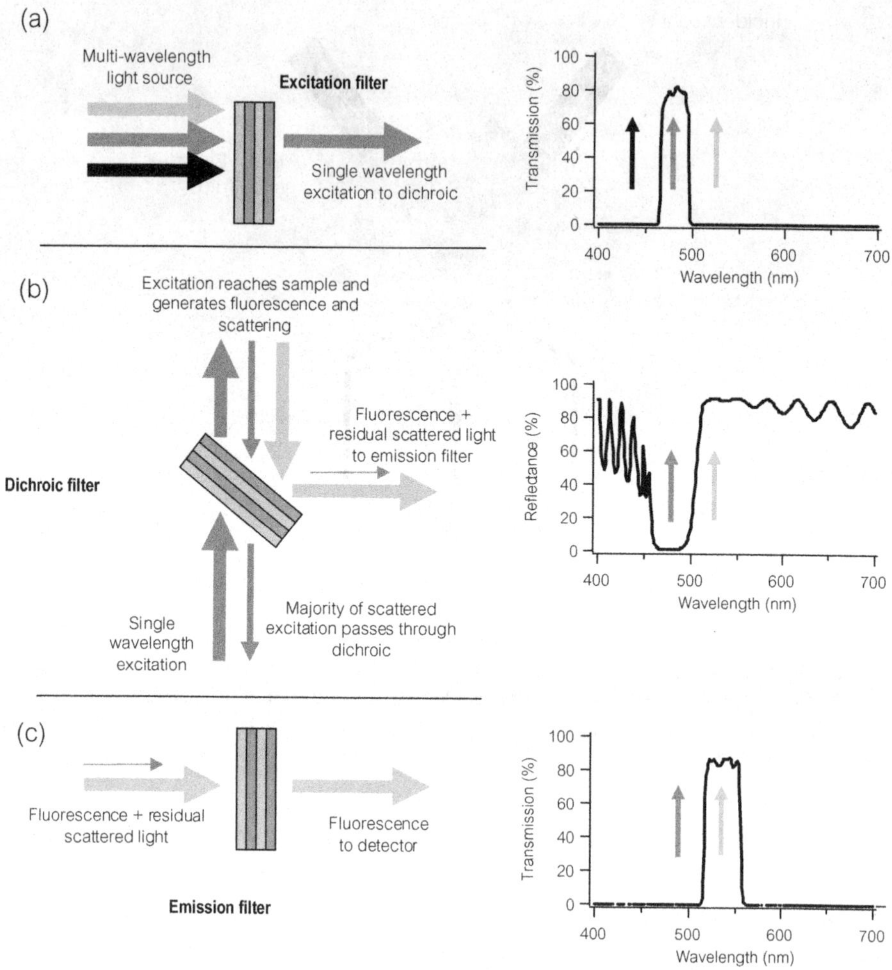

Figure 3.15 Illustration of the three filter types used in single molecule fluorescence experiments. (a) The excitation filter selects the correct excitation wavelength from a multiple wavelength light source. (b) The dichroic filter, through which the excitation light passes, reflects the returning emitted fluorescence from the sample separating it from most of the scattered excitation light which is re-transmitted. (c) The emission filter removes much of the remaining scattered light and any unwanted fluorescence. The right-hand panels illustrate the transmission characteristics of typical examples of each filter. (Curves shown, top to bottom are: Chroma Inc. filters D480/30x, 505DCLP, and D535/40m).

very thin samples (depending on the working distance/focal length of the objective) or the use of a water immersion objective. A thorough comparison of these two approaches can be found elsewhere [35].

Using either through-the-objective or prism TIR configurations the sample volume is restricted in the direction perpendicular to the sample plane. Depending on the precise geometry and the divergence of the beam at the BFP, the area of the interface that is illuminated varies, but is generally of the order of 100 μm^2. If a microscope objective collects the fluorescence signal and directs it to a 'point detector' (such as an avalanche photodiode), then the signal from quite a large area is integrated and single molecule sensitivity may be lost. Controlling the density of immobilized molecules at the interface could counteract this. However, generally the sample plane is imaged onto an array detector, such as an intensified CCD. This then forms an image of the surface at the detector and so inherent spatial sensitivity in the plane of the sample is obtained. Such an arrangement is often referred to as total internal reflection fluorescence microscopy or TIRFM. It is clear then that the small penetration depth of the exciting light above the glass/solvent interface (of the order of $\lambda/2$ nm) combined with the spatial sensitivity in the plane of the sample (technically dependant on the detector array resolution in combination with the imaging optics, but essentially diffraction limited to a few μm) fulfil the requirements for background rejection outlined at the start of Section 3.2.

3.3 Methods for discriminating signal from noise

Background signal rejection is critical for the successful implementation of single molecule detection. We have discussed the origins of the background signal: Rayleigh and Raman scattered light (from the sample, solvent and impurities) and extraneous fluorescence (from impurities). We have discussed the requirement to minimize the excitation/collection volume to reduce these background signals and described the options for the delivery and collection of light and how these achieve this. We will now consider methods to 'condition' the detected signal to remove background noise by using spectral or temporal discrimination.

3.3.1 Spectral discrimination

Spectral discrimination is the selection of certain wavelength (or energy) photons using thin-film dielectric, glass or notch filters or diffraction gratings. Diffraction gratings provide wavelength sensitivity by taking advantage of the wavelength dependent angles of diffraction of light with respect to the angle of incidence of the broad wavelength range fluorescence. Thus, monitoring the signal at different angles relative to the diffractive element (grating) provides wavelength sensitivity. Diffraction gratings are seldom used for single molecule fluorescence experiments because the transmission efficiency of these optical elements at the desired wavelength is low ($<50\%$) and the rejection of unwanted wavelengths poor. The exception is the case where it is essential to be able to 'scan' the detection wavelength in order to record single molecule fluorescence emission spectra [36]. Glass colour filters (or absorption filters—materials showing differential wavelength dependent absorption of light) are relatively inexpensive, easily cleaned, and have optical properties that are stable over long periods of time. However, they have the disadvantage of relying exclusively on absorption to reduce background signal. The amount of background noise rejection therefore depends on the thickness of the filter, which in turn affects the transmission of the fluorescence signal. Furthermore, the glasses used for these filters can themselves be fluorescent which, in the worst case, could generate a signal overlapping in wavelength with the desired fluorescence signal, effectively reducing the overall signal-to-noise ratio. In general, glass colour filters are therefore not preferred for single molecule fluorescence experiments. Holographic notch and super-notch filters are based around volume transmission gratings and operate in a similar manner to conventional diffraction gratings deviating the direction of propagation of a particular wavelength. Holographic notch filters provide significantly enhanced transmission efficiencies and much stronger rejection of unwanted wavelengths compared to conventional line gratings. Furthermore, they can be designed in a range of configurations, allowing single or multiple narrow bands of light to be transmitted or blocked. The design of many types of notch filters also results in good optical stability and long lifetimes. However, the main disadvantage of notch filters is their cost, which can be ten times higher than thin-film interference filters with similar performances that are discussed later.

Thin-film interference filters are composed of thin films of dielectric material, each approximately the thickness of the wavelength of light, layered in a stack. The reflections from the interfaces between the layers interfere constructively or destructively depending on the wavelength of the light, the thickness of the layers, and the refractive index of the layer material. Their operation is analogous to the colours produced in detergent bubbles, there the different colours observed trace contours of uniform thickness. Filters consisting of many stacked layers can be designed to pass narrow wavelength bands (from a few nanometers wide pass band to tens of nanometers wide) with transmission efficiencies of 80–95% at visible wavelengths. Auto-fluorescence of these filters is low and the blocking of wavelengths outside the pass band can be very high (several orders of magnitude lower transmission than in the pass band, typically of the order of optical density[4] 4 or 0.01%). However, like glass colour filters, interference filters also suffer a number of disadvantages. The dielectric material forming the thin films tends to

[4] Optical density is a measure of the transmittance of a material. It is given by $OD = \log_{10}(1/T)$, where T is the transmission of a component expressed as a fraction. Thus for a component with 1% transmission efficiency $T = 0.01$ and $OD = 2$.

be quite soft, which means that the filters are easily damaged and can degrade. Cleaning interference filters should be avoided as the thin coatings are easily damaged; the lifetime of this type of filter may only be 5 years (accelerated if moisture/solvent vapours are present) after which noticeable degradation of performance may occur.

In addition, the availability of materials restricts the range of wavelengths the filters can be designed to pass or block. In particular, performance of these types of filters at ultra-violet wavelengths is poor, and glass colour filters can outperform interference filters at these wavelengths. Interference filters are also designed to operate at a particular angle of incidence (as the path length through the dielectric layers will vary as a function of angle of incidence, affecting interference of light at particular wavelengths) and therefore care must be exercised in the design of the optical arrangement to ensure that light does not impinge on the filter in a range of angles (for example, one should not focus through an interference filter).

Furthermore, thin-film interference filters often suffer from a polarization sensitivity, which can produce unwanted selectivity in applications where polarization sensitivity is desired (see Section 3.4). Fortunately, a number of suppliers (Chroma Technology Corp., VT, USA for example) are now becoming familiar with the requirements of filters for single molecule applications and are constantly developing the technology. Compared to other types of filters the interference type cannot be beaten on cost; good quality filters suitable for single molecule work are available for around US$150.

Single molecule detection experiments usually incorporate three main types of thin-film interference filters, often called excitation, dichroic, and emission filters (see Figure 3.15). Excitation filters may be required when broad emission wavelength lamps are used as the excitation source, to narrow the range of excitation wavelengths used. Even when laser sources are used an excitation filter is worth incorporating to reject unwanted luminescence or other laser emission lines that might overlap the detection wavelength range of the experiment. Excitation filters are generally used at near-normal incidence with collimated beams and reflect a portion of the undesired wavelengths back towards the source.

Dichroic filters (or dichroic mirrors) are generally used at an angle of incidence of 45° and are used to separate light into two (or more) colour ranges. In single molecule instrumentation these filters enable the use of the epi-fluorescence configuration discussed in Section 3.2. The excitation light is transmitted through the dichroic and the back propagating fluorescence from the sample is reflected by the dichroic into the detection pathway (or *vice versa* depending on if the dichroic is long- or short- pass) whilst the reflected or backscattered excitation light is transmitted through the filter back towards the source and away from the detection path.

It is worth noting that regardless of the theoretical performance of a filter it will transmit/reflect a proportion of the unwanted wavelengths and so in single molecule detection it is a matter of placing different types of filters in series until sufficient blocking is achieved and the signal-to-noise ratio is acceptable (acknowledging some loss of the desired signal with each filter addition). The dichroic filter alone is rarely sufficient to provide the spectral discrimination required and therefore emission filters are placed in front of the detector(s) to reject light outside a desired pass band (or above or below a particular wavelength).

3.3.2 **Temporal discrimination**

The Rayleigh and Raman scattering processes are essentially instantaneous and therefore the background signal produced by scattering follows the temporal profile of the excitation. The time evolution of fluorescence emission on the other hand is determined by the lifetime of the excited state of the fluorophore (see Chapter 4). A fluorophore such as a typical fluorescent dye will remain in the excited state for at least a few nanoseconds before relaxing back to the ground state resulting in emission of a fluorescence photon. Thus, fluorescence emission occurs on a nanosecond timescale following excitation while Rayleigh and Raman scattering are pseudo-instantaneous. Therefore, if pulsed lasers are used, with pulse widths of femto- or picoseconds, then the scattered light signal can be discriminated from much of the fluorescence signal using time-gated detection. This type of instrumentation is complex and very expensive, and therefore not widely used and the reader is referred elsewhere for further detail [37].

3.4 **Wavelength or polarization selection optics**

A number of single molecule sensitive techniques (see Chapter 2) are based on the detection and processing of the fluorescence signal integrated over a range of wavelengths either from freely diffusing molecules or molecules immobilized on a surface. Such techniques represent a first step proof-of-principle for any single molecule fluorescence instrumentation. However, greater insights in some areas can often be obtained by more sophisticated analyses. For example, FRET (see Section 3.1 and Chapters 2, 5, and 6) can provide more detailed structural information by monitoring the fluorescence from two dyes attached to a single molecule (such as a protein or nucleic acid) which have spectrally distinct fluorescence emission characteristics and between which energy can be transferred.

Methodologies such as coincidence detection [38] and cross-correlation [39,40] spectroscopy also rely on the spectral discrimination of two, or more, fluorescence signals. In addition to spectral discrimination another informative parameter is polarization (see Chapters 2, 5, 6, and [36,41]).

Both spectral and polarization discrimination of the background subtracted signals are trivial to implement (see Figure 3.16). Standard dichroic filters are available for all common dye pairs and inexpensive polarization beam splitters are commonly available from optical component manufacturers.

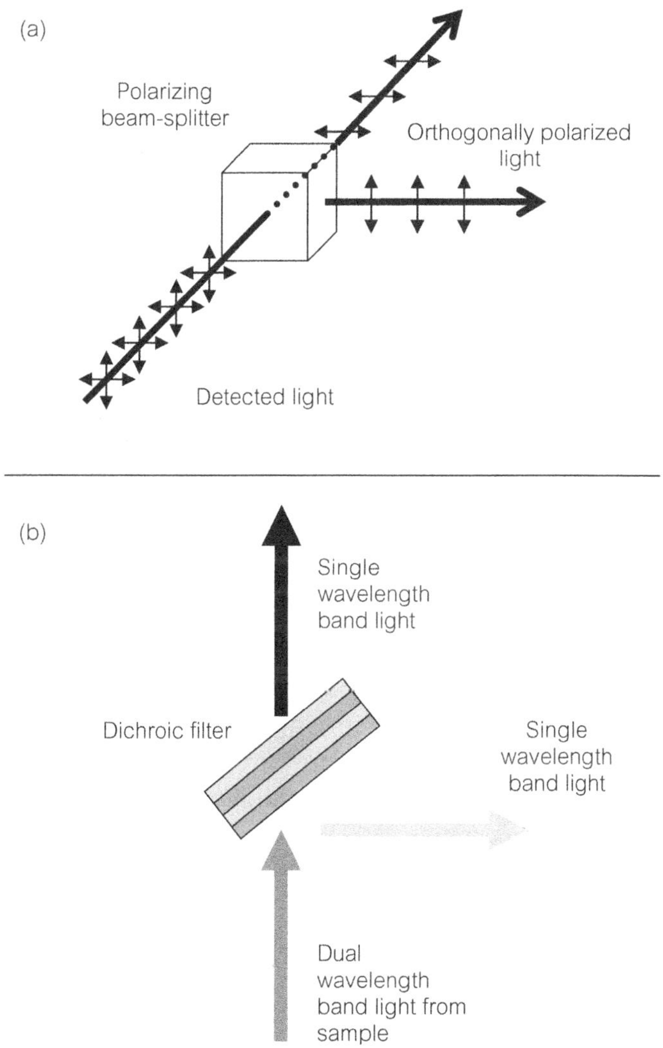

Figure 3.16 Illustrations of the use of a polarizing beam splitter to select fluorescence emission polarization (a) and a dichroic mirror to process two-colour fluorescence emission in experiments such as FRET (b).

3.5 Excitation sources

The excitation source for single molecule studies is chosen on the basis of the excitation wavelength that is required, which in turn depends on the fluorophore to be used. Since single molecule detection requires a fluorophore with a high quantum yield, by far the most commonly used are specifically designed fluorescent dyes. Although a very broad range of such dyes is available (see Chapter 4), in general, the ones chosen for single molecule fluorescence experiments have absorption bands in the blue–green region of the spectrum because of the availability of low cost laser excitation sources in that region. Single molecule detection has been achieved using lamps as the excitation source [4], however lasers with gain media such as argon ion (488 or 514.5 nm) and Nd:YAG (532 nm) provide a more stable and controllable source, are typically already polarized and have collimated beams and Gaussian intensity profiles perpendicular to the direction of propagation. Regardless of the excitation type or wavelength, the stability of the light source is critical and it is on this topic that we focus our attention.

Any fluctuations in the fluorescence intensity from an analyte caused by fluctuations in the output of the light source (or poor pointing stability of the beam) can severely affect the results of experiments that rely on photon counting statistics such as FCS, PCH, and, to a lesser extent, FRET (see Chapter 2). Poor pointing stability (i.e. variation in the direction of the laser beam) causes changes in the excitation volume in confocal arrangements and modulation of the angle of incidence in TIRF experiments. Even if a perfect light source, free from all intensity fluctuations, is incident on a detector then the output of the detector (photon counts per given time interval) is not steady but exhibits fluctuations within a Poisson distribution. This is because the quantum mechanical nature of the interaction of a photon with a detector leads to there being a statistical probability that the arrival of the photon results in an output signal.[5] Since single molecule fluorescence detection involves the counting of small numbers (<200) of photons in short time intervals (<ms), such inherent fluctuations caused by the detection process can prove to be the dominant source of noise in the experiment. However, this phenomenon can also be used as a convenient test of the stability of the instrumentation. If the scattered light or fluorescence from a concentrated sample that does not induce variations in intensity[6] is measured, then the output of the detector should follow a Poisson distribution in counts per unit time interval (see Figure 3.17). Any deviation from this (i.e. a broader distribution)

[5] See Chapter 2 for a complete discussion of this effect.

[6] A convenient method is to use a concentrated suspension of powdered milk and remove any emission filters so that the scattered excitation light is detected. It may be necessary to insert attenuating filters to reduce the scattered light intensity to a level similar to that encountered in single molecule fluorescence detection.

indicates that fluctuations are occurring over and above those induced by the stochastic nature of photon detection and is indicative either of laser intensity fluctuations or shortcomings of the instrumentation in terms of mechanical stability. This is an essential test for experiments that later rely on a statistical analysis of the distribution of photon counts from the detector (see Chapter 2).

Beam quality and mode quality also both impact on the light distribution in the excitation volume and therefore on the results of fluctuation spectroscopies that rely on theoretical descriptions of the sample volume (see Section 3.2.2 and Chapter 2). Mode quality specifically refers to the light pattern of a laser beam perpendicular to the direction of propagation. Mode quality is of particular relevance only for laser sources where the lasing cavity can generate non-axial rays (but can be used in an analogous sense to describe any light source). Most laser modes are expressed formally by two-indices following the acronym TEM (transverse electromagnetic mode). The light distribution for the TEM_{00} mode is essentially Gaussian and so most appropriate for use in these experiments (see Section 3.2.2). Higher order modes (TEM_{01} and above) all have intensity profiles that are highly non-Gaussian. The beam quality is also an important concept in optical design using lasers. Essentially, the beam quality parameter of a laser, M^2, describes how ideal the laser

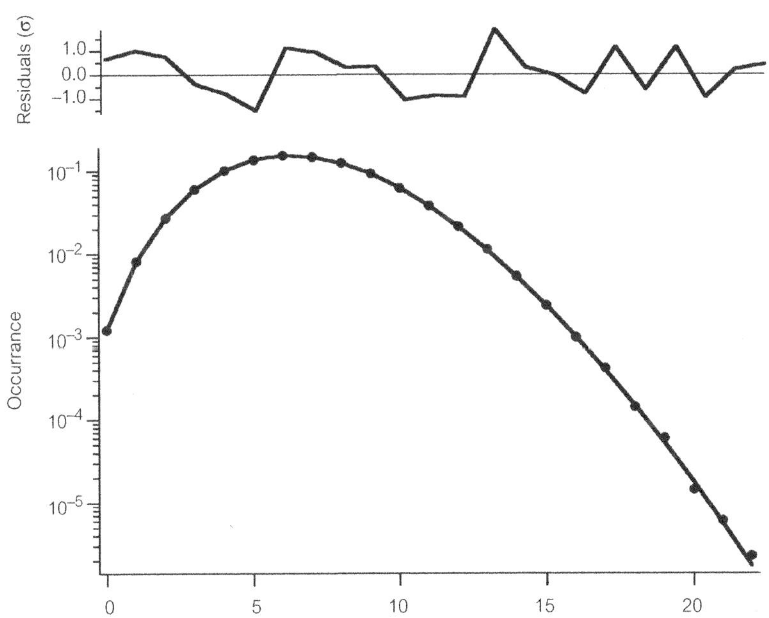

Figure 3.17 Photon count histogram of an ideal scatterer placed at the laser focus. The collected photon count distribution (circles, normalized) is fitted exactly by a Poissonian function (line) indicating that there are no fluctuations in the detected signal arising from instability of the light source or other instrumentation.

beam is (how close the beam is to a diffraction-limited Gaussian beam). Typical values for low power gas phase lasers are <1.3. Lasers with beam quality values as close to 1 as possible will mean that the output beam can be focused to as near the diffraction limit as possible and, for TEM_{00}, with a near Gaussian intensity profile. For a more detailed discussion about the origins and meanings of these parameters see [42].

Excitation of the fluorophore can be carried out using a one- or multi-photon approach as has been discussed. One-photon excitation at wavelengths suitable for most commercially available dyes is straightforwardly achieved using continuous wave (output) lasers. Since single molecule fluorescence experiments require no more than a few milliwatts of excitation power, large frame lasers are typically not required. An excellent alternative to older gas phase lasers are semiconductor lasers. These have significant advantages in that they are often easily controllable using a PC, this permits simple experimental integration and automation and they often require no water or fan assisted cooling which can cause vibrations that have serious implications for fluctuation spectroscopy. In addition they need only a low voltage single-phase power supply, are compact and relatively cheap. Small diode pumped (optically pumped semiconductor lasers—OPSL), or direct diode-emission lasers have been incorporated into instruments with single molecule sensitivity [37,43]. Very inexpensive diodes can now provide sufficient stability, spectral purity, and beam quality for single molecule applications and as more semiconductor materials become available, with band-gaps corresponding to useful wavelengths, this type of light source will no doubt be used more for single molecule spectroscopy.

In order to achieve multi-photon absorption very high intensities are required and this can be achieved at the focus of a high numerical aperture objective using short-pulsed laser sources. Typically mode-locked[7] systems such as titanium sapphire lasers are used which provide the added benefit of being tuneable over quite a broad range of wavelengths especially when used in conjunction with non-linear optical techniques such as optical parametric oscillation [42]. Modern mode-locked lasers also do not require three-phase power supplies or cooling water but are expensive and require significant space.

These pulsed lasers are also suitable for time gating to reduce background signal as discussed in Section 3.3.2 and for making time-resolved measurements, adding temporal resolution to signal amplitude, wavelength, and polarization information that can be obtained with single molecule resolution.

In Table 3.2 a non-exhaustive summary of laser light sources for single molecule spectroscopy is presented with a brief indication of their typical

[7] Mode locking refers to a method in which pico- or femto-second pulsed laser output is achieved by synchronizing the many modes inside the laser cavity [42].

Table 3.2 Summary of light sources suitable for single molecule fluorescence experiments

Type Examples	Wavelength (nm)	Dyes[a]	Laser[a]	
Continuous wave (single photon)	488	Rhodamine Green, Alexa Flour 488, GTP, FITC	Gas phase argon ion	[9,21,62]
	514	Rhodamine 6G, TMR	Gas phase argon ion	[11,63]
	532	Cy3, TAMRA	Solid state, diode pumped	[64–67]
	786	Ir132, ir125	Diode, direct emission	[43]
Pulsed (single photon)	635	DiD, Cy5	Diode laser, direct emission	[58,68,69]
Pulsed (multi-photon)	770	Yellow/green dyes	Titanium Sapphire	[70,71]
	790	Alexa Fluor 488	Titanium Sapphire	[16]
	830	Rhodamine Green	Titanium Sapphire	[16]
Conventional lamps	Visible	Cy3	Mercury lamp	[4]

[a] The flourescent dyes listed are only a sample of those that can be excited at this wavelength, as are the light sources that provide these wavelengths—this list NOT exhaustive.

applications and references to recent demonstrations of their use in single molecule fluorescence spectroscopy.

3.6 Microscope objectives for single molecule fluorescence detection

Perhaps the single most important optical element in our single molecule detection system is the microscope objective. In many experiments it is critical to the generation of a small excitation volume and in epi-fluorescence geometries it also collects the fluorescence from the sample. Microscope objectives are compound lenses, consisting of many individual elements. In general the design is intended to provide the required magnification, a small focal spot, and, as far as possible, an aberration free image with high collection efficiency. As far as single molecule fluorescence experiments are concerned, the suitability of a microscope objective can be assessed by considering the numerical aperture, the magnification, whether the objective is oil immersion or designed to operate in air, and the degree of aberration correction.

The numerical aperture (NA) of an objective describes the solid angle over which light is collected by the lens. The NA is given by,

$$\mathrm{NA} = n \sin(\mu) \tag{3.6}$$

where n is the refractive index of the imaging medium and μ is the half angle of the solid cone defined by the collected light (see Figure 3.18(a)). In microscopes with an inverted design (the objective pointing upwards), it is necessary to image through a thin coverglass. In this case when the medium between the objective and coverslip is air (see Figure 3.18(b)), the numerical aperture is limited to a value of ~1 due to refraction at the coverslip—air interface. To achieve a higher NA, an immersion objective is required. An oil immersion objective has a high refractive index oil layer between the frontmost objective lens and the coverglass (see Figure 3.18(c)) where the oil and the coverglass are generally chosen to match closely the refractive index of the objective. Objectives are available for use with a variety of immersion fluids, but most commonly with water or oil, which have refractive indices of approximately 1.33 and 1.51, respectively. Typically, in practice, water immersion objectives provide numerical apertures of up to 1.2 and oil objectives of up to 1.45. It should be remembered that whatever the NA of the objective, the NA of the system is (to an approximation) limited by the lowest

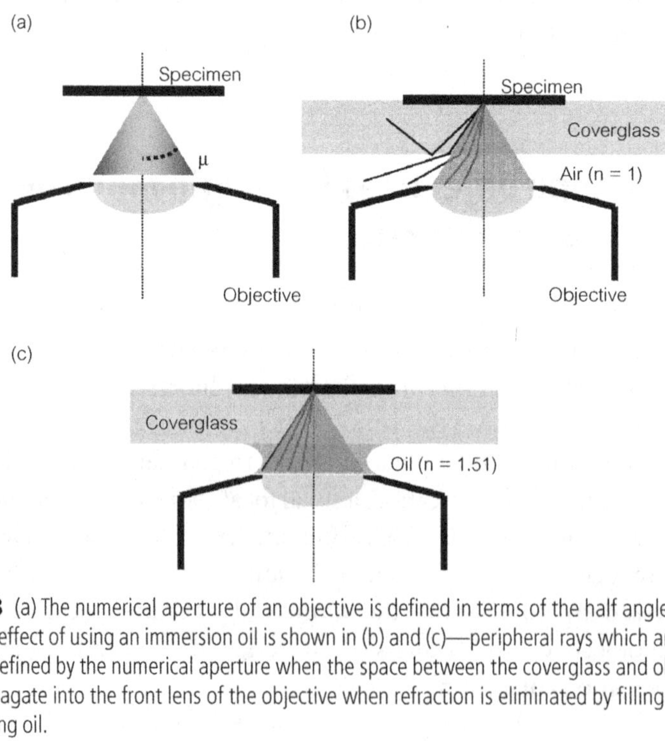

Figure 3.18 (a) The numerical aperture of an objective is defined in terms of the half angle of the cone of rays (μ). The effect of using an immersion oil is shown in (b) and (c)—peripheral rays which are refracted out of the cone defined by the numerical aperture when the space between the coverglass and objective is filled with air, propagate into the front lens of the objective when refraction is eliminated by filling the space with index matching oil.

refractive index substance between the objective and sample. Thus a 1.45 NA objective, even when used with the correct coverslip and immersion oil, would still have a reduced NA if the light were being focused (or collected) through solution, in other words if the specimen is not in contact with the coverslip. For work in aqueous conditions where the sample is not in contact with the coverslip (in diffusion experiments it is desirable to place the detection volume several μm into the solution, to prevent artefacts from molecules attached to the nearby surface), the NA is effectively limited to a value close to the refractive index of the solvent used (so ~1.33 for water). Therefore, under certain circumstances high NA objectives are of little use, although they are still crucial for through-the-objective based TIRF (see Section 3.2.3).

The NA can have a large effect on the collection efficiency of the objective. For example, NAs of 1.45, 1.3, and 0.95 correspond to 40%, 26%, and 10% of the total possible sphere of collection around an object. Thus small improvements in NA can result in significantly more photons at the detector.

Optical aberrations can degrade the quality of images, change the light distribution at the focus, reduce the resolving power, and increase the focal spot size (therefore increasing the sample volume) of an objective. The primary aberrations commonly experienced in microscopy are spherical, coma, lateral and longitudinal chromatic, curvature of field, and astigmatism [44]. Fortunately, high NA objectives are often corrected for these aberrations to a great extent and they are therefore not usually an issue from the point of view of single molecule fluorescence measurements. Generally, the choice of objective is between fluorites (abbreviated variously as Fl, Fluor, Fluar), apochromats (abbreviated as Apo), plan fluorites (Plan FL), and plan apochromats (Plan Apo). Fluorites provide better spherical and chromatic correction, apochromats provide the maximum spherical and chromatic correction and plan objectives provide the best correction of curvature of the field of view. It must, however, be emphasized that all such objectives only provide the maximum performance when used in the correct way. For example, the performance of water immersion objectives is not compromised if the object of interest is located some distance into an aqueous solution. However, the performance of oil immersion objectives is degraded significantly under similar conditions. Further, different objective/microscope manufacturers incorporate aberration correction into the microscope differently. Sometimes the aberration correction may not be incorporated into the objective but is provided by different sets of lenses in the microscope body that it is intended to be used in. Thus care must be taken in custom-built configurations.

Modern objectives are of the infinity corrected type, meaning that the image plane is at infinity (see Figure 3.19). Light collected by an infinity corrected objective is thus emitted as a collimated beam (sets of parallel rays) offering the

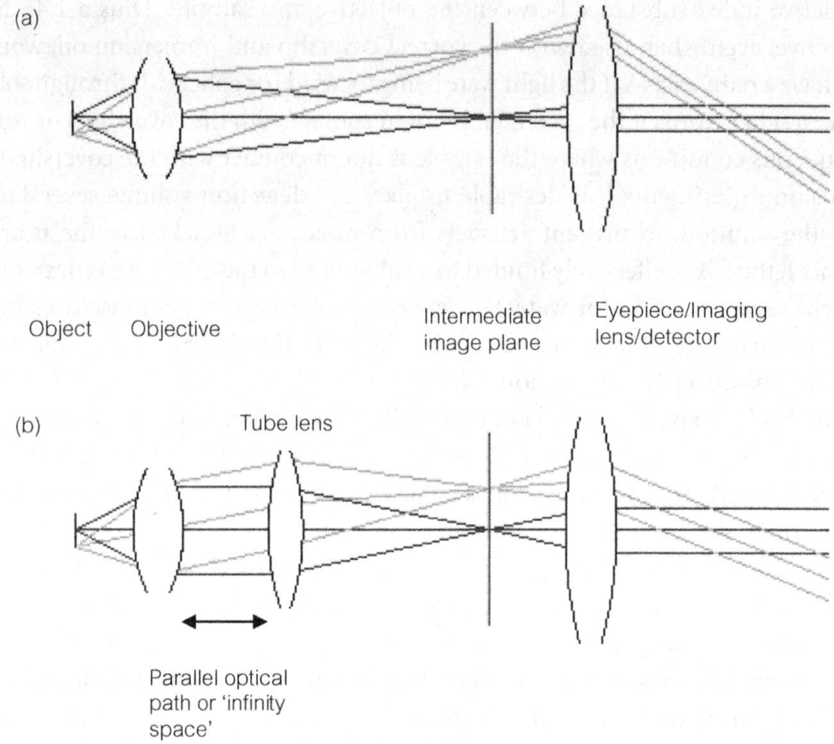

Figure 3.19 Ray diagrams of (a) finite tube length and (b) infinity corrected microscope arrangements. The intermediate image plane is shown as a vertical line. In the finite objective the intermediate image is generated directly by the microscope objective, the light converging as it emerges from the back aperture of the objective. In the infinity corrected model, light from points in the object emerge from the objective as parallel rays and a second tube lens is needed to form the intermediate image. (a) and (b) are both shown forming final images with a lens representing the microscope eyepiece. However, one of the overriding tenets of single molecule fluorescence instrumentation is to keep the number of optical components at a minimum. Thus, for example, in a point detection system (a confocal system) the detector *may* be placed at the intermediate image plane. Indeed from this point of view the finite objective may in prinicpal perform better as it is not necessary to use a tube lens. (Figure was kindly produced using Zemax (Zemax Development Corporation) by Kurt Baldwin at Avacta Ltd www.avacta.com).

advantage that a variety of optical components, such as interference filters that require a collimated beam to function optimally, can simply be placed in the collection light path (note off-axis rays may still be present, and are important in image formation, but the angular distribution of these around the optic axis is narrow). With infinity corrected optics it is necessary to place an additional image-forming lens, referred to as the tube lens, in the optical path to form an intermediate image (at the confocal pinhole or detector, for example). As a result, the magnification of the complete optical system is dependent on the tube lens focal length. In addition, the magnifications printed on infinity corrected objectives are only correct if a tube lens of appropriate focal length is used. For example, the tube lengths of microscope objectives designed by Nikon, Olympus, and Zeiss are 200, 180, and 160 mm, respectively. Thus infinity corrected objectives cannot be interchanged between microscopes without slight changes in the overall magnification. Since extra optical components are often required between the objective and tube lens (the so-called infinity space) in a single molecule fluorescence detection scheme, then the length of this part of the microscope is an important consideration. For a given pupil size (i.e. effectively the objective or tube lens aperture) the length of the infinity space clearly affects the maximum angle at which off-axis (but parallel) rays are collected (see Figure 3.19). If the tube length is too short then room is not available for additional optical components, such as a dichroic, interference filters, and polarizing elements. If it is too long then peripheral rays are lost, reducing the effective field-of-view in imaging single molecule systems. Modern objectives have been optimized for tube lengths in the range 200–250 mm which, if too short, may mean that a custom built 'compromised' instrument is needed rather than one based on a commercial optical microscope.

It is clear from the preceding discussions that the objective is at the heart of the single molecule fluorescence instrument and care must be taken to use them properly to obtain maximum performance. High numerical aperture objectives have short focal lengths of the order 1 mm. However, this is not necessarily measured from the front optical surface and it is the working distance (i.e. the separation between the front optical surface and the focal plane) of the objective that is the important parameter. Working distance is generally inversely proportional to both magnification and numerical aperture and can be as small as a few hundreds of microns, which has implications for the thickness of the sample substrate or expensive damage to the front optical surface of the objective! The choice of glass coverslip is also important. The glass slides used in oil immersion objectives should be of the correct thickness and refractive index for the particular objective (the objective manufacturer should provide specifications). Similarly, the choice of immersion oil, in particular the refractive index, should be made carefully for optimum performance. Low fluorescence oils that are non-drying and have differing viscosities at various temperatures (making handling easier and preventing the oil simply running off an inverted objective) are available from a variety of suppliers.

Crucial in the design of the instrument is the focusing control for the microscope objective. For TIRF applications it is necessary to place the centre of the depth of focus of the objective (the objective focal plane) near the glass water interface. For diffusion confocal and two-photon designs it is desirable to always place the focal plane away from the surface (to prevent surface adsorption artefacts) and the same depth into the solution for each repeat experiment, otherwise aberrations cause the size of the focal volume to change as a function of the depth of the focus into the solvent. Thus, accurate and stable control of the position of the objective is essential. In commercial microscopes such control may be provided by adjustment of the sample stage height. However, in our experience it is desirable to have a control resolution of around 100 nm for ease and repeatability of focusing. In this case mechanical stages are often not sufficient and piezoelectric single-axis translation stages designed for objectives are invaluable. They can provide long travel, high resolution, and stability (through closed-loop operation). Focusing of the microscope can be achieved in a number of ways. If a commercial microscope is being used, a viewing port (either fitted with binoculars or an inexpensive CCD camera) can be used and the microscope can be focused onto surface features such as dust or scratches on the glass. In the case of surface immobilized samples one may even be able to focus onto the single molecules. In the case of two-photon or confocal systems, once focused onto the glass surface, the objective focus can then be advanced into the solution a known and repeatable amount. These practical aspects are discussed further in Section 3.9.

Table 3.3 summarizes a range of microscope objectives suitable for single molecule fluorescence experiments along with some references giving examples of the single molecule applications in which they have been used.

Table 3.3 Summary of some of the microscope objectives suitable for single molecule fluorescence experiments and examples of their most suitable (although not exclusive) application

	Numerical aperture	Magnification	Type	Most suitable applications[a]
Olympus	1.65	100	Apo[b], oil	TIRF [72]
	1.45	60	Plan Apo, oil	TIRF [62]
	1.4	100	Plan Semi-Apo, oil	Confocal, MP [65,71]
	1.20	60	Plan Apo, water	Confocal, cell imaging, MP [16]
Zeiss	1.45	100	Plan Fluar, oil	TIRF [9]
	1.4	100/63	Plan Apo, oil	Confocal, MP
	1.3	100/63/40	Fluar, oil	Confocal, MP [73]
	1.2	63/40	C Apo, water	Confocal, cell imaging, MP [21]
Nikon	1.4	100/60	Plan Apo, oil	Confocal, MP [49]
	1.3	100/40	Plan Fluor, oil	Confocal, MP [11]
	1.2	60	Plan Apo, water	Confocal, cell imaging, MP

MP: multi-photon; TIRF: total internal reflection fluorescence.

[a] All objectives listed in this table are suitable for prism based TIRF, those also suitable for through-the-objective TIRF are highlighted TIRF.

[b] The Olympus 1.65 NA objective is designed for demanding TIRF applications. This objective requires quartz coverslips and special immersion oil.

3.7 Detectors for single molecule fluorescence experiments

The choice of detector is another critical stage in the development of a single molecule fluorescence experiment. In general the detector should fulfil the following criteria:

(1) high quantum efficiency over the spectral range of interest,

(2) good linearity of quantum efficiency over the spectral range of interest,

(3) sufficient time response for the application,

(4) low noise (i.e. low average 'dark' count), permitting single-photon detection.

We will restrict our discussions to detectors suitable for work in the visible and near-infrared part of the optical spectrum: photomultiplier tubes (PMTs), avalanche photodiodes (APDs), and electron multiplying charge coupled devices (EMCCDs). These detectors may be considered in two categories. PMTs and APDs are single point detectors, the single output of which is proportional to the integrated light intensity impinging on the detector area. Such detectors can be used for fluctuation spectroscopy using light collected from a point within the sample or the sample could be raster scanned to record and image (see Section 3.9). CCDs are imaging detectors that contain an array of detector elements and the object plane of the sample can be imaged directly onto the detector surface. The suitability of each of these types of detector for a given single molecule spectroscopy application depends on the way the devices work and therefore a brief discussion of their operation and application will be provided here.

3.7.1 Single point detectors

A PMT produces a small output current burst when a single incident photon impinges on the photocathode by a process of electron multiplication within a dynode chain. The basic layout of a typical PMT is illustrated in Figure 3.20. The photon strikes the photocathode, is absorbed and generates a photoelectron with a certain quantum efficiency determined by the photocathode material [42]. When a photoelectron is produced it is accelerated towards the first dynode by the electric field created by the focusing electrode. The photoelectron strikes the first dynode and results in the generation of secondary electrons (again with a certain efficiency), each of which is accelerated in an electric field and strikes a subsequent dynode, generating further secondary electrons. The process of amplification continues down the dynode chain until the anode collects the

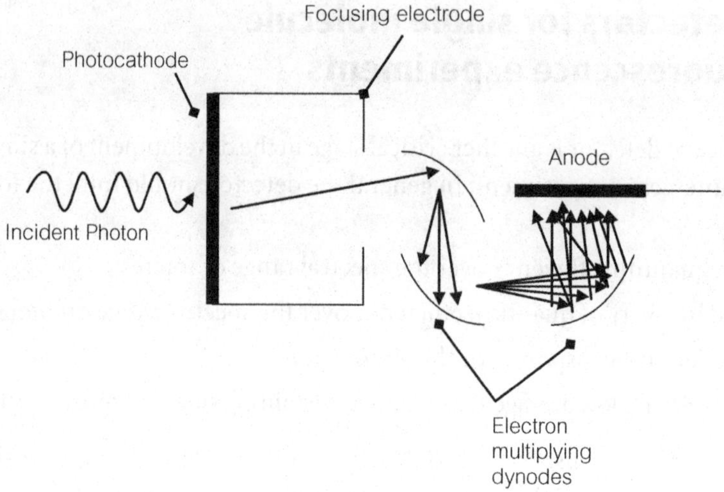

Figure 3.20 Illustration of the basic principle of operation of a photomultiplier tube (see text for description of the proceses involved).

electrons emitted by the final dynode and an external circuit detects this current. PMTs often suffer from poor linearity in output current over a large range of intensities, often have low quantum efficiencies (typically <25%) and also a narrow spectral response. Dark current, generated when no light is falling on the photocathode, due to thermionic emission of electrons, is a source of noise in PMTs that can be reduced by cooling (thermoelectrically or with liquid nitrogen). Multiplication noise (i.e. the variation in output signal for any given single photon impinging on the photocathode due to the chain of probabilities involved in the amplification process) is also an issue with PMTs at very low light levels. Advantages of PMTs are their low cost, good time response and fast transit times. Table 3.4 provides some examples of the use of PMTs in single molecule detection schemes. We note however that the PMT is now less commonly used due to the improved performance of, in particular, avalanche photodiodes.

The avalanche photodiode generates a large detectable signal when a photon strikes the device at the diode junction [42] (and in operation is in many way analogous to a PMT). A single pair of charge carriers (electron and hole) may be generated by the absorption of the photon and this is amplified by the avalanche breakdown effect [42] giving a large detectable signal. The device is biased by high voltages (up to several kV) that provide a field that accelerates the charge carriers generated at the semiconductor interface such that they are given sufficient energy to generate further charge carriers by impact ionization with the atoms in the semiconductor lattice. These secondary charge carriers are themselves accelerated leading to the generation of more charge carriers in

Table 3.4 Examples of detectors suitable for single molecule fluorescence studies and a summary of their applications in the literature

Detector type	Application
PMT	Multi-photon diffusion FCS/PCH [73] Single-photon diffusion detection / FCS [11]
APD	Multi-photon diffusion FCS/PCH [73] Single-photon diffusion FRET [49] 3 colour FRET [65] Lifetime measurements [69]
ICCD/EMCCD	Stoichiometry determination [62] GFP imaging in cells [9] Nucleotide kinetics [67] Combined TIRF/AFM [72]
CMOS array	Parallel single molecule detection/FCS [64]
No gain CCD	Demonstration of use of unmodified microscope [4]

FRET: fluorescence resonance energy transfer; FCS: fluorescence correlation spectroscopy; TIRF: total internal reflection fluorescence; PCH: photon counting histogram; ICCD: intensified charge coupled device; EMCCD: electron multiplying charge coupled device; CMOS: complimentary metal oxide semiconductor; AFM: atomic force microscope.

a cascade, resulting in an avalanche of charge carriers, or a burst of current and a voltage spike (depending on the method used), from the device. The bias voltage poises the diode near this critical breakdown point, such that a single photon can be detected.

APD detectors are very widely used for single molecule fluorescence spectroscopy. Compared with PMTs, APDs display excellent linearity, spectral response across the visible range, and quantum efficiencies in excess of 60% (in the red part of the spectrum). For example, the SPCM-AQR [45] series of APDs produced by Perkin Elmer Optoelectronics (Quebec, Canada) provide all this in a single cigarette-pack sized box requiring no high-voltage supply and with very low dark currents with no active external cooling. The output of such a device is a TTL[8] pulse for each photon detected which can be input directly into a range of data collection hardware (see Section 3.8). APDs are point detectors and because of the small size of their active areas (~100 μm diameter) they offer the advantage that they can also function as the confocal 'aperture' in epi-fluorescence microscopes, removing the need for a pinhole and hence increasing the overall detection efficiency. In addition, the time resolution provided by APDs is very high, making them suitable for time-resolved techniques. Single-photon

[8] A 'TTL pulse' is a transistor–transistor logic compatible pulse. TTL is a standard for a particular type of integrated circuit that is almost ubiquitous among counting electronics.

counting APDs have been used in single molecule fluorescence correlation spectroscopy [46], time-resolved single molecule measurements [47,48] using the time correlated single-photon counting technique and fluorescence resonance energy transfer measurements [49].

Point detectors with small active areas such as APDs can be mounted in a commercial microscope immediately after the tube lens. However, care must be taken to ensure that all of the light collected by the objective is focused onto the active area, which requires high-resolution stages to be employed to position the detector. In the case of larger area detectors such as PMTs a further lens is required in order to avoid saturating a small area of photocathode, which leads to poor detector performance. Both detector types can easily be damaged by exposure to high intensity light when both powered and un-powered. It is important to note that in the case of single-photon counting modules such as the Perkin Elmer SPCM-AQR, saturation can occur as light levels impinging on the detector increase. This however appears as a *loss* of the output signal. Care must be taken to avoid misinterpreting this fall in output signal, so that more light is not permitted to fall on the detector, causing damage.

3.7.2 **Imaging detectors**

Imaging detectors comprise a two-dimensional array of typically micron scale detector elements that can each be addressed by read-out circuitry so that an image of the sample can be acquired. A suitable optical arrangement must be used to image the object onto the plane of the detector array (for example, matching the intermediate image size—see Figure 3.19—to the imaging array size). CCDs are by far the most common imaging arrays; in these each detector element is formed from a charge storage device. Incident photons generate charge carriers in each element which are accelerated and are stored using an applied potential. The amount of charge stored in this 'well' will thus be proportional to the integrated light intensity that has fallen on that element of the array. Readout of the array is achieved by movement of the charge from each element of the array to the next, either on an individual basis or line by line, as is illustrated schematically in Figure 3.21. In this simple illustration the charge in the bottom row of the array is moved down into an output register, and the rows above are all moved down by one. The contents of the output register are then moved left to right and the charge in each element of the register is amplified as it is read out to provide the output current signal. The read out process is a limiting factor on the frame rate of CCD cameras. This is determined by the speed of the electronics that move the charge, the time required to clear out residual

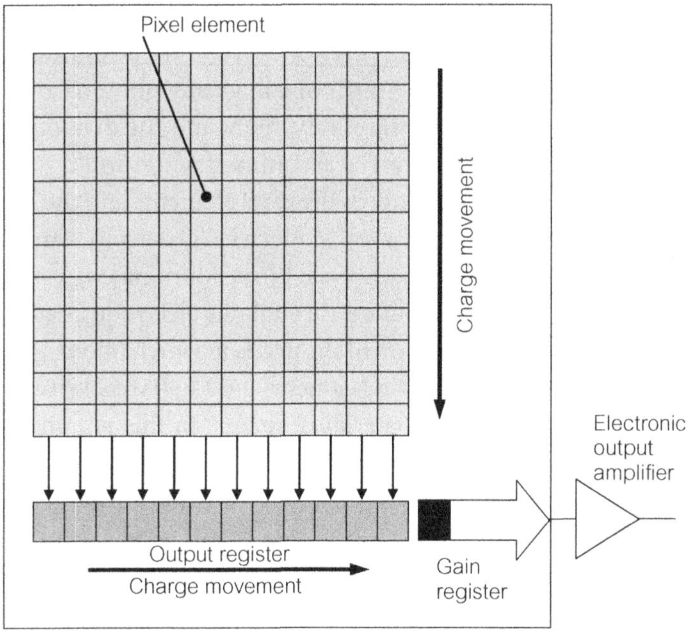

Figure 3.21 Schematic representation of the structure of a typical CCD. Elements are read out sequentially by first moving charge downwards line by into an output register using a series of electrodes. This register is then read out one element at a time and the signals amplified by an external electronic current amplifier. In the EMCCD systems on-chip gain is provided before external amplification to lift even very low signals well above the read out noise floor.

charge between exposures, and the speed of reading the signal from the output register. Much CCD development is focused on increasing the readout rate, for example, other geometries of array and register structures can be used, such as frame transfer devices [42]. Other techniques such as binning can also speed up data acquisition. Binning is the process of combining the charge from two or more pixels, which has a three-fold advantage. First, the size of the pixel array is effectively reduced which increases the frame rate. Second, better noise performance is achieved because read out noise is reduced (because the noise introduced by the read out electronics is fixed per read out event and not proportional to the amount of charge being read out) which is a limiting source of noise in CCD technology. Third, since CCDs typically have arrays of 512×512 pixels or more, very large data files are produced which, depending on the frame rate, might cause data handling problems further downstream. By decreasing the effective resolution of the array binning can reduce the volume of data significantly.

To be suitable for single molecule detection a CCD should fulfil a number of requirements: its spatial and temporal resolution should be appropriate to the application, the quantum efficiency (QE) of converting photons arriving at the CCD surface to charge generation should be high, and the dynamic range (i.e. the range of input intensities that can be accommodated) should be large. CCD QE is defined by the physical structure of the pixel elements and the electrodes as well as the semiconductor material used. Modern CCD detectors are available in front and back illuminated configurations. Front illuminated systems require the incident photons to be transmitted through an electrode structure on the front face of the device into the region of the pixels in which photo-generation of charge occurs. Thus, front illuminated cameras tend to have a virtually zero QE in the UV because the electrode material is opaque in this region of the spectrum, moderate QE of 10–40% in the visible and higher QE (>40%) in the near infra-red. Back illuminated CCD designs have greatly improved QE because the electrode structure is located on the backside of the sensitive pixel elements (with respect to the direction of incoming photons). In the absence of the front electrode, the number of incident photons reaching the semiconductor material increase and QE of up to 50% in the near-UV and greater than 80% in the visible to infra-red region can be achieved. The quantum efficiency also tends to be a strong function of temperature and so all high-sensitivity CCD cameras are equipped with cooling as optimal operation temperatures can be as low as −90°C.

The basic CCD detector is not suitable for fast frame rate, low light level applications because of the frequency dependent read out noise (from electronic circuit amplification) and gain must be introduced in order to render these devices capable of single molecule detection. Intensified CCDs (ICCD) and electron multiplying CCDs (EMCCD) are the two most common systems that are used. ICCDs were the first development of CCDs intended to extend the detection sensitivity to near single incident photon levels. They generally combine devices called micro channel plates (MCP), essentially an array of small PMTs, onto the front of the CCD array. ICCDs are however complex, expensive, prone to noise (especially from cross talk between MCP elements on adjacent pixels), and have a finite lifetime. An alternative, EMCCD technology, is used in many of today's low light level cameras, although development of low light level cameras is rapid meaning that the state-of-the-art is constantly changing. The EMCCD uses an additional register between the output register and the output amplifier, called the gain register (see Figure 3.21). High potentials applied to the gain register provide the stored charge with sufficient energy that they can cause impact ionization as they move through the gain register, generating further charge carriers and hence providing an overall multiplication of the output charge. The overall gain provided by the EMCCD compared with a standard CCD is therefore several orders of magnitude and thus the output current from a few incident photons (or even single photons with adequate cooling) can be well above the read out noise floor (from the external current amplifier). Since the electronic read out noise is also frequency dependent, due to the characteristics of the read out circuitry, both lower light levels and higher read out speeds are possible with EMCCD detectors.

The time resolution of typical EMCCDs is a function of a number of variables including readout speed and signal-to-noise ratio, which depends on quantum efficiency and read out noise. In the context of single molecule fluorescence experiments exposure times of 10 ms (i.e. a frame rate ~100 Hz) are routinely achieved, whilst through the use of binning and signal optimization, exposure times as short as several ms might yield good data. In general though, better temporal performance is obtained from point detectors such as APDs, but of course these lack the potential to form images in a simple manner. Improving the temporal resolution of CCDs is an area of current development by the CCD manufacturers.

EMCCDs and ICCDs are available from a range of manufacturers including the iXon from Andor Technology (Northern Ireland, UK) and the Cascade from Roper Scientific (Arizona, USA). Both of these devices have proven single molecule fluorescence detection capabilities and specifications such as quantum efficiency and pixel array size are quite comparable between devices but subtle differences regarding temporal resolution, noise levels, and software exist, which will inform the choice for any specific application. Modern CCD cameras can be coupled to commercial microscopes without modification and are supplied with basic image analysis software. Many systems are available with software development kits that enable custom control software to be developed for integration into custom single molecule instrumentation. Computer control is generally provided through fast, propriety internal expansion cards (fast USB interfaces are also being introduced), which in a practical sense limit the control software environment to Microsoft Windows. One important point to consider when designing the detector/data handling aspects of a single molecule fluorescence experiment is that very large data sets can be generated. Gigabytes of data are rapidly produced by 100 Hz frame rates of a 512 × 512 pixel CCD image and therefore it may be necessary to spool images in real time to hard disk; this aspect of the instrument deserves as much consideration as selection of the camera.

Provided in Table 3.4 is a list of recent applications of the aforementioned detectors in single molecule experiments.

3.8 Acquisition cards and software

Data acquisition hardware for CCD imaging instrumentation is generally proprietary and therefore will not be discussed in detail here. The three common forms of data capture in single point, single molecule experiments are: time series recording (fluorescence burst counting), autocorrelation (for FCS), and time correlated single photon counting (TCSPC) for time-resolved or fluorescence lifetime measurements. In this section we will review the operation of the hardware to implement these detection schemes and refer the reader to reported examples of their use (see Table 3.5).

Collecting the raw data, such as the TTL pulses that correspond to fluorescence bursts detected by an APD, as a time series recording permits the greatest flexibility in subsequent analysis. These forms of analysis may include calculation of the autocorrelation or cross-correlation functions of one or more signals [50,51], PCH/FIDA [52] [22,53], burst detection [11], or time trajectory analysis from immobilized molecules [54–56]. Recording the time trace of pulses from a detector can be performed by any input card that can operate in multi-channel scaling (MCS) mode. An example is the MCA series of PC expansion cards from Fast ComTec GmbH (Oberhaching, Germany), which are able to count TTL pulses from APD modules directly without external signal pre-processing. Output signals from PMT detectors generally require the use of an additional discriminator before the MCS card, which then provides a TTL pulse each time a pulse with amplitude greater than a pre-defined threshold is detected. In the case of the Fast ComTec MCA-2 card two digital inputs are provided enabling, for example, FRET data from two APD modules to be collected simultaneously. If more inputs are required, for example dual colour, dual polarization measurements, then it is also possible to externally trigger two or more of these boards to provide

Table 3.5 Some examples of data acquisition schemes for single molecule fluorescence measurements

Application	Data acquisition hardware	Detector type used	References
Single-photon detection, PCH and FCS	MCS type	APD, PMT	[11,16,43,74]
Time resolved	PCI based TCSPC expansion board	APD	[58,69,75]
Hardware FCS	Multiple Tau Digital correlators	APD, PMT	[6,16,21,73]

FCS: fluorescence correlation spectroscopy; PCH: photon counting histogram; TCSPC: time correlated single photon counting; MCS: multi-channel scalar; APD: avalanche photodiode; PMT: photo-mutiplier tube; PCI: peripheral component interconnect.

synchronization of data collection. Similar cards are available from a variety of manufacturers and it is only important to ensure that the maximum specified data input rates are in excess of the maximum data rates from the detector, that sufficient binning time resolution is provided for the application and that software development kits are available for user customization and system integration.

Autocorrelation (see Chapter 2) can be performed on raw time series data collected by MCS cards such as those described earlier, although there are a number of inefficiencies associated with post-processing autocorrelations (see Chapter 2), not least that long data acquisitions have to be performed and the analysis has to be carried out before one can tell if the experiment has been at all successful. Hardware digital correlators [46] (e.g. the ALV-5000 series, ALV GmbH, Germany) can take the digital output from APD modules and directly perform an auto- or cross-correlation and display the result in real time. The method of operation of one particular hardware correlator was covered in detail in Chapter 2 Section 2.4.2.

Time correlated single photon counting (TCSPC), which provides the capability to measure fluorescence lifetimes [57,58] is a conceptually more complex technique, but relatively easily implemented using a PC expansion card. In the TCSPC technique, the time delay between an excitation laser pulse and the detection of a single fluorescence photon from the sample is determined and a histogram of counts over a range of delay times is gradually built up from many excitation/detection cycles to form the observed fluorescence decay (see Chapter 2, Section 2.7.4). The measurement of the delay time must be made with a resolution much better than the fluorescence decay time and this typically implies a time resolution of at least 100 ps. The TCSPC technique has the disadvantage that this statistical counting approach requires quite long data acquisition times which, in single molecule experiments, may result in problems when photobleaching (or diffusion rates) may limit the total possible observation time (number of photons available) from a fluorophore. In single molecule TCSPC measurements where few photons are detected, the accuracy with which fluorescence lifetimes, or changes in lifetime, can be determined may be poor (see Chapter 2). Data capture boards that are capable of recording the time trace of detected photons, as well as the arrival time of the fluorescence photon relative to an excitation pulse, provide useful opportunities for post-processing of data as well as measurements of lifetime, if they are feasible. Such boards (designed with single molecule TCSPC in mind) are available from among others PicoQuant (Berlin, Germany) and Beckler and Hickl (Berlin, Germany) and can take their input directly from the digital TTL output of APD modules.

Single molecule instrumentation generally requires the integration of a number of devices into a single system that requires control, data collection, and analysis by a PC of very large data sets. Many hardware manufactures are able to supply dynamic linked libraries (DLLs) that allow easy programming of control and communication software in Microsoft Visual C or Microsoft Visual BASIC. Alternatively, dedicated data analysis and experiment control software, such as LabView (National Instruments, USA) and Igor Pro (Wavemetrics, USA), produce excellent results without advanced levels of programming knowledge, providing that these applications are capable of interfacing with the hardware that has been selected.

3.9 Realizing single molecule instrumentation

In this section we provide some practical details of simple single molecule instruments. We note that in a number of ways these instruments may not provide optimum solutions for all studies (we simplify the designs to two-colour detection for basic FCS or FRET measurements, with no polarization sensitivity or time-resolved capability) thus, while instruments constructed with little deviation from these guides will function well, they do not suit every application. Instead of providing detailed schematics for the wide range of possible techniques we hope to provide insight into the basic layout of single molecule instruments as well as to point out many of the important aspects in their design, construction, and use. We outline two configurations: one suitable for diffusion FRET, FCS, or PCH and scanning confocal spFRET measurements of surface immobilized systems and the other a TIRFM suitable for single colour, single molecule experiments or two-colour immobilized spFRET.

3.9.1 Commercial systems

Commercial systems from manufactures such as Nikon, Olympus, and Zeiss now exist with many design aspects directed towards single molecule studies. Of particular note is the Confocor series of fluorescence correlation microscopes available from Carl Zeiss [59]. Such commercial systems are excellent solutions for many labs. The cost of these systems can, however, be prohibitive. Furthermore, modifications to these microscopes may be necessary, particularly in the light of the rapid rate of emergence of 'new' single molecule techniques. (e.g. it is doubtful a commercial system would be able to carry out the measurements outlined in [9,60] without significant modification).

3.9.2 **Design for a scanning confocal microscope**

Description

Shown in Figure 3.22 is the layout of a simplified scanning confocal system. The instrument is suitable for both single molecule fluorescence studies of Alexa 488 dye-labelled molecules and FRET studies for the dye pair Alexa 488–Alexa 594 (see Chapter 4—of course the configuration is immediately suited to other spectrally similar dyes and is easily adaptable for other dye pairs). We now tour the instrument shown in Figure 3.22 and refer to the labelled components in this Figure with *italics*. Light from a 488 nm laser (*Laser*) passes through a laser line filter (*LLF*) to remove any other lasing wavelength (this laser is intra-cavity doubled being pumped by infra-red radiation) and a quarter wave plate (*1/4*, zero order quarter waveplate) to circularly polarize the laser light to reduce photoselection of molecules with a particular dipole orientation. The circularly polarized, collimated beam, is guided by a kinematic platform mounted mirror (*M1*) through an iris (*I1*) adjusted to let the beam just pass through but to minimize secondary beams due to reflections from surfaces. A second mounted mirror (*M2*) then guides the beam to a 'spatial filter' assembly. The spatial filter consists of a lens-pinhole-lens system (*L1-P1-L2*). The spatial filter serves to create an as true to a near-Gaussian beam profile as is practically possible. This is necessary as the laser output may consist of ring artefacts (from scattering by dirt on the laser output lens for example) or proportions of undesirable transverse modes. The first lens (*L1*, plano-convex, 25 mm diameter, 50 mm focal length) focuses the light to a diffraction-limited spot. A pinhole is then placed precisely at the focus (*P1*, 15 μm diameter). If the pinhole is matched with the lens to the size of the diffraction-limited beam (equation 3.1), then the light propagating through the pinhole will have a clean near-Gaussian intensity profile. It is then only necessary to collimate this diverging beam with a second lens (*L2*, plano-convex, 25 mm diameter, 50 mm focal length), producing a near-perfect Gaussian beam. In reality, such an arrangement can be awkward to set up and we have found excellent results can be obtained by deliberately choosing an inappropriately small pinhole. In this way a diffraction pattern from a circular aperture (the pinhole) is produced (observed on the other side of the pinhole). This consists of a characteristic bright central spot and a concentric ring pattern of decreasing intensity (an Airy disc pattern). A second spatial filter (an iris, *I2*) can be placed to block out all light except the central spot. However, such a configuration has two important consequences. First, the beam can no longer necessarily be focused to the diffraction limit by the microscope objective; in effect the pinhole will be imaged onto the sample plane and this therefore limits the minimum spot size. However, this will not be

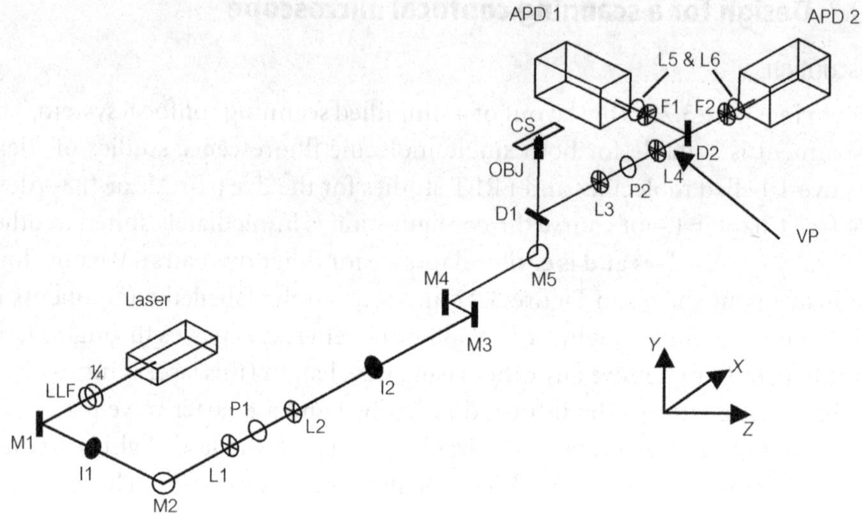

Part	Description (optical components)
Laser	Laser (e.g. 488 nm Sapphire OPSL, Coherent Inc.)
1/4	Zero order quarter waveplate (WPQ05M-488, Thorlabs, USA)
LLF	Laser line filter (e.g. HQ487/15x, Chroma Technology Corp., USA)
M1–M5	Protected aluminium mirrors, 25 mm diameter, ~5 mm fused silica substrate (PF10-03-G01, Thorlabs, USA)
I1–2	Adjustable iris (ID25/M, Thorlabs, USA)
L1–L6	50 mm focal length, 25 mm diameter plano-convex lenses, broadband anti-reflection coated. (LA1131-A, Thorlabs, USA)
P1	15 μm diameter pinhole (P15S, Thorlabs, USA)
D1	Shortpass dichroic mirror, custom piece (e.g. 488DSCX, Chroma Technology Corp., USA)
OBJ	Microscope objective ×100, 1.45 NA infinity corrected oil immersion (Plan Fluar, Zeiss, Germany)
CS	Coverslip or sample chamber (e.g. Lab-Tek II chambered coverglass, Nalge Nunc International, USA)
P2	50 μm diameter pinhole (P50S, Thorlabs, USA)
VP	Removable viewing mirror/optics for focusing
D2	Longpass dichroic mirror (e.g. 565DCLP, Chroma Technology Corp., USA)
F1	Donor channel emission filter (e.g. HQ525/50M, Chroma Technology Corp., USA)
F2	Acceptor channel emission filter (e.g. HQ620/30M, Chroma Technology Corp., USA)
APD1–2	Avalanche photo diode detectors. (e.g. SPCM-AQR-15, Perkin Elmer Optoelectronics, USA)

Figure 3.22 Schematic of the simple confocal microscope that produced much of the data not otherwise referenced in this text. Specific examples of components are given for a configuration to measure diffusion spFRET or FCS (autocorrelation or cross-correlation) for the dye pair Alexa Fluor 488 and Alexa Fluor 594 (Moelcular Probes Inc., USA). These examples represent, in most cases, an arbitrary choice of supplier and are provided only to enable the user to see the specifications of the parts used. (Figure was kindly produced using Zemax (Zemax Development Corporation) by Kurt Baldwin at Avacta Ltd www.avacta.com).

limiting for this configuration because the size of the pinhole in the objective focal plane is given by the ratio of the focal lengths of the collimating lens, *L2*, and the objective (image size = pinhole size × objective focal length/*L2* focal length), which is still close to the diffraction limit. Second, the light intensity profile of the central spot (properly the Airy disk) in the diffraction pattern is not perfectly Gaussian (although it is a close approximation, see [28,61] for a further discussion). Despite these shortcomings we have found this configuration to be easy to align, inexpensive and to produce artefact-free FCS and PCH data suggesting a near three-dimensional Gaussian PSF. With a good quality laser system the spatial filter assembly may in fact be totally redundant and similarly good results may be obtained with or without it.

The spatial filter may also be used to expand the beam (in order to vary the input beam diameter into the microscope objective, see Section 3.6). In a system where the pinhole size (*P1*) is matched to the lens (*L1*) spot size, then the ratio of the focal lengths of the collimating and focusing lenses give the factor of expansion. In our case the pinhole is smaller than the spot size. So the collimated beam diameter can be calculated, to an approximation, from equation 3.1 and in this case the collimated spatially filtered beam diameter is of the order 3–4 mm in diameter, which is smaller than the back aperture diameter of the microscope objective (around 5–6 mm).

After the spatial filter, the beam is then guided by three kinematic platform mounted mirrors (*M3–5*). The last mirror is oriented to divert the beam vertically to allow an inverted configuration with respect to the microscope objective. After *M5* the beam passes through a kinematic mounted dichroic mirror (*D1*, mounted at 45°) creating an episcopic arrangement. The beam then passes through the microscope objective (*OBJ*) and is focused into the sample.

The collected fluorescence then emerges from the objective's back aperture and is reflected from dichroic *D1* into the detection path. Confocallity is provided by the lens (*L3*, plano-convex, 25 mm diameter, 50 mm focal length), pinhole (*P2*, 50 μm diameter), lens (*L4*, plano-convex, 25 mm diameter, 50 mm focal length) arrangement. The lens–pinhole combination was chosen empirically (see Section 3.2) and once again was found to provide a three-dimensional PSF in combination with the other elements of the system by carrying out control FCS experiments (see Chapter 2). Lens *L4* provides collimation. The collected light then proceeds to a second dichroic mirror (*D2* mounted at 45°). The light propagating in the two detection channels then proceeds to two emission filters (*F1* and *F2*) and then two lenses (*L5* and *L6*, plano-convex, 25 mm diameter, 50 mm focal length) which focus the light onto the avalanche photo diode detectors (*APD1/2*, SPCM-AQR-15, Perkin Elmer Optoelectronics, USA).

Mechanical arrangement and alignment

All components can be arranged on a standard optical table, 25 mm grid spacing of 6 mm diameter metric threaded holes for ease of arrangement. All optical components were mounted in standard steel or aluminium 'post and bases' unless noted below. Translation along the optic axis (along *x*, see Figure 3.22) was provided for *L1* and *L2* to allow accurate focusing and collimation by low-resolution single-axis stages (e.g. MS1/M, Thorlabs, USA). Mirrors were mounted in kinematic mirror mounts (e.g. KMS/M, Thorlabs, USA). *P1* was mounted in a two-axis stage (*y–z*, for example, ST1XY-A, Thorlabs, USA) to allow alignment. *I2* was mounted in a two-axis stage (*y–z*, for example, ST1XY-A, Thorlabs, USA) to allow alignment in order to properly remove higher order fringes in the diffraction pattern. *D1*, *OBJ*, and the sample stage (holding the coverslip *CS*) were all attached to a large post (for example, XT95, Thorlabs, USA) rigidly attached to the table. *OBJ* was mounted in a coarse mechanical focusing tube (for example, SM1V10, Thorlabs, USA) and also in a high-resolution piezoelectric stage (e.g. MIPOS100, piezosystem jena, Germany) to provide translation for fine focusing (in *y*). The sample stage (supporting the coverslip) was fixed (in *y*). In this way focusing is provided by vertical motion of the, comparatively, low mass objective (possible due to the use of infinity corrected optics). To provide a scanning capability, the sample stage (fixed in *y*) incorporated a high-resolution closed-loop two-axis (*z* and *x*) scanning stage (for example, P620 family, Physik Instrumente, Germany). *L3* and *L4* were mounted in low-resolution single-axis stages (e.g. MS1/M, Thorlabs, USA). *P2* was mounted in a two-axis stage (*y–z*, for example, ST1XY-A, Thorlabs, USA) to allow alignment. *APD1/2* were mounted on three-axis stages (e.g. MT1/M, Thorlabs, USA).

Alignment proceeds in the same way to any other optical instrument. Components are placed in series, starting at the light source. Components should not deviate the beam (unless this is the intention). Where necessary beams should be deviated through angle of only 90° or 45° and kept parallel (or perpendicular) to the optical bench where possible. Alignment for various components can be achieved by observing the transmitted beam on a card before and after the component is inserted, and by replacing optics (in the mounts) with an aperture (to check the beam passes through the centre of the optic) followed by a mirror (to check that the beam is at normal incidence). Alignment of pinholes can usually be achieved by maximizing the transmitted intensity and by checking for symmetry in the intensity profile of the transmitted spot / diffraction pattern. Collimation can be achieved by observing the beam at far distance (as is practically possible) or using a 'shear plate' and making small adjustments (along *x* only) of the collimation lens. Initial alignment of the confocal system is made significantly easier as a consequence of the dichroic *D1* being imperfect. The sample is replaced with either a clean coverslip or mirror. The excitation light is reflected strongly and mainly propagates back through the dichroic *D1* along the path of the excitation light. The mirror or coverslip can then easily be placed in the focal plane by observation of this reflected beam on the iris (*I2*) and ensuring that the return and incident beams trace the same path and are of the same diameter. A small proportion of the reflected light is also reflected off the dichroic (although this is in fact non-ideal). This reflected light is also collimated if the mirror/coverslip is at the objectives' focal plane. This collimated, reflected beam can now be used to set up the confocal pinhole assembly (*L3, P2*, and *L4*) by alignment such that maximal transmission is achieved. In this way, confocality of the pinhole and objective focus is ensured. Alignment of the dichroic *D2* and the APD detectors can then be trivially achieved by maximizing the detected signals either using a fluorescent dye as the sample (if the concentration is sufficient enough emission will also leak into the longer wavelength channel), or by increasing the laser power and aligning with the detected scattered/reflected light.

Sample introduction and routine focusing

It is clear from Figure 3.22 that the design does not incorporate conventional binocular (or CCD camera) observation of the sample. Such an arrangement could easily be added, however we have found that quick focusing can be achieved by observing the diffraction pattern of reflected excitation light after *P2*. A new sample (glass coverslip and, typically, aqueous solution) is introduced and an observation mirror is placed into the beam path after *L4* (indicated in Figure 3.22 by *VP*). Maximizing the observed intensity in the diffraction pattern at *P2* is trivial by eye. This then provides a point of reference at which the objective focal plane is placed on the glass–water interface. Using the fine closed-loop control on the objective focusing stage then allows one to place the objective's focal plane any depth into the solution above the glass–water interface. Thus in diffusion FRET or FCS measurements the sample volume can be placed at a known and repeatable distance into the solution. The sample volume will thus be highly reproducible in size and profile. This is necessary to avoid surface absorption effects generating artefacts in free random diffusion fluctuation measurements. Typically, we place the objective sample plane 10 μm above the glass–coverslip interface.

Experimental parameters and methods

As has been discussed in Section 3.6 the choice of sample coverslip is crucial. The coverslip material and thickness must be matched to the optical requirements of the objective used. The actual form of the chamber (bare coverslip, closed flow chamber or open chamber) is unimportant except with regard to the actual experiment being conducted (i.e. evaporation, solvent and possibly a desire to reduce photobleaching through flow of the sample—see Chapter 2).

Laser excitation power is inconvenient to measure as the light emerging from the objective rapidly diverges. As a result power is typically measured while the beam is collimated, before the objective, and corrected for the transmission efficiency of the objective (provided by the manufacturer). Such a procedure is approximate but few single molecule measurements depend on knowing the absolute excitation power precisely. In our laboratory an excitation power of around 30 μW is used for diffusion measurements (for Alexa Fluor 488) and a lower power (of the order of 1 μW) for scanning confocal surface immobilization studies, although these figures do vary depending on the particular experiment.

For many studies of biological samples, temperature control is necessary. In our laboratory we have obtained the best results with simple recirculated water heating/cooling using a common laboratory water bath. A hollow sample stage is constructed from an efficient thermal conductor and water passed through to regulate the temperature of the sample placed on it. If an immersion objective was used then temperature control would be inaccurate as a large temperature gradient is generated by the mass of the objective acting like a heat sink. Thus it is essential that the objective is also heated or cooled with an appropriate jacket. The temperature in our configuration is then monitored by a calibrated thermocouple placed close to the laser focus directly in the solution. Feedback can then be provided to the water bath if desired. Many common recirculated water baths use pulsed pumps (as opposed to a steady flow): we have not however experienced problems with these pump designs in terms of vibrations but advise caution. Peltier heating/cooling devices, with the sample placed directly on the device, should be avoided due to the small expansion and contraction of these devices during operation. They could of course be used in combination with 'heat pipes' however it may well be necessary to still use recirculated water to actively cool the 'hot' side of the Peltier.

3.9.3 Design for a total internal reflection fluorescence microscope

Description

Figure 3.23 shows the layout of a simplified total internal reflection fluorescence microscope (TIRFM). We briefly tour the configuration here, before expanding on experimental details such as alignment, laser power, and optomechanical considerations. The instrument we describe is suitable for single molecule fluorescence studies of Alexa 488 dye-labelled molecules and FRET studies for the dye-pair Alexa 488–Alexa 594 (see Chapter 4—of course the configuration is easily adaptable and is immediately suited to spectrally similar dyes).

Light from a 488 nm laser (*Laser*) passes through a laser line filter (*LLF*) to remove any other lasing wavelengths (this laser is intra-cavity doubled and pumped by infra-red radiation) and a quarter wave plate (1/4, zero order quarter waveplate) to circularly polarize the laser light (see a discussion of the polarization states of the evanescent field generated in TIRF in Section 3.2.3). The circularly polarized, collimated beam, is guided by a kinematic platform mounted mirror (*M1*) through an iris (*I1*) adjusted to let the beam just pass through but to minimize secondary beams due to reflections from surfaces. A second mounted mirror then guides the beam to a 'spatial filter' assembly. The spatial filter consists of a lens-pinhole-lens system. The spatial filter assembly is similar to that described in Section 3.9.2. In this case the pinhole is matched with the lens to the size of the diffraction-limited Gaussian beam. The spatial filter is also used in this instrument to significantly expand the beam. In this configuration the beam is later focused onto the back focal plane (BFP) of the objective lens (Figure 3.13), resulting in collimated light emerging (at some angle) from the objective (see later). In this case the diameter of the beam that is focused onto the BFP is then related to the diameter of the resulting collimated beam from the objective and therefore the area of the sample surface that is illuminated. Adjustment of this beam diameter is provided by the iris (*I2*). After the spatial filter the beam is then guided by two kinematic platform mounted mirrors (*M3–4*). The expanded collimated light is then focused onto the BFP of the objective (*OBJ*) through the dichroic (*D1*) by the lens (*L3*). Note that this is non-ideal use of the dichroic as the light is converging, however, the focal length is long so the angles involved are not great. The converging beam is collimated by the objective and emerges at some angle (illustrated schematically, but not accurately in the top inset in Figure 3.23). The particular angle is related to the numerical aperture of the objective and the position in the BFP that the light is focused to (see Section 3.2). If the light is focused to the back focal point (i.e. the centre of the BFP) then the light is not deviated and emerges parallel to the axis y. As the position of the focus is moved (in any direction) away from the back focal point (but kept in the BFP so moved in the z–x plane) then the light deviates at an increasing angle. The maximum deviation is reached near the periphery of the objective's back aperture (and limited by the effective numerical aperture). Note that the size of the focused spot in the BFP will not only affect the diameter of the resulting collimated beam but will also affect the degree of collimation that results. This is one reason for the beam diameter control provided by *I2*. Indeed reflections and optical aberration will effectively result in a number of rays incident on the sample substrate at a variety of angles. Details of the optomechanical components

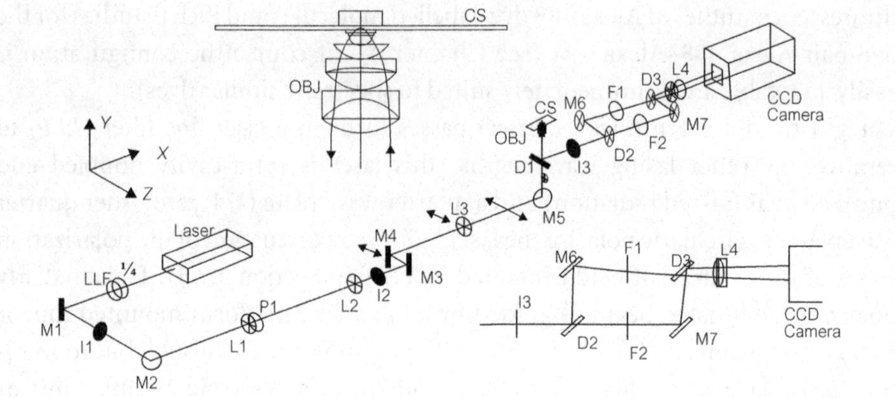

Part	Description (optical components)
Laser	Laser (e.g. 488 nm Sapphire OPSL, Coherent Inc. UK)
1/4	Zero order quarter waveplate (WPQ05M-488, Thorlabs, USA)
LLF	Laser line filter (e.g. HQ487/15x, Chroma Technology Corp., USA)
M1–M7	Protected aluminium mirrors, 25 mm diameter, ~5 mm fused silica substrate (PF10-03-G01, Throlabs, USA)
I1–3	Adjustable iris (ID25/M, Thorlabs, USA)
L1	25 mm focal length, 25 mm diameter plano-convex lens, broadband anti-reflection coated. (LA1951-A, Thorlabs, USA)
P1	15 μm diameter pinhole (P15S, Thorlabs, USA)
L2	150 mm focal length, 25 mm diameter plano-convex lens, broadband anti-reflection coated. (LA1433-A, Thorlabs, USA)
L3	200 mm focal length, 25 mm diameter plano-convex lens, broadband anti-reflection coated. (LA1708-A, Thorlabs, USA)
D1	Shortpass dichroic mirror, custom piece (e.g. 488DSCX, Chroma Technology Corp., USA)
OBJ	Microscope objective ×100, 1.45 NA infinity corrected oil immersion (Plan Fluar, Zeiss, Germany)
CS	Coverslip or sample chamber (e.g. Lab-Tek II chambered coverglass, Nalge Nunc International, USA)
D2	Shortpass dichroic mirror (e.g. 555DCSX, Chroma Technology Corp., USA)
F1	Acceptor channel emission filter (e.g. HQ620/30M, Chroma Technology Corp., USA)
F2	Donor channel emission filter (e.g. HQ525/50M, Chroma Technology Corp., USA)
D3	Longpass dichroic mirror (e.g. 565DCLP, Chroma Technology Corp., USA)
L4	150 mm focal length, 25 mm diameter plano-convex lens (acromatic doublet), broadband anti-reflection coated. (AC254-150-A1, Thorlabs, USA)
CCD Camera	Back illuminated EMCCD (Andor iXon, Andor Technology, Northern Ireland)

Figure 3.23 Schematic of the simple total internal reflection fluorescence microscope that produced much of the data not otherwise referenced in this text. Specific examples of components are given for a configuration to measure spFRET for the dye pair Alexa Fluor 488 and Alexa Fluor 594 (Moelcular Probes Inc., USA). These examples represent, in most cases, an arbitrary choice of supplier and are provided only to enable the user to see the specifications of the parts used. Inset: Detail of the arrangement delivering red and green images to different sides of the CCD and of the peripheral illumination scheme at the objective (figure was kindly produced using Zemax (Zemax Development Corporation) by Kurt Baldwin at Avacta Ltd www.avacta.com).

used to vary the incident angle of the primary collimated beam are given in the next sub-section.

If the emerging beam impinges on the glass coverslip/solvent interface at an angle greater than the critical angle for total internal reflection, then a return beam will be observed leaving the objective on the opposite side of the back aperture, assuming the objective is also focused onto the interface (see Figure 3.23, inset top). In the main figure this beam is simply dumped (to a black body), in a later section we will describe briefly how this beam can be used to provide auto-focusing and auto-TIRF angle feedback signals.

Fluorescence from the sample (along with scattering) emerges from the back aperture of the objective as a collimated beam. This light is then reflected from the dichroic (*D1*) onto the detection optics (Figure 3.23, inset, bottom). The light then passes through an adjustable iris (*I3*). The purpose of this iris is to alter the apparent image size (or apparent field of view) formed later on the camera (closing the aperture reduces the maximum angle at which off-axis rays are collected efficiently, therefore effectively reducing the field of view). Thus for a fixed magnification system one can match the image size to the CCD size. This can of course also be adjusted by matching the required field of view size and overall magnification to produce the correct sized image on the CCD. More usefully, in this system one can use the iris to ensure that in two-colour mode both images (red and green) can fit onto the CCD, but allows a slightly larger area to be observed efficiently when one colour operation is used.

The collimated beam now passes through a shortpass dichroic (*D2*), which therefore transmits the donor colour image and reflects the acceptor colour image. The transmitted green donor signal proceeds onto an emission filter (*F2*), a mounted mirror (*M7*), and then is reflected off a suitable longpass dichroic (*D3*) and brought close to the acceptor signal beam. The acceptor signal beam having been guided by a mounted mirror (*M6*) is emission filtered (*F1*) and passed through the dichroic *D3*. The two spatially separated signals are then focused onto the CCD by the achromatic doublet lens *L4*, which is the tube lens for this system. The overall magnification is then given by the ratio of the objective focal length and tube lens focal length, so is approximately $\times 130$. Thus, if the CCD is placed at the focus of the lens *L4* then the maximum field of view that can be observed is around 60 μm^2 due to the magnification of the system (the CCD has 512×512 pixels each 16×16 μm so an active area ~8×8 mm).

Mechanical arrangement and alignment

All components were arranged on a standard optical table, 25 mm grid spacing of 6 mm diameter metric threaded holes, for ease of arrangement. All optical components were mounted in standard steel or aluminium 'post and bases' and using standard optical mounts with no special requirements—unless noted below. Translation along the optic axis (along *x*, see Figure 3.23) was provided by low-resolution single-axis stages (e.g. MS1/M, Thorlabs, USA) for *L1* and *L2* to allow accurate focusing and collimation. *P1* and *I2* were mounted in a two-axis stage (*y*–*z*, for example ST1XY-A, Thorlabs, USA) to allow alignment. *M4* and *L3* were mounted on a base plate on a single axis translation stage (in *z*). As the beam from *M3* was set to be incident at 45° on *M4* and the beam from *M4* was set to pass through the centre of the lens *L3*, at normal incidence, then translation of the stage moves *M4* precisely along the *z* axis (movement indicated by double-headed arrows). This then translates the focal point of the lens *L3* across the objective's BFP without altering the position of the focus of *L3* out of the BFP. In this way an angle of incidence of the collimated light from the objective can be selected (see a further discussion of alignment and achieving TIRF in the following paragraph). Lens *L3* was further mounted in a removable mount (for initial alignment, described in the following paragraph). *D1*, *OBJ*, and the sample stage (holding the coverslip *CS*) were all attached to a large post (e.g. XT95, Thorlabs, USA) rigidly attached to the table. *OBJ* was mounted in a coarse mechanical focusing tube (e.g. SM1V10, Thorlabs, USA) and also in a high-resolution piezoelectric stage (e.g. MIPOS100, piezosystem jena, Germany) to provide focusing (in *y*). The lens *L4* was mounted in a focusing tube (e.g. SM1V10, Thorlabs, USA) and the CCD camera was mounted on a large low-resolution three-axis stage (e.g. MT1/M, Thorlabs, USA). The sample stage (holding the coverslip/sample chamber) was mounted in an inexpensive single-axis stage with open-loop piezoelectric actuator (e.g. 07TES507, Melles Griot, USA). This allowed interrogation of different areas of the sample surface (after an area is exhausted due to photobleaching) without the need for refocusing.

Laser and spatial filter alignment proceeds as for the confocal system described earlier. Specific to the TIRFM lens *L3* is focused onto the BFP of *OBJ* by observing the emerging beam from the objective at a distance and then moving either lens *L3* (in *x*) or the objective (in *y*), until the emerging beam is collimated. Initial alignment of the detection optics is once again made significantly easier as a consequence that the dichroic *D1* is imperfect. The sample is replaced with either a clean coverslip or mirror. With the beam emerging un-deviated from the objective the lens *L3* is removed (thus a collimated beam enters the objective back aperture, centred on the optic axis and is focused to a near-diffraction-limited spot). The excitation light is reflected strongly (rather the Airy disc pattern of the focused beam is imaged) and mainly propagates back through the dichroic *D1* along the path of the excitation light. The mirror or coverslip can then easily be placed in the focal plane by observation of this reflected beam at the iris (*I2*), as for the confocal system. A small proportion of the reflected light is also reflected off the dichroic (although this is in fact non-ideal). This reflected light is also collimated if the mirror (or coverslip–air interface) is at the objective's focal plane. This collimated reflected beam could now be used to coarsely position many of the remaining detection optics. A new sample should now be placed onto the microscope, we find 200 nm fluorescent beads, such as FluoSpheres (Molecular Probes, USA), applied to a coverslip with the excess rinsed off, results in an excellent bright test sample. The objective should then be focused (e.g. using the reflected light observed on iris *I2*). Note that adjusting the objective position when focusing the excitation light will alter the relative position of the BFP with respect to the focus of *L3* (when replaced), we have not however encountered problems with this configuration. For initial alignment an iterative process can be followed, for routine focusing the objective is never moved more than a few hundred micrometers. The long focal length of *L3* works in our favour by making these changes insignificant. Lens *L3* is now reinserted resulting in an un-deviated collimated beam emerging from the top of the objective and with the objective focused onto the glass/air interface. The stage holding *L3* and *M4* can now be translated (in *z*) and one can observe the angle of the light emerging from the objective change. The angle can then be adjusted until total internal reflection occurs: monitored either by a card looking for the return spot, or until the transmitted beam is no longer seen. If one further adjusts the position of *L3* and *M4* one should observe intensity dependence in the return spot that follows the calculations in Figures 3.9 and 3.12.

It is now instructive to note the effect of small adjustments of the objective focus. If one observes the return spot on a card whilst adjusting the focus slightly (noting the start position!), then the return spot will be observed to translate across the card. In this way the position and intensity of the return spot can be used in an electronic feedback system incorporating a quadrant photodiode to maintain focus and TIRF angle alignment, respectively. This might be useful in maintaining alignment in experiments that require long observation times.

If the instrument is placed in total internal reflection and the sample is in focus it should now be possible to observe the image (near the focus of *L4*) by placing a card in the detection path. In this case, the image should be a pattern of bright spots (the fluorescent beads). It should now be possible to arrange the remaining optical elements and to form this image on the CCD.

Whilst no special procedures are necessary to form an image on the CCD, high quality image formation is in fact not trivial. More precisely, formation of good aberration free images is demanding. The confocal system described in the previous section is a rather simple optical system and indeed optical manipulation of single collimated beams from point sources and then focusing to points is surprisingly forgiving. Image formation, however, is far more demanding and in simple optical system such as this aberrations can dominate if care is not taken. This is particularly relevant in the FRET TIRFM as different spectral ranges are separated, optically transformed, and then recombined onto the same CCD. It is thus easy to introduce differential degrees of aberrations that may cause relative stretching (for example) of the final red and green images. If this is not too significant (no more than a few pixels) it may be possible to correct for in software. It is then no coincidence that, for example, the optical paths for the red and green signals (Figure 3.23, inset bottom) are identical, although it may be necessary to add aberration correction to one of the paths (donor or acceptor) with additional lenses.

The test sample of bright fluorescent beads is again useful for initial alignment and checks. It should be found that the beads are sufficiently bright that even at low detector gains sufficient fluorescence is detectable in both colour images—one can therefore check that the images transpose, or use the images as calibration data to provide an offset (or at least quantify the problem) for later single molecule images. The reader should note that imaging of single molecules on a surface is, in this respect, a forgiving imaging application.

Sample introduction and routine focusing

Unlike the confocal system, sample introduction and focusing is rather trivial as the image from the surface can be observed on the CCD. Once the microscope is focused (by making small adjustments to the objective position using its closed-loop piezoelectric stage) the sample can then be translated to a nearby area for measurement (as the area initially used for focusing will be bleached). Electronic shutters or attenuators should be incorporated into the optical path to limit photobleaching of the sample.

Experimental parameters and methods

As has been discussed in Section 3.6, the choice of sample coverslip is crucial. The coverslip material and thickness must be matched to the design requirements of the objective used. The actual form of the chamber (bare coverslip, closed flow chamber or open chamber) is unimportant except with regard to the actual experiment being conducted (i.e. evaporation and solvent). Note that with different substrates and solvents, different critical angles for TIR will exist (see Table 3.1). Laser excitation powers (which can be easily measured for the propagating beam un-deviated by the objective) will vary somewhat on the requirements of the experiment (a balance must be struck between signal-to-noise and bleaching lifetime) and is generally of the order of several milli-watts. Temperature control can be provided in the same manner as that described for the confocal system in the previous section.

References

[1] **Sathy, P, Philip, R, Nampoori, VPN,** and **Vallabhan, CPG,** Photoacoustic observation of excited singlet-state absorption in the laser-dye rhodamine 6g. *Journal of Physics D-Applied Physics* 27 (1994) 2019–2022.

[2] **Eisenberg, D,** *The Structure and Properties of Water*, Clarendon Press, Oxford, 1969.

[3] **Fang, XH** and **Tan, WH,** Imaging single fluorescent molecules at the interface of an optical fiber probe by evanescent wave excitation. *Analytical Chemistry* 71 (1999) 3101–3105.

[4] **Adachi, K, Kinosita Jr, K,** and **Ando, T,** Single-fluorophore imaging with an unmodified epi-fluorescence microscope and conventional video camera. *Journal of Microscopy* **195** (Pt 2) (1999) 125–132.

[5] **Pawley, JB,** *Handbook of Biological Confocal Microscopy*, Plenum Press, New York, 1995.

[6] **Hess, ST, Huang, SH, Heikal, AA,** and **Webb, WW,** Biological and chemical applications of fluorescence correlation spectroscopy: A review. *Biochemistry* **41** (2002) 697–705.

[7] **Haustein, E** and **Schwille, P,** Ultrasensitive investigations of biological systems by fluorescence correlation spectroscopy. *Methods* **29** (2003) 153–166.

[8] **Axelrod, D, Burghardt, TP,** and **Thompson, NL,** Total internal-reflection fluorescence. *Annual Review of Biophysics and Bioengineering* 13 (1984) 247–268.

[9] **Mashanov, GI, Tacon, D, Knight, AE, Peckham, M,** and **Molloy, JE,** Visualizing single molecules inside living cells using total internal reflection fluorescence microscopy. *Methods* **29** (2003) 142–152.

[10] **Paige, MF, Bjerneld, EJ,** and **Moerner, WE,** A comparison of through-the-objective total internal reflection microscopy and epifluorescence microscopy for single-molecule fluorescence Imaging. *Single Molecules* 2 (2001) 191–201.

[11] **Osborne, MA, Balasubramanian, S, Furey, WS,** and **Klenerman, D,** Optically biased diffusion of single molecules studied by confocal fluorescence microscopy. *Journal of Physical Chemistry B* **102** (1998) 3160–3167.

[12] **Pawley, JB,** *Fundamental Limits in Confocal Microscopy*, Plenum Press, New York, 1995.

[13] **Wilson, T,** in **JB Pawley** (Ed.), *Handbook of Biological Confocal Microscopy*. Plenum Press, New York, 1995, pp. 167–182.

[14] **Bestvater, F, Spiess, E, Stobrawa, G, Hacker, M, Feurer, T, Porwol, T,** *et al.*, Two-photon fluorescence absorption and emission spectra of dyes relevant for cell imaging. *Journal of Microscopy-Oxford* **208** (2002) 108–115.

[15] **Foldes-Papp, Z** and **Rigler, R,** Quantitative two-color fluorescence cross correlation spectroscopy in the analysis of polymerase chain reaction. *Biological Chemistry* **382** (2001) 473–478.

[16] **Heinze, KG, Rarbach, M, Jahnz, M,** and **Schwille, P,** Two-photon fluorescence coincidence analysis: Rapid measurements of enzyme kinetics. *Biophysical Journal* **83** (2002) 1671–1681.

[17] **LeCaptain, DJ** and **Van Orden, A,** Two-beam fluorescence cross-correlation spectroscopy in an electrophoretic mobility shift assay. *Analytical Chemistry* **74** (2002) 1171–1176.

[18] **Li, HT, Ren, XJ, Ying, LM, Balasubramanian, S,** and **Klenerman, D,** Measuring single-molecule nucleic acid dynamics in solution by two-color filtered ratiometric fluorescence correlation spectroscopy. *Proceedings of the National Academy of Sciences of the United States of America* **101** (2004) 14425–14430.

[19] **Schwille, P, Haupts, U, Maiti, S,** and **Webb, WW,** Comparison of one- and two-photon excitation for intracellular applications of fluorescence correlation spectroscopy. *Biophysical Journal* 74 (1998) A36.

[20] **Rigler, R**, **Mets, U**, **Widengren, J**, and **Kask, P**, Fluorescence correlation spectroscopy with high count rate and low background-analysis of translational diffusion. *European Biophysics Journal* 22 (1993) 169–175.

[21] **Hess, ST** and **Webb, WW**, Focal volume optics and experimental artifacts in confocal fluorescence correlation spectroscopy. *Biophysical Journal* **83** (2002) 2300–2317.

[22] **Kask, P**, **Palo, K**, **Ullmann, D**, and **Gall, K**, Fluorescence-intensity distribution analysis and its application in biomolecular detection technology. *Proceedings of the National Academy of Science of the United States of America* **96** (1999) 13756–13761.

[23] **Perroud, TD**, **Bo, HA**, **Wallace, MI**, and **Zare, RN**, Photon counting histogram for one-photon excitation. *Chemphyschem* **4** (2003) 1121–1123.

[24] **Perroud, TD**, **Huang, B**, **Wallace, MI**, and **Zare, RN**, Photon counting histogram for one-photon excitation (vol 4, pg 1121, 2003). *Chemphyschem* **4** (2003) 1280–1280.

[25] **Perroud, TD**, **Huang, B**, **Zare, RN**, Effect of Bin Time on the Photon Counting Histogram for One-Photon Excitation. *Chemphyschem* **6** (2005) 905–912.

[26] **Axelrod, D**, Total internal reflection fluorescence microscopy in biology. *Methods in Enzymology* **361** (2003) 1–33.

[27] **Axelrod, D**, Selective imaging of surface fluorescence with very high aperture microscope objectives. *Journal of Biomedical Optics* **6** (2001) 6–13.

[28] **Hecht, E** and **Zajac, A**, *Optics*, Addison-Wesley, Tokyo, 1982.

[29] **Lipson, SG**, **Lipson, H**, and **Tannhauser, DS**, *Optical Physics*, Cambridge University Press, Cambridge, 1995.

[30] **Lakowicz, JR**, *Principles of Fluorescence Spectroscopy*, Plenum Press, New York, 1983.

[31] **Forkey, JN**, **Quinlan, ME**, and **Goldman, YE**, Protein structural dynamics by single-molecule fluorescence polarization. *Progress in Biophysics and Molecular Biology* **74** (2000) 1–35.

[32] **Wakelin, S** and **Bagshaw, CR**, A prism combination for near isotropic fluorescence excitation by total internal reflection. *Journal of Microscopy-Oxford* **209** (2003) 143–148.

[33] **Forkey, JN**, **Quinlan, ME**, **Shaw, MA**, **Corrie, JET**, and **Goldman, YE**, Three-dimensional structural dynamics of myosin *V* by single- molecule fluorescence polarization. *Nature* **422** (2003) 399–404.

[34] **Stout, AL** and **Axelrod, D**, Evanescent field excitation of fluorescence by epi-illumination microscopy. *Applied Optics* **28** (1989) 5237–5242.

[35] **Conibear, PB** and **Bagshaw, CR**, A comparison of optical geometries for combined flash photolysis and total internal reflection fluorescence microscopy. *Journal of Microscopy* **200** (2000) 218–229.

[36] **Ying, LM** and **Xie, XS**, Fluorescence spectroscopy, exciton dynamics, and photochemistry of single allophycocyanin trimers. *Journal of Physical Chemistry B* **102** (1998) 10399–10409.

[37] **Muller, R**, **Zander, C**, **Sauer, M**, **Deimel, M**, **Ko, D-S**, **Sibert, S**, **arden-Jacob, J**, **Deltau, G**, **Marx, NJ**, **Drexhage, KH**, **Wolfrum, J**, Time-Resolved identification of single molecules in solution with a pulsed semiconductor diode laser. *Chemical Physics Letters* **262** (1996) 716–722.

[38] **Li, HT**, **Ying, LM**, **Green, JJ**, **Balasubramanian, S**, and **Klenerman, D**, Ultrasensitive coincidence fluorescence detection of single DNA molecules. *Analytical Chemistry* **75** (2003) 1664–1670.

[39] **Van Orden, A**, **Fogarty, K**, and **Jung, J**, Fluorescence fluctuation spectroscopy: A coming of age story. *Applied Spectroscopy* **58** (2004) 122A–137A.

[40] **Dittrich, PS, Muller, B,** and **Schwille, P,** Studying reaction kinetics by simultaneous FRET and cross-correlation analysis in a miniaturized continuous flow reactor. *Physical Chemistry Chemical Physics* **6** (2004) 4416–4420.

[41] **Ha, T, Laurence, TA, Chemla, DS,** and **Weiss, S,** Polarization spectroscopy of single fluorescent molecules. *Journal of Physical Chemistry B* **103** (1999) 6839–6850.

[42] **Wilson, J** and **Hawkes, JFB,** *Optoelectronics: An Introduction*, Prentice Hall Int., New York, 1989.

[43] **Soper, SA** and **Legendre Jr., BL,** Single molecule detection in the near-IR using continuous-wave diode laser excitation with an avalanche photon detector. *Applied Spectroscopy* **52** (1998) 1–6.

[44] **Welford, WT,** *Aberrations of the Symmetrical Optical System*, London, Academic Press, 1974.

[45] **Li, LQ** and **Davis, LM,** Single photon avalanche diode for single molecule detection. *Review of Scientific Instruments* **64** (1993) 1524.

[46] **Wohland, T, Rigler, R,** and **Vogel, H,** The standard deviation in fluorescence correlation spectroscopy. *Biophysical Journal* **80** (2001) 2987–2999.

[47] **Neuweiler, H, Schulz, A, Bohmer, M, Enderlein, J,** and **Sauer, M,** Measurement of submicrosecond intramolecular contact formation in peptides at the single-molecule level. *Journal of the American Chemical Society* **125** (2003) 5324–5330.

[48] **Knemeyer, JP, Herten, DP,** and **Sauer, M,** Detection and identification of single molecules in living cells using spectrally resolved fluorescence lifetime imaging microscopy. *Analytical Chemistry* **75** (2003) 2147–2153.

[49] **Schuler, B, Lipman, EA,** and **Eaton, WA,** Probing the free-energy surface for protein folding with single-molecule fluorescence spectroscopy. *Nature* **419** (2002) 743–747.

[50] **Elson, EL,** Quick tour of fluorescence correlation spectroscopy from its inception. *Journal of Biomedical Optics* **9** (2004) 857–864.

[51] **Krichevsky, O** and **Bonnet, G,** Fluorescence correlation spectroscopy: The technique and its applications. *Reports on Progress in Physics* **65** (2002) 251–297.

[52] **Muller, JD, Chen, Y, Gratton, E,** in **R Rigler** and **ES Elson** (Eds), Photon Counting Histogram Statistics. *Fluorescence Correlation Spectroscopy: Theory and Applications*. Springer, Berlin, 2001, pp. 410–437.

[53] **Chen, Y, Muller, JD, So, PTC,** and **Gratton, E,** The photon counting histogram in fluorescence fluctuation spectroscopy. *Biophysical Journal* **77** (1999) 553–567.

[54] **Rhoades, E, Gussakovsky, E,** and **Haran, G,** Watching proteins fold one molecule at a time. *Proceedings of the National Academy of Sciences of the United States of America* **100** (2003) 3197–3202.

[55] **Haran, G,** Single-molecule fluorescence spectroscopy of biomolecular folding. *Journal of Physics-Condensed Matter* **15** (2003) R1291–R1317.

[56] **Haran, G,** Noise reduction in single-molecule fluorescence trajectories of folding proteins. *Chemical Physics* **307** (2004) 137–145.

[57] **Bohmer, M** and **Enderlein, J,** Fluorescence spectroscopy of single molecules under ambient conditions: Methodology and technology. *Chemphyschem* **4** (2003) 793–808.

[58] **Bohmer, M, Wahl, M, Rahn, HJ, Erdmann, R,** and **Enderlein, J,** Time-resolved fluorescence correlation spectroscopy. *Chemical Physics Letters* **353** (2002) 439–445.

[59] **Jankowski, T** and **Janka, T,** in **R Rigler** and **ES Elson** (Eds.), Confocor 2 – The Second Generation of Fluorescence Correlation microscopes. *Fluorescence Correlation Spectroscopy: Theory and Applications*. Springer, Berlin, 2001, pp. 331–345.

[60] **Mashanov, GI, Tacon, D, Peckham, M,** and **Molloy, JE,** The spatial and temporal dynamics of pleckstrin homology domain binding at the plasma membrane measured by Imaging single molecules in live mouse myoblasts. *Journal of Biological Chemistry* **279** (2004) 15274–15280.

[61] **Pedrotti, FL** and **Pedrotti, LS,** *Introduction to Optics*, Pentice-Hall Int., London, 1993.

[62] **Abe, K, Kaya, S, Hayashi, Y, Imagawa, T, Kikumoto, M, Oiwa, K,** *et al.*, Correlation between the activities and the oligomeric forms of pig gastric H/K-ATPase. *Biochemistry* **42** (2003) 15132–15138.

[63] **Meseth, U, Wohland, T, Rigler, R,** and **Vogel, H,** Resolution of fluorescence correlation measurements. *Biophysical Journal* **76** (1999) 1619–1631.

[64] **Gosch, M, Serov, A, Anhut, T, Lasser, T, Rochas, A, Besse, PA,** *et al.*, Parallel single molecule detection with a fully integrated single-photon 2X2 CMOS detector array. *Journal of Biomedical Optics* **9** (2004) 913–921.

[65] **Hohng, S, Joo, C,** and **Ha, T,** Single-molecule three-color FRET. *Biophysical Journal* **87** (2004) 1328–1337.

[66] **Joo, C, McKinney, SA, Lilley, DMJ,** and **Ha, T,** Exploring rare conformational species and ionic effects in DNA Holliday junctions using single-molecule spectroscopy. *Journal of Molecular Biology* **341** (2004) 739–751.

[67] **Nishizaka, T, Oiwa, K, Noji, H, Kimura, S, Muneyuki, E, Yoshida, M,** *et al.*, Chemomechanical coupling in F-1-ATPase revealed by simultaneous observation of nucleotide kinetics and rotation. *Nature Structural and Molecular Biology* **11** (2004) 142–148.

[68] **Vallee, RAL, Tomczak, N, Kuipers, L, Vancso, GJ,** and **van Hulst, NF,** Effect of solvent on nanoscale polymer heterogeneity and mobility probed by single molecule lifetime fluctuations. *Chemical Physics Letters* **384** (2004) 5–8.

[69] **Bohmer, M,** Time-resolved confocal scanning device for ultrasensitive fluorescecne detection. *Review of Scientific Instruments* **72** (2001) 4145–4152.

[70] **Ide, T, Takeuchi, Y, Aoki, T,** and **Yanagida, T,** Simultaneous optical and electrical recording of a single ion- channel. *Japanese Journal of Physiology* **52** (2002) 429–434.

[71] **Ide, T, Takeuchi, Y,** and **Yanagida, T,** Development of an experimental apparatus for simultaneous observation of optical and electrical signals from single ion channels. *Single Molecules* **3** (2002) 33–42.

[72] **Sarkar, A, Robertson, RB,** and **Fernandez, JM,** Simultaneous atomic force microscope and fluorescence measurements of protein unfolding using a calibrated evanescent wave. *Proceedings of the National Academy of Sciences of the United States of America* **101** (2004) 12882–12886.

[73] **Eid, JS, Muller, JD,** and **Gratton, E,** Data acquisition card for fluctuation correlation spectroscopy allowing full access to the detected photon sequence. *Review of Scientific Instruments* **71** (2000) 361–368.

[74] **Chirico, G, Olivini, F,** and **Beretta, S,** Fluorescence excitation volume in two-photon microscopy by autocorrelation spectroscopy and photon counting histogram. *Applied Spectroscopy* **54** (2000) 1084–1090.

[75] **Vallee, RAL, Tomczak, N, Kuipers, L, Vancso, GJ,** and **van Hulst, NF,** Single molecule lifetime fluctuations reveal segmental dynamics in polymers. *Physical Review Letters* **91** (2003) art. no.-038301.

FOUR

Preparation of samples for single molecule fluorescence spectroscopy

4.1 Introduction

We have seen in Chapters 2 and 3 that the measurement and analysis of the intensity of fluorescence emission from single molecules can readily be achieved. The missing component of the discussion so far is the production of biological molecules or their complexes that are labelled in such a manner as to report on the process of interest. Consequently, before any single molecule measurements are undertaken it is necessary to consider what dye, or dye pair, should be used together with the location and method of their conjugation to the biomolecule of interest. In principle, the labelling procedure should be trivial as it is relatively easy to derivatize both nucleic acids and proteins with a wide variety of different moieties using established protocols. Such methods are usually adequate when preparing biomolecules labelled with a single fluorophore at one or more sites. However, the production and purification of samples in which each biomolecule is labelled with two different fluorophores at two specific sites (a requirement of fluorescence resonance energy transfer 'FRET' experiments) can provide a formidable challenge that must be overcome before many of the techniques and methods described previously are undertaken.

In this chapter we will briefly discuss the photophysical properties of commonly used dyes and, by reference to biomolecular systems studied by single molecule FRET in the literature, discuss the properties of commonly used dye pairs. We will then describe, in detail, the labelling and purification methods used in our laboratories to generate milligram quantities of a single chain protein labelled with two different dyes. Finally, we will discuss a variety of methods by which biomolecular systems have been tethered at or close to a surface, a necessary prerequisite for performing many of the experiments described in other chapters.

4.2 **Dye selection**

4.2.1 **Photophysical considerations of dye selection**

In this section we review a number of important concepts that affect the choice of dyes for single molecule fluorescence applications. We introduce some concepts such as quantum yield, quenching, photobleaching, and blinking, many of which have already been highlighted in the context of single molecule fluorescence experiments and analysis in Chapters 2 and 3. We also refer the reader to a number of sources that provide further information on some or all of these topics [1–4].

The fundamental absorption and emission properties of dyes are often represented in an energy level diagram that shows the various electronic and vibrational energy levels that may exist in a molecule, together with the pathways that exist between these various distinct states (the 'Jablonski' diagram). A simplified Jablonski diagram that does not include effects due to solvent quenching, intermolecular quenching or FRET (all of which will be covered later) is shown in Figure 4.1. The Jablonski diagram elegantly depicts molecular electronic and vibrational energy levels illustrating the phenomenon of light absorption and emission. At equilibrium, a fluorophore is likely to exist in the lowest vibrational energy (V_0) level of the molecular ground state (S_0). Upon absorption of a photon, whose energy closely matches an electronic transition in the fluorophore (we consider only single-photon absorption in this discussion), the fluorophore is excited to a

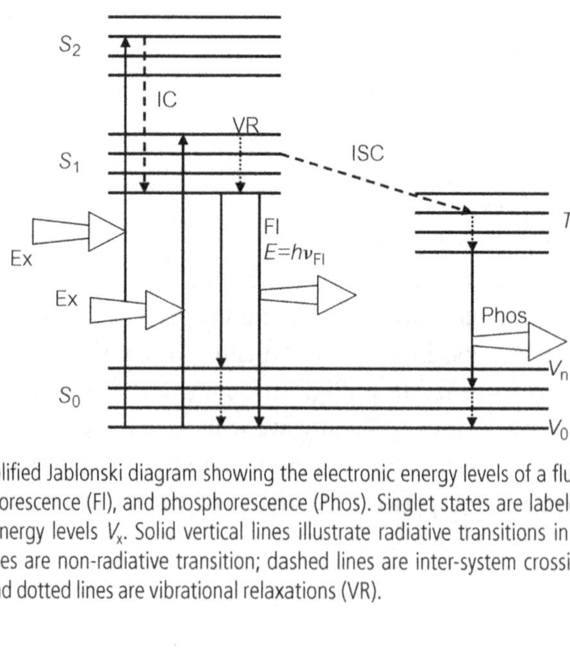

Figure 4.1 Simplified Jablonski diagram showing the electronic energy levels of a fluorophore, illustrating excitation (Ex), fluorescence (Fl), and phosphorescence (Phos). Singlet states are labeled S_x, triplet states T_x and virbrational energy levels V_x. Solid vertical lines illustrate radiative transitions in the direction of trhe arrows, broken lines are non-radiative transition; dashed lines are inter-system crossing (ISC) and internal conversion (IC), and dotted lines are vibrational relaxations (VR).

high vibrational energy level in the first or second electronic singlet states (S_1 or S_2), a process which occurs essentially instantaneously ($<10^{-15}$ seconds). Rapid relaxation ($\sim 10^{-12}$ s) then occurs to the lowest vibrational energy level of the first singlet state (S_1), a process termed internal conversion (IC). Molecules in this equilibrated excited state can then lose energy through two primary mechanisms (for this simplified discussion): fluorescence involving the emission of a photon and relaxation to the ground state, or inter system crossing (ISC) to the first triplet state (T_1). Molecules in the first triplet state can then eventually relax to the ground state (note ISC to a higher triplet level can occur followed by IC to T_1), the rate constant for this transition, termed phosphorescence is $\sim 10^{-7}$ s, much slower than for the fluorescence relaxation transition S_1–S_0 ($\sim 10^{-9}$ s). Energy may also be transferred back to S_1 and result in delayed fluorescence with relaxation to S_0. For single molecule fluorescence we are most concerned with fluorescence due to radiative relaxation from S_1 or ISC to the triplet state T_1.

Fluorescence

Fluorescence is caused by relaxation of an electron from $S_{x>0}$ to S_0 through emission of a photon. A shift of the emitted photon's wavelength towards the red end of the spectrum generally occurs, indicating that energy has been lost. Examination of the Jablonski diagram illustrates the primary cause of this shift: loss of energy in the excited state due to relaxation (IC) and vibrational relaxation (VR) within a state to the lowest vibrational energy level of the first singlet state (S_1). The resulting wavelength shift (the Stokes shift) is commonly observed in spectra, such as that shown in Figure 4.2. The relatively small stokes shift of

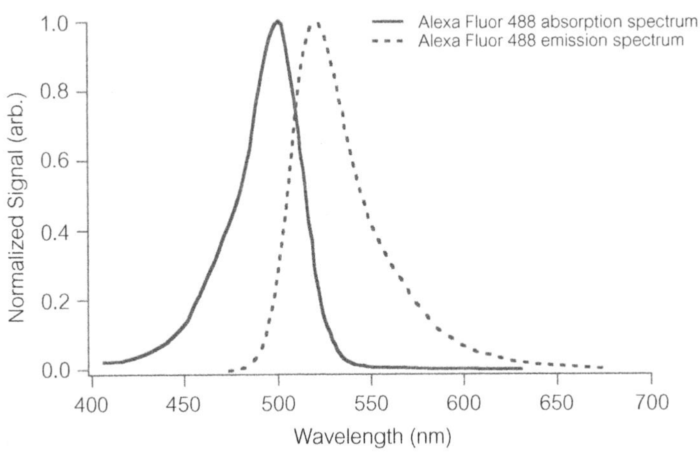

Figure 4.2 Normalized absorption (solid line) and fluorescence emission spectra (dashed line) for the dye Alexa Fluor 488 (see text).

typically <20 nm (which is always similar for common, strongly visibly fluorescent dyes) has a number of consequences for single molecule spectroscopy. In particular, it means that it is generally not possible to find two-dye molecules that have disparate emission signals (so as to allow resolution of their emission signals) when exciting fluorescence with the same wavelength excitation source. Alternatives for this dual-channel single molecule detection were discussed in Chapters 2 and 3. It is also essential that the Stokes shift is sufficient to allow efficient separation of the majority of the fluorescence emission from the excitation light in order to reduce the background signal from scattered light (see Chapter 3).

Figure 4.2 shows the absorption spectrum (solid line) for the dye Alexa Fluor 488. The absorption spectrum gives an indication of the relative amount of light that is absorbed at different wavelengths and is often presented with the optical density on the ordinate. If the extinction coefficient of the dye (with units of $cm^{-1}M^{-1}$) at some wavelength is known then the concentration (molarity) of the solution can be determined. Absorption spectra are recorded by monitoring the change in the amount of absorbed light as a function of wavelength. A common alternative to measurement of the absorption spectrum is the measurement of an excitation spectrum (as this can often be performed on the same instrument used to measure the fluorescence). In this case the detected wavelength is fixed at a region where fluorescence emission exists and the excitation wavelength is scanned. In this way a plot of the relationship between the excitation wavelength and the relative amount of fluorescence is obtained. In most common cases the excitation and absorption spectra for a molecule will be indistinguishable in shape and so both give an indication of the relative emission intensity at a given excitation wavelength (although in some cases the absorption spectrum may indicate wavelengths where absorption occurs but relaxation to the ground state is non-radiative). For common dyes the characteristics of the emission spectrum are independent of the excitation wavelength. Thus the position of the maxima in the excitation/absorption spectra of a dye compared to the available excitation wavelengths is not crucial, however for the single molecule instrument it is an important consideration, in order to maximize the fluorophore brightness.

A high quantum yield is another important requirement for single molecule measurements. The fluorescence quantum yield is defined as the ratio of the number of photons emitted to the number of photons absorbed. Quantum yields less than 1 can occur as not every relaxation from a radiatively generated excited state necessarily leads to radiative emission. The quantum yield is thus a direct measure of the brightness of a dye molecule. In the next section we will briefly consider other pathways.

It should be noted that the maximum number of fluorescence photons emitted per second from a fluorescent molecule is actually limited by the lifetime of the excited state of the molecule. The quantum yield is often taken to mistakenly imply that if more emission photons are needed one can simply increase the excitation intensity without limit. The absorption of a photon and the generation of equilibrated excited states are essentially instantaneous processes ($\sim 10^{-12}$ s) and the transition time to the ground state (via photoemission or some other mechanism) is even more rapid ($\sim 10^{-15}$ s). The molecular excited state, however, can persist for much longer before relaxation ($\sim 10^{-9}$–10^{-8} s) and so this sets a limit on the maximum emission rate from the molecule. Consequently, once saturation of the excitation – emission cycle has been reached, increasing the excitation power will not then increase the brightness, or number of photons per second, from a molecule. Indeed it may introduce, or exaggerate other, undesirable photo induced effects, some of which will be discussed in the next section.

Bleaching, blinking, quenching, and the triplet state

Ideally, dyes selected for single molecule fluorescence spectroscopy should have: a high extinction coefficient at the laser wavelength which is available to the researcher, a high fluorescence quantum yield and display emission that is spectrally distinct from the excitation wavelength, is steady (or at least does not turn off and on, 'blinking') and which persists for a long period of time. Common fluorophores, unfortunately, fall somewhat short in many of these categories and in this section we briefly outline some of the more common problematic photophysical effects.

All fluorescent molecules are prone at some stage to the effect of irreversible photobleaching. The exact mechanism of photobleaching is poorly understood but is mainly thought to be due to photo-oxidation [5,6] and is therefore strongly affected by laser power. The effect can be reduced by the use of oxygen scavenging cocktails introduced to the solvent and by reduction in the excitation intensity. In any experiment one must therefore balance the required observation time, the signal-to-noise and the importance of any artefacts that may be manifested due to photobleaching (e.g. the 'zero' peak in FRET and reduced apparent diffusion rates in FCS, see Chapter 2). Typical fluorophores chosen for single molecule fluorescence work generally emit of the order 10^5–10^6 photons before photo destruction. If the excited state lifetime is of the order of 5 ns (and we assume saturation) then the typical single fluorophore may only be fluorescent for approximately 1 ms. It is clear from this that single molecule experiments are rarely conducted near the point of saturation of the excitation–emission cycle.

If the quantum yield for ISC is high saturation may also occur, as the triplet lifetime is much longer than the singlet excited state lifetime [6,7]. Molecules that show large ISC are therefore generally not very bright. The lifetime of the triplet state can be reduced through quenching with the ground triplet state of molecular oxygen [6–8]. In this manner the likelihood of the onset of saturation is decreased, although, as has been discussed, somewhat frustratingly photo-oxidation has been suggested as a primary cause of photobleaching. Nevertheless, it has been possible to assemble cocktails of quenchers and scavengers which, when combined with care in setting the excitation laser power, produce good results both in terms of reduced triplet yield and reduced photobleaching.

As well as mechanisms that cause a persistent reduction in fluorescence or even its total elimination, a number of effects can also lead to intermittent reduction or loss of fluorescence. These phenomena are grouped under the term 'blinking'. Reductions in the apparent fluorescence level from a dye molecule (quenching) are often associated with bi-molecular interactions (which result in a reduction of the fluorescence quantum yield) or alternatively quenching is often due to an intramolecular conformational change or photo-activated states [9], or by some transient inter-fluorophore excited state interaction (excimer formation). Complex blinking behaviour (the intermittent quenching of fluorescence) has been observed in a number of dye molecules used for single molecule fluorescence work [9] but was particularly problematical in the early studies of strongly visibly fluorescent natural protein chromophores (auto-fluorescent proteins) [10], in particular the green fluorescent protein (GFP) and the yellow fluorescent protein (YFP), which we discuss as examples. Proteins such as GFP are particularly useful for single molecule studies as, despite their large size, they can be tagged to a target protein by genetic manipulation (making a so-called fusion protein) obviating the need to derivatize the purified target protein with functionalized fluorophores. This makes them particularly suitable for *in vivo* cellular studies. A number of studies have, however, noted significant fluorescence dynamics apparent in the fluorescence trajectories of isolated GFP on both fast (sub-millisecond) and slow (millisecond – second) timescales [11–15]. The intermittency in fluorescence in these molecules has been assigned to a number of causes including slow interconversion in the excited state between different isomers, charge states, photobleaching, spectral diffusion (changes, particularly shifts, in the absorption or emission spectra of single dyes) [10] and pH coupled conformational changes [16]. The problems associated with these autofluorescent proteins, which elegantly represent the problems associated with many such intra-molecular effects in other types of potential single molecule fluorophores, as well as a discussion of the potential usefulness of GFP can be found in a recent review [10]. It should be noted that the causes of many of these complications are associated with the relatively large size and complexity of these autofluorescent molecules (238 amino acids for GFP), however different isomerization states and local environmental effects may also exist in small chromophores. Simple dye molecules can be particularly affected by their local environment, such as the close proximity of nucleotide bases or amino acid side groups that are known to quench other types of fluorophores (tryptophan – histidine quenching for example). This 'quenching' may either be transient, occurring in the excited state (collisional quenching), or slow; inter-molecular complex formation (which might also be reversible) that results in a non-fluorescent ground state molecule (static quenching). Small chromophores can, for example, show significant heterogeneity in their emission characteristics (which clearly have a number of consequences for a variety of measurements). In one example Hou and co-workers [17] measured the emission spectra of individual molecules of the dye Nile Red in PVA and PMMA polymer films, and observed dramatic spectral variation between individual molecules. A figure from this paper is presented and discussed in Chapter 2, Section 2.7.7.

The complexity and high degree of situation specific behaviour of all the effects that can modulate dyes fluorescence means that it is difficult to anticipate the expected phenomenon *a priori*. As a result one must take extreme care to test for and eliminate any effects that might provide a trivial explanation for changes in the fluorescence signal that might be observed in a given experiment: thorough controls are always necessary.

One form of quenching that is of particular relevance to the experiments described elsewhere in this book is quenching of a dye (called the donor) through resonance energy transfer to a second molecule (the acceptor). In the case where the quenching molecule (the acceptor) is itself fluorescent, this is often termed fluorescence resonance energy transfer (FRET). The case of a fluorescent quencher (acceptor) is illustrated in the Jablonski diagram shown in Figure 4.3.

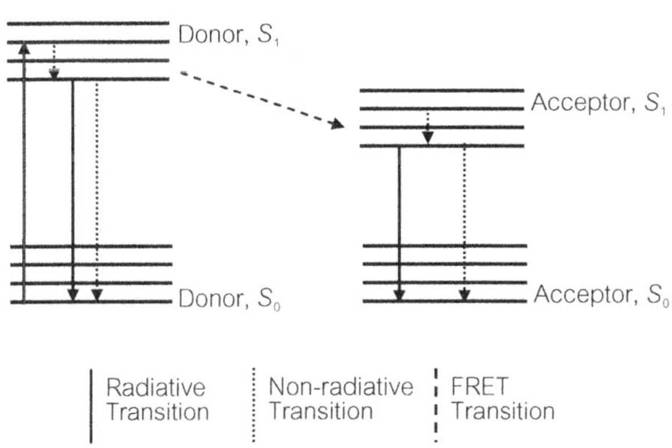

Figure 4.3 Jablonski type diagram showing the simplified energy-level arrangement for a donor–acceptor FRET pair. Intersystem crossing has been ignored, triplet states of both molecules have been omitted and the possibility of direct radiative excitation of the acceptor is neglected.

We see that a radiative transition (absorption of a photon) can generate an excited state in the donor molecule, this energy can then be lost through three mechanisms: emission of a photon (donor fluorescence), non-radiative relaxation (a quenching mechanism, such as those discussed earlier) or non-radiative energy transfer to the acceptor, generating an excited state in the acceptor. The acceptor may then return to the ground state either through radiative emission (acceptor fluorescence) or non-radiative relaxation.

A thorough discussion of the theory of the mechanism of energy transfer from the donor to the acceptor, first determined by Förster [18] using a classical electromagnetic description, can be found elsewhere [4,18] and a discussion of many of the practical aspects of FRET, in the context of single molecule measurements, can be found in Chapter 2, Section 2.5. Of most importance (and relevant to the current discussion) is that the energy transfer is non-radiative (although energy transfer due to absorption of an emitted photon is possible, it is not described by FRET) nor is a coupled excited state or other complex produced. The key is the distance over which the effects occur, in the case of FRET this is generally in the 1–10 nm range; at distances of less than 1 nm the formation of coupled excited states can occur, and while this may still result in apparent energy transfer (excitation at the donor absorption maximum, emission at acceptor wavelengths) this is not described by the theory of FRET. A common analogy used to explain FRET is the classic physics lab experiment illustrating the action of coupled oscillators using identical pendula. If one pendulum is set to swing (the analogy being that an excited state is generated in the donor molecule through absorption of a photon) then, if conditions are right, this energy will be transferred completely to the second coupled (resonant) pendulum (the acceptor). In the case of this mechanical demonstration the energy will be transferred repeatedly backwards and forwards between the two oscillators (until all energy is lost to friction). In the analogy the excited state in the second oscillator (the acceptor) will decay via the emission of an acceptor photon. Note that energy transfer back to the donor is generally not possible (a deviation from the pendulum analogy) as a significant loss in energy occurs upon transfer to the acceptor, creating a barrier for energy transfer back to the donor. For those wishing to relive their school physics lessons, and apply the pendulum analogy, we reference a novel computer-based Java applet that simulates coupled pendula [19]. We can extend this analogy by considering the particular system of the pair of dye molecules in resonance, by considering a classical electromagnetic description of light and matter (although we note that a real understanding of the mechanism of FRET is beyond the scope of this text, requiring an understanding of quantum electrodynamics). In a classical electromagnetic description light can be considered an oscillating electric field, interaction with the donor induces an oscillating dipole in the donor molecule. The oscillating dipole generates an oscillating electric field. If the acceptor molecule is sufficiently close then this oscillating electric field from the donor can induce an oscillating dipole in the acceptor molecule at the resonant frequency. This results in loss of energy in the donor and the potential for radiative relaxation of the acceptor resulting in the emission of light.

As for the coupled pendula, the acceptor and donor dye must be matched in some way if FRET is to occur. A discussion of the requirements for FRET for a particular pair of dyes, and how this may be quantified, can be found in Chapter 2, Section 2.5. The most important parameter that is discussed in that section is the Förster distance, R_0. This distance is the scalar separation between the two dyes at which the efficiency of transfer from the donor to the acceptor is 50%. Typical values are of the order of 50 Å for visibly fluorescent dye pairs suitable for FRET. The particular utility of FRET is that this length scale is similar to the size of many biological systems (proteins, membranes etc.) and that the Förster theory demonstrates that the dependence of the transfer efficiency goes as the inverse sixth power of the scalar separation between the dyes. Thus even small changes in the distance (e.g. due to conformational changes) are easily measurable, for example a 5 Å change in separation from 54 to 49 Å for a dye pair with $R_0 \sim 54$ Å, results in a change in FRET efficiency from 50% to 64%, which is easily measurable. This utility can also be a disadvantage of FRET and systems must be designed very carefully to ensure that a measurable differential is likely to be seen from any structural change that is to be probed. For example, at separations larger than ~100 Å, regardless of the distance change, FRET is useless. Similarly, at distances less than ~20 Å efficiency changes are insignificant.

Dyes for single molecule spectroscopy

The range of dye molecules available for single molecule studies is increasing all the time. Development is occurring in chromophore design to optimize most of the desirable qualities that have been discussed in earlier sections. High quantum yields and extinction coefficients at the common laser wavelengths are desirable (typically values of >0.1 and 20,000 $cm^{-1}M^{-1}$, respectively [5] are encountered). Similarly, low triplet quantum yield, long times before photobleaching and a lack of any blinking due to photo-induced isomerization or other effects are all advantageous. Table 4.1, whilst not exhaustive, summarizes a variety of dyes that are often used in single molecule investigations, collated from numerous sources in the literature. In addition to the characteristics mentioned earlier, a number of special properties are necessary for dyes chosen to act as FRET pairs (see Chapter 2, Section 2.5). Some examples of these dye pairs and their properties and applications are provided in Table 4.2. It should be noted that, for all of the examples in these tables, the exact properties that a particular dye will display

Table 4.1 Photophysical properties of some common dyes with potential for single molecule fluorescence studies

Fluorophore	λ_{ex}^{1p} (nm)	λ_{em}^{1p} (nm)	QY	ε ($cm^{-1}M^{-1}$)	SS (nm)	τ_f (ns)	λ_{ex}^{2p} (nm)	λ_{em}^{2p} (nm)	Reference
FITC	495	520	0.7	73,000	25	—	947	530	[87–89]
FAM	495	520	0.7	83,000	25	—	—	—	[88]
TMR	554	585	0.2–0.5	95,000	31	2.1	849	570	[5,90,91]
R6G	530	556	—	105,000	26	—	—	—	[92]
Cy2	489	506	—	—	17	—	905	520	[87,93]
Cy3	550	570	0.14	150,000	20	~1	1032	578	[3,87,90]
Cy5	650	670	0.15	250,000	20	~1	—	—	[3,5,90]
Cy5.5	675	694	—	250,000	19	—	—	—	[3]
Cy7	743	767	0.02	250,000	24	~0.8	—	—	[3,90,94]
ECFP	458	472	0.4	26,000	14	—	—	—	[94]
EGFP	395,470	509	0.8	30,000	39	3.2	—	—	[5,90,94]
EYFP	514	527	0.6	84,000	13	3.7	—	—	[90,94]
DsRed	532	582	0.29	22,500	50	2.8	—	—	[90]
Bodipy Fl	504	510	—	70,000	6	—	920	526	[87,88]
Bodipy R6G	528	547	—	70,000	19	—	—	—	[88]
AF488	495	520	0.5–0.9	80,000	25	—	985	530	[5,87]
AF546	554	570	—	112,000	16	—	1028	582	[87,95]
AF555	555	565	—	150,000	10	—	—	—	[3]
AF594	590	617	—	92,000	27	—	1074	619	[87,95]
AF633	632	647	—	100,000	15	3.2	—	—	[3,90]
AF647	650	665	—	240,000	15	—	—	—	[3]
AF660	663	690	—	130,000	27	—	—	—	[3]
AF680	679	702	—	180,000	23	—	—	—	[3]
AF700	702	723	—	190,000	21	—	—	—	[3]
TR	596	620	0.5	85,000	24	—	1108	616	[5,87]
Bodipy TMR	544	570	—	56,000	26	—	—	—	[88]
Atto488	501	523	0.8	90,000	22	3.1	—	—	[96]
Atto532	532	553	0.9	115,000	21	3.8	—	—	[96]
Atto565	563	592	0.9	120,000	24	3.4	—	—	[3,96]
Atto594	601	627	0.85	120,000	26	—	—	—	[96]
Atto633	629	657	0.64	130,000	28	3.2	—	—	[96]
Atto700	700	719	0.25	120,000	19	3.2	—	—	[96]

It is important to note that the exact characteristics of a particular dye will be very dependant on the precise molecular environment/solvent, as such these values are given only as a guide. The list was compiled from manufacturers web sites and published work. Abbreviations: TMR (tetramethylrhodamine), Cy*x* (cyanine dye *x*, Amersham Biosciences, UK), AF*xxx* (Alexa Fluor dye *xxx*, Invitrogen Ltd., UK), R6G (rhodamine 6G), TR (texas red), FITC+FAM (fluorescein derivatives). ECFP, EGFP, EYFP (enhanced—cyan, green and yellow fluorescent protein, respectively, Clontech Laboratories, Inc., USA). Bodipy is a trademark of Invitrogen Ltd., UK. Atto*xxx* is a trademark of ATTO-TEC GmbH, Germany. λ_{ex}^{1p}, λ_{em}^{1p} and λ_{ex}^{2p}, λ_{em}^{2p} are the one- and two-photon absorption and emission maximum, respectively. QY is the fluorescence quantum yield, SS the Stokes shift, ε the molar extinction coefficient and τ_f the fluorescence lifetime. Further information regarding two-photon measurements can be found in Chapter 3.

Table 4.2 Dye pairs used in single molecule fluorescence resonance energy transfer experiments

System	Donor	Acceptor	R_0 (Å)	Labelling strategy	Reference
Protein (CI2)	TMR	Cy5	53	Ligation of a peptide corresponding to residues 1–39 of CI2 (N-terminally labelled with TMR NHS ester) to the C-terminal fragment (residues 40–64). A sulphydryl reactive derivative of Cy5 then conjugated to cysteine 40 (introduced to facilitate ligation)	[37]
Protein (AK)	AF488	TR	49	Conjugation of a pair of inserted cysteine residues with maleimide functionalized dyes	[85]
Helicase binding to DNA duplex	Cy3	Cy5	~60	Donor and acceptor incorporated in phosphoramidite form into their respective strands of DNA during oligonucleotide synthesis	[63]
Protein (Syntaxin 1a), complex with SNARE proteins	AF488	AF594	54	Single-step double labelling of pairs of cysteine residues introduced at specific sites	[97]
Polyproline rods	AF488	AF594	54	Each proline rod contained an amino- and carboxy-terminal glycine and cysteine residue, respectively. Each rod sequentially labelled with suphydryl reactive AF488 maleimide followed by the amine reactive AF594 succinimidyl which in this case is specific for the N-terminus	[98]
Duplex DNA	TMR	Cy5	53	Oligonucleotides synthesized containing either a thiol modifier C6 or an aminomodifier C6 dT (or both) at various positions. Each oligonucleotide then conjugated with either maleimide or hydroxysuccinimide derivatized dyes	[28]
Protein (CspTm)	AF488	AF594	54	Introduction of cysteine residues at the N- and C-termini. The sulphydryl groups then conjugated with maleimide derivatized dyes in a two-step procedure (label with donor, isolate, label with acceptor and then purify double labelled proteins (donor–acceptor) from other species	[44]

Table 4.2 *(Continued)*

System	Donor	Acceptor	R_0 (Å)	Labelling strategy	Reference
Coil–coil peptide GCN4 tethered to a surface	R6G	TR	37–51	Each monomer identical but derivatized with either R6G or TR (using a carboxysuccinamide ester) at their N-terminus whilst on solid phase. Cysteine residues and polynucleotide tag at each C-terminus allows disulphide cross-linking and immobilization onto positively charged amino-silanized glass, respectively	[34]
DNA reconstituted into a nucleosome	Cy3	Cy5	60	Oligonucleotide primers containing either an aminolink-dC or an aminolink-dT were conjugated with hydroxysuccinimide esters of Cy5 and Cy3 respectively. After purification, these oligonucleotides used as primers in a polymerase chain reaction to generate a double labelled 164 bp nucleosome positioning fragment	[27]
Single ribozyme (RNA) tethered to a surface	Flu	Cy3	53	5′ end of RNA labelled with donor by *in vitro* transcription. Acceptor annealed to the 3′ end of the ribozyme by hybridization of a DNA oligonucleotide (derivatized at its 5′ and 3′ ends with biotin and Cy3). Biotinylated DNA oligonucleotide binds tightly to streptavidin-coated overslip	[99]
Ribozyme tethered to a surface in complex with its RNA substrate	Cy3	Cy5	–	3′ end of ribozyme extended to allow the hybridization of a DNA oligonucleotide derivatized with acceptor and biotin (for binding onto streptavidinated glass substrate). 5′ end of ribozyme then annealed to donor-labelled substrate	[30]
DNA hairpin tethered to a surface at its apex	Cy3	Cy5	54	Each hairpin consisted of a 30 bp complimentary region (stem) connected by a $(dT)_5$ loop. A DNA oligonucleotide shorter than the full length hairpin was synthesized in which the middle dT of the loop contained at C6-biotion moiety for immobilization and the donor was conjugated by an amino-modified dT on the partially double-stranded stem. The hairpin (which allows only one site for dT incorporation) was completed by DNA polymerase with dTTP-Cy5 substituted for dTTP	[29]
RNA polymerase and DNA	TMR	Cy5	–	Single cysteine mutants of RNA polymerase subunit labelled with TMR maleimide. Double-stranded DNA Fragment synthesized and labelled with Cy5	[54]

Table 4.2 *(Continued)*

System	Donor	Acceptor	R_0 (Å)	Labelling strategy	Reference
Catabolite acitivator protein and DNA	TMR	AF647	64	Mutant of CAP containing a surface-exposed cysteine labelled with TMR maleimide. The consensus DNA sequence to which this protein binds was labelled with Alexa 647 NHS-ester via an amino-modifier C6 dT	[17]

Abbreviations: TMR (tetramethylrhodamine), CI2 (chymotrypsin inhibitor 2), Cy*x* (cyanine dye *x*), AK (adenosine kinase), AF*xxx* (Alexa Fluor dye *xxx*), CspTM (cold-shock protein from *Thermotoga maritima*), R6G (rhodamine 6G), TR (texas red), Flu (fluorescein), bp (base pair).

in a particular molecular environment might vary tremendously and so we offer this information as a guide only. Notable by their absence from these tables are chromophores that fluoresce in the ultra-violet, especially naturally fluorescent amino acids (such as tryptophan). This may seem surprising as changes in the intrinsic fluorescence of proteins is used extensively to monitor many processes throughout biochemistry and biophysics. However, these fluorophores have photophysical properties that are not amenable to single molecule detection. For example, tryptophan possesses a low extinction coefficient (often $<5000\ cm^{-1}M^{-1}$ at wavelengths where it is exclusively excited) and a low quantum yield (often <0.05) when incorporated into peptides/proteins in aqueous solutions [20,21]. Perhaps most significantly, the extinction coefficient and quantum yield are also highly variable and extremely environmentally sensitive (e.g. see [20,21]) much more so than the visible fluorescent dyes presented in Tables 4.1 and 4.2; further these chromophores tend to photobleach rapidly. For these reasons single molecule studies with intrinsic protein fluorescence are difficult (see for one attempt at a pseudo-single molecule study of a multi-tryptophan protein [22]). Single molecule experiments using the intrinsic UV fluorescence of proteins might be possible by careful selection of tryptophan location and the use of higher quantum yield tryptophan analogues or different solvents (solvents with lower dielectric constants than water can result in massively enhanced fluorescence from internal tryptophan residues, when it is possible to use such solvents). However, a second important reason for the lack of such studies in the literature is the performance of the current generation of optics and light detectors suitable for single molecule spectroscopy (where the total detection efficiency is often an order of magnitude less for the entire instrument at UV compared to visible and IR wavelengths where it is already low). The development of instrumentation with improved detection efficiency in the UV, however, may soon allow such studies to be feasible (see Chapter 3).

4.2.2 Other dye considerations

As well as photophysical properties, the chemical and steric features of the dye should also be considered when selecting dyes(s) for single molecule studies. For example, despite its widespread use, fluorescein and its derivatives display pH sensitive fluorescence [23]. Other properties to consider are the solubility of the dye in aqueous buffers and the effect of conjugation of these large dyes on the physicochemical properties of the biomolecule of interest. Steric freedom of the dye molecules is also assumed when the orientation factor κ (used in FRET studies) is taken to be 2/3 (see Chapter 2). Newer generation dyes such as the Alexa series (Invitrogen Ltd., UK) have been designed to not only have superior optical properties but also to obviate many of these problems. For example, many of these dyes are highly pH insensitive, water soluble and are conjugated to the biomolecule, without quenching of fluorescence, via a flexible relatively long (usually ~5 carbon aliphatic chain) linker.

4.3 Labelling of biomolecules

As can be seen from Table 4.2, nucleic acids, proteins or heterogeneous complexes of one or both of these, have been the most commonly investigated biomolecules using single molecule fluorescence techniques. As no intrinsic naturally occurring fluorophores with suitable photophysical properties for single molecule experiments are found in either nucleic acids or polypetides (see Section 4.2.1), it is necessary to introduce such moieties into these biomolecules [5,24,25]. This can be done in three ways:

(1) biosynthesis of the biomolecule to contain a fluorophore (e.g. production by bacteria)

(2) chemical synthesis of a labelled biomolecule

(3) chemical modification of a biomolecule produced by either synthetic or biosynthetic methods.

The method employed is dependent on whether the target molecule is a nucleic acid or polypeptide and in the case of proteins, the number of amino acids in the target.

4.3.1 Nucleic acids

Both ribonucleic and deoxyribonucleic acid polymers are almost exclusively synthesized using the same solid phase phosphoramidite chemistry utilized in the synthesis of oligonucleotide primers for the polymerase chain reaction (PCR) [26]. Nowadays, the reliability of automated synthesis methods is such that even oligonucleotides containing up to ~200 base pairs (bp) can be made routinely at little cost (*n.b.* most manufacturers recommend more stringent purification methods for oligonucleotides >60 bp which can significantly increase the cost). Fluorophores in phosphoamidite form are usually introduced into the oligonucleotide during the synthesis or can be conjugated, after synthesis, to either the 5′ or 3′ end of the oligonucleotide or at internal positions, to pyrimidine (thymine or cytosine) derivatives introduced at specific positions.

As we shall see later, primary aliphatic amines and sulphydryl groups (SH) are the functional groups in biomolecules that are most commonly used for conjugation to dyes. A reactive amine or thiol group can be introduced to either the 5′ or 3′ end of the oligonucleotide *via* an aliphatic carbon linker chain of varying length covalently linked to the terminal phosphate group. To label bases within the nucleic acid, it is necessary to incorporate a modified thymine or cytosine base. These bases have been modified so that a six carbon linker terminating in an amine is attached to the C5 of the pyrimidine ring.

As well as single molecule fluorescence studies, oligonucleotides labelled with fluorophores are used in many different applications, such as quantitative PCR. Consequently, labelled oligonucleotides are readily available from commercial manufacturers but the range of dyes available from each company varies. This laboratory has obtained labelled DNA oligonucleotides from both MWG-Biotech, UK and IBA, Germany and obtained excellent results from both.

The ability to readily synthesize and purify relatively long sequences of nucleic acids containing a dye (or another moiety such as biotin, for example) at a particular location makes such systems relatively straightforward to assemble. If a pair of dyes are to be incorporated at distal ends of a long segment of duplex DNA (e.g. to measure the large conformational change of DNA that occurs upon nucleasome formation [27]), then PCR can be performed using template DNA and forward and reverse primers that have been derivatized with the appropriate fluorophores. Another important property of nucleic acids is the ability of one oligonucleotide to anneal to another of complementary sequence. This is advantageous when undertaking FRET experiments on nucleic acids as each dye can be incorporated into different oligonucleotides, each of which has been purified to homogeneity by standard HPLC or electrophoresis techniques and then hybridized to one another. This technique has been used in the construction of DNA duplexes of constant length in which the dyes are separated by different distances (allowing the distance dependence of FRET to be quantified [28,29]). The ability to synthesize long sequences of nucleic acids labelled at a precise location followed by hybridization to another labelled nucleotide allows a wide variety of experimental designs. The versatility of nucleic acids in this respect is exemplified by the study of the docking of the P1 duplex into the folded core of the *Tetrahymena* ribozyme (an RNA enzyme) [30]. In this study, the 5′end of the ribozyme (created biosynthetically) was annealed to its oligonucleotide substrate labelled with donor. The 3′ end of ribozyme was extended to allow the hybridization of a DNA oligonucleotide derivatized not only with acceptor but also with biotin which, as we shall see, can be used to tether biomolecules to a surface.

The development of automated solid phase synthesis of nucleic acids and chemical strategies to incorporate fluorophores at any position along their length, together with the wide array of molecular biological techniques available for their modification has resulted in a system highly amenable to single molecule fluorescent studies. By contrast, proteins, to which we now turn our attention, are rarely synthesized *in vitro* and thus a different approach needs to be taken.

4.3.2 **Proteins**

Proteins perform a dazzling array of functions in biology that include structural, enzymatic, regulatory, and mechanical roles to name but a few examples. Proteins have their structure and function determined by the amino acid sequence and the precise manner in which the polypeptide chain spontaneously folds into a native conformation. They represent a fascinating area for single molecule biophysical research. However, by contrast to nucleic acids, methodologies for the solid-phase synthesis of proteins are still in their infancy and the construction of systems amenable for single molecule fluorescent studies can still represent a significant hurdle to overcome.

As discussed in Section 4.3, there are three routes by which fluorophores can be incorporated into biomolecules. These are: (1) biosynthesis, (2) chemical synthesis, and (3) chemical modification. For reasons described later, chemical modification is by far the most widely adopted approach. Accordingly, we shall only briefly discuss the direct biosynthetic and synthetic approaches before describing, in detail, the chemical modification of proteins.

Biosynthesis

Fluorophores can be directly incorporated into biological systems by the cellular machinery that translates the nucleic acid into a sequence of amino acids (the ribosome). The vague term 'biological system' is used deliberately to include three very different approaches: *in vitro* translation systems, the insertion of amino acids analogues using amino acid auxotrophs and the production of proteins fused to fluorophores formed solely from proteins, such as the green fluorescent protein (GFP) family. *In vitro* translation systems have been used for many years to introduce modified or unnatural amino-acids into the polypeptide chain of the protein under study. In this method, the translational machinery of a cell is reconstituted in a test tube. The nucleic acid to be translated (RNA) or transcribed and translated (DNA) is designed to contain a nonsense codon (a triplet that does not code for an amino acid) at the point in the sequence where the modified amino acid is to be introduced upon translation. The tRNA molecule with an anti-codon complimentary to the nonsense codon is then chemically modified with the desired unnatural amino-acid which is then incorporated into the nascent polypeptide chain during translation. Unfortunately, this process usually produces only small amounts of protein and the chemical loading of tRNA, which in cells is carried out enzymatically by a specific tRNA synthetase for each tRNA species, can only be performed on a small scale. Zhang *et al.* [31] overcame both of these problems by using a directed evolution approach to select a tRNA–tRNA synthetase pair. The evolved tRNA synthetase aminoacetlyates the tRNA with the desired amino acid *in vivo* which is then incorporated into the nascent polypeptide in response to the amber nonsense codon TAG. Using this approach these workers incorporated m-acetyl-L-phenylalanine at a specific location into the Z domain protein. This keto-containing amino acid was then selectivity modified with hydrazide-derivatized fluorophores *in vivo* and the labelled protein then purified using standard techniques.

Close analogues of the amino acid tryptophan can be inserted into a polypeptide simply by adding the desired amino acid to the medium in which a tryptophan-deficient auxotroph of *E.coli* is cultured. However, this approach cannot be used in single molecule studies as these tryptophan analogues (which can be covalently attached to $tRNA^{Trp}$ and hence incorporated into the nascent chain) still have unsuitable photophysical properties.

GFP is a single chain protein first isolated from the *Aequorea Victoria* jellyfish that forms a highly visible intrinsic fluorophore by rearrangement of a Ser-Tyr-Gly sequence upon folding to a near native state. This non-toxic protein has become ubiquitous throughout biochemistry and cell biology because of its ability to act as an easily detectable marker: by fusing the gene for GFP in frame with a gene of interest, it is possible to ascertain the level of expression and the cellular location of the protein encoded by this gene, using simple fluorescence microscopy. Many variants now exist with altered spectral properties so it is possible to perform FRET experiments using proteins to which GFP and BFP (blue fluorescent protein), for example, have been fused. This approach was used by Philipps and co-workers [32] to establish an *in vivo* screening system for the selection of mutated proteins with increased thermodynamic stability relative to the wild type protein. To do this, BFP and GFP were fused in frame to the N and C-termini of an immunoglobulin V_L domain. Efficient FRET was only observed in fusions where the V_L domain was fully folded bringing the two reporters in close proximity.

By countering the problematic effect of rapid GFP photobleaching (see Section 4.2.1), Mashanov *et al.* [33] were able to observe the slow binding and release of single peckstrin homology domains at the plasma membrane of mouse myoblasts. This work is reviewed in detail in Chapter 6.

Chemical synthesis

We have discussed how the development of solid phase synthesis techniques has revolutionized nucleic acid research, similarly methods that allow the synthetic construction of peptides and proteins could potentially revolutionize many areas of structural biology. For peptides and small proteins (<40 amino acids in length), solid phase synthesis is straightforward and allows conjugation of the fluorophore either whilst on the resin or after cleavage. This approach was taken by Talaga *et al.* [34] in their study of the folding of a disulphide cross-linked heterodimer (each monomer consisted of 44 amino acids) of the two-stranded coiled coil from GCN4. However, automated solid phase synthesis cannot be used to efficiently generate proteins >50 amino acids in length due to the accumulation of erroneous sequences and the decreasing yield of the coupling steps due to aggregation and folding of the newly formed peptide chain. Such problems can be obviated by using cysteine-directed native chemical ligation or conformationally assisted ligation (see [35] for an excellent review of both of these methods) of pairs of synthetic peptides (total synthesis) or ligation of a synthetic peptide with recombinantly expressed protein fragments (semi-synthesis). Both of these methods involve the synthesis of the N-terminal portion of the protein that terminates in a thioester (COSR). A requirement for native chemical ligation is that the peptide corresponding to the C-terminal portion of the protein has a cysteine residue at its N-terminus. This then allows rapid reversible thiol exchange followed by irreversible intramolecular rearrangement linking the peptides by a peptide bond. The ligation rate of the peptides was found to be enhanced dramatically when the peptides self-assembled and folded to the native state, bringing the two ends into close proximity. This enhancement was such that it was possible in some cases to perform native chemical ligation in the absence of a cysteine residue at the start of the C-terminal peptide (termed conformationally assisted ligation, [36]). This technique was used to synthesize the 64 amino acid chymotrypsin inhibitor 2 labelled with a pair of fluorophores suitable for FRET analysis [37]. In this case the N-terminus was labelled with TMR whilst the N-terminal peptide (residues 1–39) was still on the resin. The self-associated peptides ligated rapidly due to the presence of a cysteine residue at the N-terminus of the C-terminal fragment (this residue is methionine in the wild type protein). The sulphydryl group of this cysteine was then used for conjugation of the protein with disulphide activated Cy5 [37]. Despite these successes these approaches remain limited in their applicability as they either require the insertion of a cysteine residue directly C-terminal to the ligation point or the peptides need to self-associate to form a native-like structure. Even in this case, the range of amino acid pairs that efficiently couple is severely limited. These problems can be circumvented by a new approach that places the directing thiol group on an auxiliary nitrogen protecting group. This technique, termed 'auxiliary directed native chemical ligation' [38], is still in its infancy but may allow more flexibility in the location and nature of ligation sites.

Chemical modification

Despite the potential of all the techniques described earlier, the method that has been the most widely adopted is chemical modification of a protein purified from a bacterial expression system. All proteins consist of chains of amino acids joined together by formation of an amide bond between the α-carboxyl group of one amino acid and the α-amino group of the next amino acid. Each amino acid consists of an amino-group, a carboxyl group, a hydrogen atom and a 'side-chain' or 'R' group all directly bonded to an α-carbon. As there are twenty different, chemically diverse side-chains it should be relatively easy, in principle, to introduce fluorophores at specific sites within proteins by reaction with these side-chains. However, many amino acids contain either unreactive aliphatic chains or carboxylic acids which have a low reactivity. Other side-chains such as arginine and tryptophan can only be modified in conditions that are not compatible with protein stability. Consequently, most methods for conjugating proteins with extrinsic fluorophores utilize the reactivity of primary amines (i.e. the α-amino group of the N-terminus and the ε-amine of the lysine side-chain) or the thiol side-chain of cysteine.

Proteins are frequently conjugated to a wide variety of compounds using amine reactive functional groups such as isothiocyanates and N-hydroxysuccinamide esters as lysine residues are frequently found on the surface of proteins. For many studies the number of labels per protein molecule is not critical. For example, in immunology and cell biology fluorescence imaging techniques (where partitioning of a protein or compound is followed by addition of a labelled antibody marker whose epitope is on the biomolecule of interest) it was found that the brightest antibodies had dye-to-protein ratios of 4–12 : 1 [39]). However, for most single molecule studies it is highly desirable (indeed a requirement for spFRET) that one or at most two (in the case of spFRET experiments) dye molecules are attached at a specific location on the protein. Unfortunately, the relative frequency of lysine residues in proteins is such that many of these residues can be present on the surface of a protein even as small as 100 amino acids [40]. Even if all but one of the lysine residues were to be replaced with arginine (which is possible though it would be time consuming and a relatively expensive undertaking), this labelling would still be unspecific due to the presence of an N-terminal amino-group on every protein. In theory, as only the free base (unprotonated) form of the amine is reactive and the pKa of the N-terminal amino group is significantly lower than that of the amino-group on the lysine side-chain (pH 8.0 and 10.0 respectively) it should be possible to specifically label the N-terminus of a protein by performing the conjugation at neutral pH. However, complete site specific labelling of the N-terminus is rarely achievable in practice.

Whilst, on average, 6% of the amino acids in proteins are lysine, only 2% are cysteine residues [40]. This means that a 100 amino acid protein may contain only a small number of cysteine residues. Removal/introduction of cysteine residues (so that only one is present) combined with the use of very specific sulphydryl reagents can thus allow a protein to be labelled at a single known site with high efficiency. Furthermore, by using standard mutagenesis kits or PCR based methods, it is relatively easy to generate a pseudo wild-type protein without any cysteine residues followed by a series of mutants each containing a single cysteine residue at different locations. This protein engineering approach has proved to be the method of choice for producing proteins suitable for single molecule fluorescence studies. As we shall see later, this technique has been extended to selectively label a single protein with two different dyes at different locations.

4.3.3 **Chemistry of fluorophore derivatives**

In order to covalently link a dye to either a nucleic acid or polypeptide it is necessary to derivatize the fluorophore with a functional group that is reactive towards specific functional groups that occur naturally in these biomolecules or that have been introduced in their synthesis. As we have seen, despite their different chemical structures both nucleic acids and proteins are almost always conjugated to fluorophore derivatives containing functional groups that react primarily with either amines or thiols. We shall now briefly describe the chemical properties of these functional groups.

Amine reactive conjugates

Dyes derivatized to contain an isothiocyanate or an N-hydroxysuccinamide ester (NHS-ester) have been used extensively to label amine-containing compounds. However, newer generation dyes [3] (e.g. Cy dyes (Amersham Biosciences, UK) and the Alexa Fluor series (Invitrogen Ltd., UK)) can only be purchased as succinimidyl esters due to their greater chemical stability when stored. Tables 4.1 and 4.2 show some new generation dyes that are extensively used in single molecule experiments because of their superior photophysical properties. Consequently, even though the isothiocyanate derivative of fluorescein (FITC), for example, is still widely used in labelling antibodies for confocal fluorescence miscroscopy studies in cell biology and immunology, their properties will not be discussed further here.

NHS-esters react with primary amines to form a carboxamide and N-hydroxy-succinamide (see Figure 4.4). NHS-esters show reasonable reactivity and high selectivity towards aliphatic amines but can also react with aromatic amines, histidines and tyrosines, but at a significantly lower rate. In proteins, this practically limits the reactive groups to either the α-amino group of the N-terminus, or the ε-amine of the lysine side-chain. In an aqueous environment the optimum pH for this reaction is 8.0–9.0. However, it should be noted that the rate of hydrolysis of succinimidyl esters increases with increasing pH (but is slow when $<$pH 9) [41].

Sulphydryl reactive conjugates

Haloacetamides and maleimides both readily react with sulphydryl groups (thiols) to yield thioether products as shown in Figure 4.4. These reagents can also react with the free base (i.e. unprotonated) form of aliphatic amines. However, as the reaction of haloacetamides or maleimides with thiols proceeds rapidly at pH 7, where most aliphatic amines are protonated, these reagents in practice, show high specificity for thiols. Pairs of thiols can oxidize to form disulphide bridges, which depending on the solvent accessibility and location of each thiol can either be inter- or intra-molecular in their nature. Labelling of such residues can still be achieved by reduction of the disulphides by action of dithiothreitol (DTT), β-mercaptoethanol or Tris(2-carboxyethyl)phosphine (TCEP). DTT

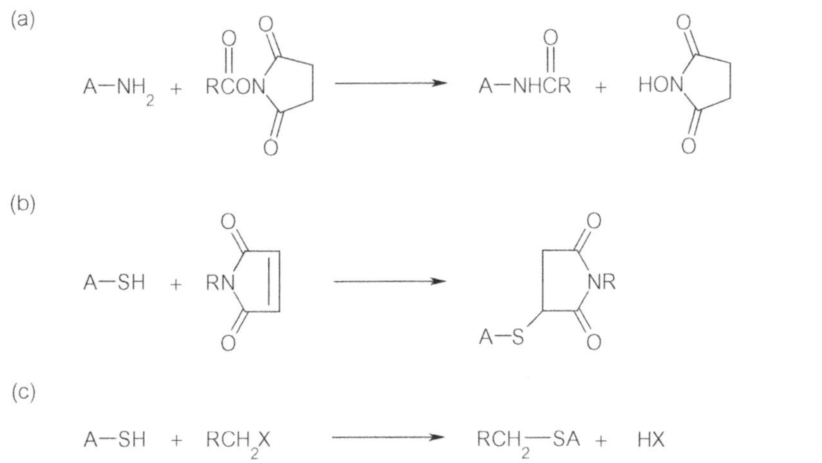

Figure 4.4 Reaction summary for the primary amine of a protein (A) with an NHS ester (a) and a sulphydryl group of a protein with a maleimide (b) or iodoacetamide (c). In all cases R denotes the fluorophore and variable length carbon linker.

and β-mercaptoethanol both contain free thiols that can compete for dye with the biomolecule to be conjugated. Thus these agents need to be removed before the thiol reactive dye derivative is added. Whilst dialysis and gel filtration procedures are typically employed, we have found ion-exchange followed by a rapid dialysis step to provide excellent results. Unlike DTT and β-mercaptoethanol, TCEP does not contain any free thiols and it is therefore not necessary to remove the reducing agent prior to the labelling reaction, a step that often leads to oxidation of the free thiols back to disulphides. However, in our experience performing these reactions in the presence of TCEP nevertheless decreases the labelling efficiency significantly. Above pH 8 hydrolysis of the maleimide may compete with the desired reaction but, nevertheless, maleimides are becoming the reagents of choice for the specific labelling of sulphydryl groups.

Haloacetamides, of which the most reactive is the iodoacetamide derivative, may also react with methionine, histidine or tyrosine residues in proteins, but usually only if free thiol groups are unavailable. A further disadvantage of haloacetamides is their instability in light, especially when in solution. The use of haloacetamides is decreasing, possibly as a result of these slight drawbacks. For example, both the Alexa Fluor (Invitrogen Ltd., UK) and the Cy dye series (Amersham Biosciences, UK) are only available as maleimide derivatives.

4.4 Doubly labelling single protein molecules for FRET studies

The generation of protein molecules that are labelled with two different fluorophores in different specific locations probably poses the biggest problems to the researcher described in this chapter. As we have seen, production of duplex DNA specifically labelled with a FRET dye pair is facilitated by the ability to produce and label each DNA oligomer separately followed by hybridization. Heterogeneous protein/DNA and protein/protein complexes can also be assembled in a similar manner relatively easily as each fluorophore can be conjugated to a separate component of the complex and purified to homogeneity. Assembly of the complex being studied then forms the FRET system. However, this strategy cannot be used when the FRET dye pair are conjugated to the same molecule. A logical method perhaps, would be to perform sequential conjugations using dyes derivatized with differing functional groups. The only readily labelled functional groups present in proteins for this purpose are the amino-groups of lysine and the N-terminus and the sulphydryl groups of cysteine. However, such an approach rarely works due to the frequency of lysines or the difficulties of finding conditions whereby the N-terminal group can be selectively conjugated. Instead, a much simpler approach has been widely adopted whereby two cysteines residues are introduced into the protein to be labelled. Labelling of the protein with each dye is then performed either sequentially (if the sulphydryl groups show different reactivity) or in a single step. The success of these methods depends on the ability to resolve, by standard chromatography methods, unlabelled protein from singly and doubly labelled protein. As many of the newer dyes contain charged groups to increase their solubility, it is possible to use high performance ion-exchange as well as hydrophobic interaction chromatography to purify each species.

If one of the two -SH groups is less reactive then, by performing sequential reactions, it is possible to specifically label a site with either the donor or acceptor. The Haas group have described in detail a protocol for performing site specific double labelling of proteins and, before describing the labelling methods used in our laboratory, we shall briefly describe the general protocol suggested by Haas [25].

4.4.1 General protocol for site specific double labelling of single proteins

1. As the double labelling of a protein is a time consuming and expensive process, it is necessary to carefully consider where the cysteines (and hence the dyes) should be introduced into the protein. Sets of donor/acceptor sites should be chosen that cause minimal perturbation to the native state or the physicochemical properties of the protein. For this reason charged residues and sites that may allow non-covalent (e.g. hydrophobic) interactions between the dye and protein are not considered. Furthermore, the donor/acceptor sites must allow rotational freedom for the dyes. In this regime, the assumption that the value for the orientation factor $\kappa = 2/3$ is most likely to be valid (see Chapter 2 for a discussion of the relevance of the orientation factor in FRET studies). Sites are further screened so that only residues that have $\geq$30% of their surface area accessible to the solvent (and therefore the modification reagent) are considered. The solvent accessible surface area of a protein on a per residue basis can be calculated using software such as DSSP [42] and the atomic co-ordinates of the three dimensional structure of the protein contained in the protein databank file [43] of a protein with a solved structure. This identifies a small set of residues that are suitable for labelling.

2. Mutants containing a single cysteine is created for every putative site. The reactivity of each of these sulphydryl groups is measured using stopped-flow and monitoring (by absorbance) the production of thiolate ions released upon

reaction of DTNB (5,5′-dithiobis-(2-nitrobenzoic acid) with free thiol groups. The effect of different reaction conditions upon the rate constant of each of these is then investigated.

3. Pairs of sites are then selected which show large differences in their reaction rates (thus allowing sequential labelling) and are positioned such that the conformational change or structural element to be characterized can be monitored, with particular attention placed on the absolute distances between pairs of sites and the expected distance change upon conformational reconfiguration.
4. After constructing the two-cysteine residue mutant, labelling can be performed with the first dye under conditions optimized in point 2 to favour the formation of a protein labelled at a unique site. Unless the other cysteine is completely buried, four species are present in solution: unlabelled protein, two singly labelled species (one at a much higher concentration) and a doubly labelled species. The two singly labelled species are purified from the other products by ion-exchange or hydrophobic chromatography techniques. The second, less reactive, site is then used to conjugate the second dye. Complete and efficient labelling of this site is achieved by performing the reaction in conditions where the protein is unfolded (in chemical denaturants such as urea or guanidinium chloride). The extent of labelling and the degree of purification of each species can be monitored by ElectroSpray-Ionization mass spectrometry (ESI-MS) at each stage of the process.

This protocol provides a generic approach to obtain a highly homogenous site specifically double-labelled protein (as long as the protein can be refolded from denaturant!). However, in many cases this detailed protocol is not needed. If each dye does not interact with the surface of the protein and is able to freely rotate, then the heterogeneity of the doubly labelled sample is not important as a protein labelled with a donor (at site one) and an acceptor (at site two) should have identical spectroscopic properties to one where the location of the donor and acceptor are swapped. If the two -SH groups show similar reactivity, as is usually the case when N- and C-terminal cysteines are introduced into a protein, doubly labelled protein (either donor–acceptor or acceptor–donor) can be obtained by reacting a sub-stoicheiometric quantity of the first dye. This yields a mixture of unlabelled protein, single labelled protein (at either site) or doubly labelled protein, which can be separated by chromatography. Addition of an excess of second dye then drives the labelling to completion. Any remaining unreacted dye is then removed by size exclusion chromatography.

This technique has been used to generate cold-shock protein labelled with Alexa 488 and Alexa 594 at its N- and C-termini [44]. However, it is unlikely that any one protocol will suit every protein that a researcher may wish to study. For example, the labelling procedure of Im9 that follows is essentially a hybrid of the two methods described earlier.

4.4.2 **Labelling of Im9: a case study**

The colicin immunity protein Im9 is a four-helix bundle protein studied intensively by the Radford group in Leeds, UK as a model for protein folding [45–52]. In order to monitor the conformation of this protein at the single molecule level by FRET it is necessary to conjugate two-dye molecules at specific locations. One partially buried cysteine residue (C23) is present in the wild-type Im9 sequence. This residue was found, by using the inexpensive test reagent, N-Ethylmaleimide followed by ESI-MS, to show low reactivity. A second solvent-exposed cysteine was thus introduced by mutation of serine 81 to allow site specific labelling of Im9 with donor (Alexa 488) and acceptor (Alexa 594) dyes. Im9S81C was overexpressed, purified to homogeneity and its identity verified by ESI-MS as described previously [50].

Reduction of cysteine residues

The mass spectrum of unlabelled Im9S81C prior to any labelling steps revealed that a significant fraction of the protein had formed intramolecular disulphide bridges. The presence of disulphides could also be detected by using pre-packed high performance anion-exchange resins such as Resource Q or MonoQ columns (GE Healthcare, UK). To maximize the yield of labelled protein it was therefore necessary to reduce these bonds prior to attempting conjugation with dyes. To do this protein was dissolved (~5 mg/ml) in 50 mM Tris, 7 M urea, and 4 mM DTT (dithiothreitol). The solution was left at room temperature for 1.5 h. Application of an aliquot of the solution to the anion-exchange column confirmed that full reduction of the disulphide had occurred. The solution was diluted 10 times with 50 mM Tris, 4 mM DTT (pH7.5), dialysed, and then freeze dried.

Conjugation of donor

The protein was dissolved (~3 mg/ml) in 100 mM Tris (pH 7.3). 50 μl of a 20× solution of Alexa Fluor 488 C_5 maleimide (Invitrogen Ltd., UK) in DMSO was added to the protein solution (950 μl) giving a molar ratio 0.7 : 1 dye:protein. The solution was stirred at room temperature whilst protected against light for 40 min. After the reaction, the solution was purified by anion-exchange chromatography to separate single-labelled species from non-labelled species. The resultant elution profile is shown in Figure 4.5. Fractions containing singly labelled dye were pooled, dialysed and then freeze-dried.

Conjugation of acceptor

950 μl single-labelled protein was dissolved (~3 mg/ml) in 100 mM Tris, 9 M urea (pH 7.3). 50 μl of a 20× solution of Alexa Fluor 594 C_5 maleimide in DMSO was added to the protein solution (a dye : protein molar ratio of 3 : 1). The solution was stirred at room temperature, protected against light for 4h. After the reaction, the solution was purified by anion-exchange chromatography to separate double-labelled species from single-labelled species (see Figure 4.6). The fractions containing doubly labelled protein were pooled, dialysed, and concentrated by a Centriprep® centrifugal concentrator (Millipore, UK).

Removal of free dye

A convenient and rapid method to assess the extent of labelling is to compare an absorbance spectrum of the protein-dyes conjugate to that expected from a 1 : 1 mixture of the dyes (the small contribution from the absorbance of the protein at the absorption maximum wavelengths of the dyes is ignored). The absorbance spectrum of the doubly labelled protein obtained following the protocol mentioned earlier is shown in Figure 4.7 and revealed that excess donor was present even after ion-exchange and dialysis (compare the measured spectrum, Figure 4.7 dashed line, to that calculated for a 1 : 1 mix of the two dyes, Figure 4.7, solid line).

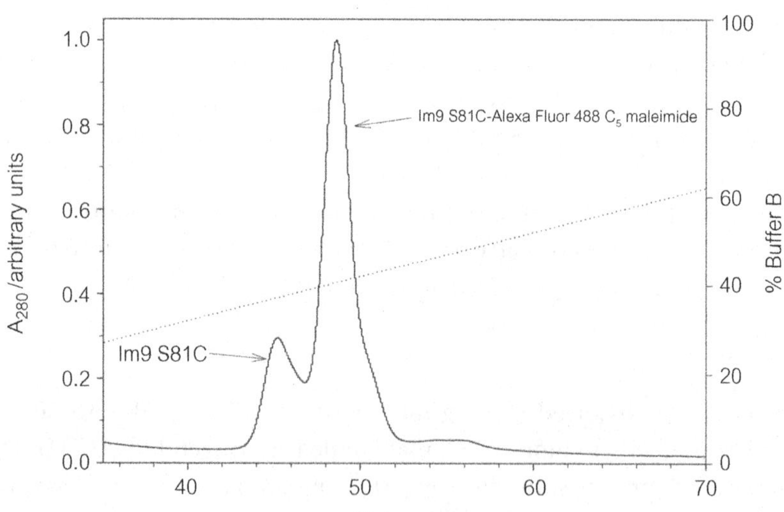

Figure 4.5 Chromatogram showing the separation of unlabelled protein (Im9 S81C) from the same protein conjugated to Alexa Fluor 488 C_5 maleimide using anion-exchange chromatography. Column: Resource Q 6 mL (GE Healthcare, UK), buffer A: 50 mM Tris.HCl, 4 mM DTT (pH7.5), buffer B: 50 mM Tris, 4 mM DTT and 1 M NaCl (pH7.5).

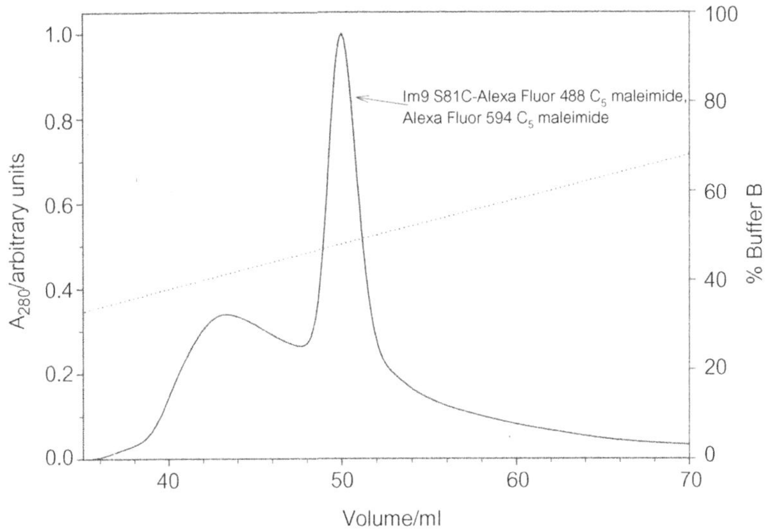

Figure 4.6 Chromatogram showing the separation of singly labelled protein (Im9 S81C-Alexa Fluor 488 C_5 maleimide) from the same protein conjugated to both Alexa Fluor 488 C_5 maleimide and Alexa Fluor 594 C_5 maleimide using anion-exchange chromatography. Column: Resource Q 6 ml (GE Healthcare, UK), buffer A: 50 mM Tris.HCl, 4 mM DTT (pH 7.5), buffer B: 50 mM Tris 4 mM DTT and 1 M NaCl (pH7.5).

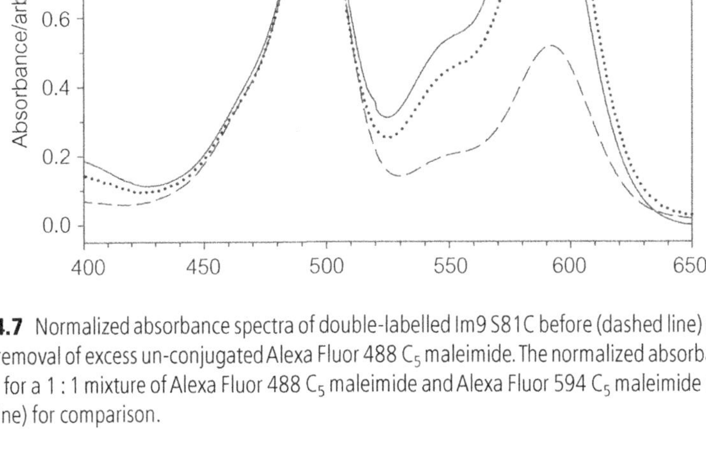

Figure 4.7 Normalized absorbance spectra of double-labelled Im9 S81C before (dashed line) and after (dotted line) removal of excess un-conjugated Alexa Fluor 488 C_5 maleimide. The normalized absorbance spectrum expected for a 1 : 1 mixture of Alexa Fluor 488 C_5 maleimide and Alexa Fluor 594 C_5 maleimide is shown (continuous line) for comparison.

In this example the donor was conjugated to the protein after the acceptor. To remove unreacted donor dye that is apparently weakly associated with the protein, the concentrated solution was applied to a Superdex Peptide gel filtration column (Amersham Biosciences, UK). The absorption spectrum of this additionally purified sample (Figure 4.7, dotted line) shows near-perfect agreement with the simulated curve for a 1 : 1 mixture (Figure 4.7, solid line). The collected solution was aliquoted and snap-frozen.

4.5 Optimizing biochemical systems for single molecule fluorescence studies

4.5.1 Buffer considerations

There are two main areas of concern with regard to buffer preparation: first the minimization of spurious transient background signals from highly fluorescent (or strongly scattering) contaminants and second the selection of buffer components to prevent unwanted solution conditions (maintaining pH and minimizing non-specific surface absorption, for example). We briefly discuss these two areas in this section.

Fluorophores used in single molecule measurements are required to have high quantum yields in the visible range of the electromagnetic spectrum, so that when combined with modern instrumentation for the detection of single molecule fluorescence, data with a high signal to noise is obtained. Experiments on biological systems are typically performed in buffered aqueous solutions where the salt concentration is typically nine orders of magnitude greater than that of the analyte (mM and pM respectively). At these concentrations, the presence of even small percentages of impurities can increase the 'noise' level (or perhaps more properly the rate at which background events which are indistinguishable from those due to the analyte, occur) to unacceptable levels. In addition, studies on phenomena such as protein folding often require additional dissolved components in the buffer, for example, concentrations of denaturant, such as urea, can be as high as 8 M. In order to minimize background signal the highest purity of solute that is available must be used. Further, one must ensure that any glassware, disposable containers or liquid handling components are equally clean. We have found that buffers prepared using Fluka BioChemika Ultra (formerly Microselect) grade reagents dissolved into ultrapure water (deionized water at >17 MΩ/cm resistance, filtered through 0.22 μm membrane filter before use) gives little background signal (in FRET studies in our laboratory less than 1% of background events that occur are above the threshold, see Chapter 2). The purity of urea stocks can be particularly variable; indeed despite the use of 'best' reagents a large number of additional background events can often be seen with the addition of high concentrations of urea to buffers. The increase in background events, using the instrumentation described in Chapter 3, is especially appreciable in the lower wavelength donor detection channel, suggesting either particulate contaminants scattering large amounts of excitation light or, more likely, small molecule fluorescent contaminants. These can be removed, or at least reduced in frequency, by repeated re-crystallization of the urea through gradual cooling of a saturated solution in hot ethanol followed by washes with cold ethanol.

Low concentrations of surfactants are an important component in single molecule studies because of their ability to prevent surface adhesion of proteins [44], which can be very significant with the low concentration of protein and the large surface area of any sample holder. This can result first in errors in serial dilutions of stocks to picomolar concentrations and second to unwanted surface induced conformational changes of the biomolecules being studied. The ability of surfactants to prevent surface adhesion, followed by possible protein denaturation, can be demonstrated by titrating the detergent Tween 20 (SigmaUltra, Sigma, UK) into identical protein solutions. Figure 4.8 shows the results of such a titration for a fixed concentration (~400 pM) of the doubly labelled protein Im9S81C described earlier (raw data on the left, proximity ratio histograms constructed as described in Chapter 2 on the right). When little detergent is present (Figure 4.8 (a and b)) the data is dominated by occasional bursts located predominantly in the green or donor channel (possibly from denatured proteins or proteins aggregated after surface interactions). This is reflected by an apparent low FRET efficiency population in the histogram. As the detergent concentration is increased, the burst frequency increases significantly and the emergence of a peak at high FRET efficiency, corresponding to native protein is seen (Figure 4.8 c–f). It should be noted that these histograms and data are not ideal as the protein concentration is high (e.g. clear overlap between events is occurring). Indeed 400 pM is around an order of magnitude higher than might be used in a 'real' experiment, but provides an effective demonstration for this point. Further, it is essential that the detergent is added to the higher concentration stock and not after serial dilution to the final concentration (~50 pM) as surface absorption (and then dissociation upon detergent addition) may already have perturbed the system under study.

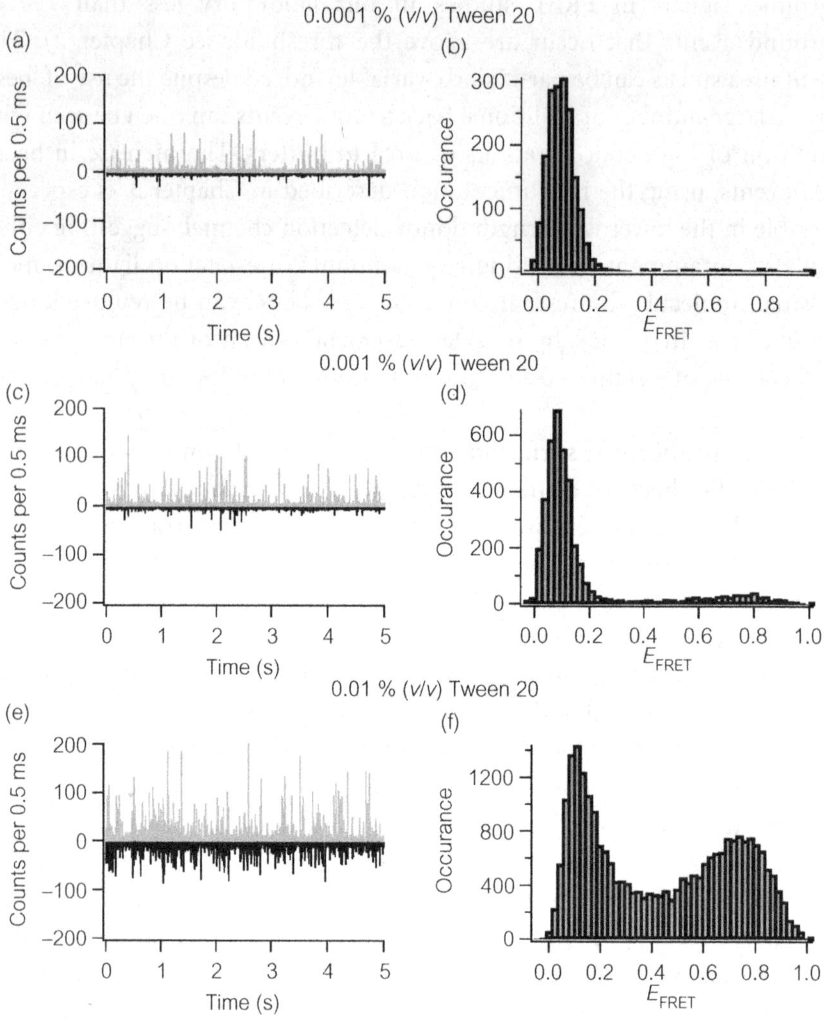

Figure 4.8 Surfactants prevent non-specific absorption of protein to the sample container. The raw data (left, donor channel grey, acceptor channel black) and the calculated proximity ratios (right, threshold = 40 counts, see Chapter 2 for the method) both show the increase of a freely diffusing species with a relatively high (presumably natively folded) proximity ratio upon titration of Tween 20 (0.0001% to 0.01%, top to bottom) into 400 pM Im9 S81C labelled with Alexa 488 and Alexa 594 C5 maleimide, 50 mM sodium phosphate pH 7.0.

4.5.2 **Minimizing the 'zero peak'**

As we discussed in Chapter 2, a 'zero peak' is often observed when analysing data obtained by diffusion spFRET methods. The origin of this peak is likely to be a consequence of acceptor photobleaching as such a molecule displays a high donor signal but low acceptor signal resulting in a proximity ratio close to zero. This peak may have a large area and so overlap with a population of FRET-active molecules that have a low proximity ratio can occur, making quantitative analysis difficult. Several methods are commonly used to minimize this artefact. One particularly simple solution is to prevent re-sampling of molecules which are likely to contain a photobleached acceptor by the use of flow from a large volume of stock analyte solution (feasible at 50 pM concentrations). Addition of oxygen scavengers such as L-carnosine or 1, 4-diazabicyclo[2.2.2]octane (DABCO), for example, can be very effective in reducing photobleaching by limiting photo-oxidation of the acceptor dye molecules, reducing the zero peak. As discussed, the photobleaching lifetime also decreases with increasing intensity of the excitation light and consequently reduction of laser power can greatly minimize photbleaching of both the donor and acceptor. This is illustrated in Figure 4.9, where the data shows the burst traces (left) and resultant proximity ratio histograms (right) for the same solution of spFRET labelled Im9 (Im9S81C) at three laser powers: 40 μW (a and b), 80 μW (b and c) and 120 μW (d and e) in a diffusion–confocal experiment. The excitation intensity was measured before the microscope objective (see Chapter 3). The increase in the magnitude of the 'zero peak' compared to the protein peak with increasing laser power is clear. In this case then a balance must be struck between signal-to-noise and the influence of any zero peak. Concurrent use of all three of the photobleaching reduction methods suggested (oxygen scavengers, flow and reduced excitation laser power) is particularly effective in reducing this troublesome artefact.

Recently, methods applied to the analysis of data from novel multiparameter experiments [7,53,54] have demonstrated a way to 'test' for active acceptor after the measurement of the proximity ratio. This therefore allows almost complete removal of inactive acceptor molecule influenced data which therefore eliminates the zero peak altogether, although at the expense of much more complicated instrumentation. This method is discussed further in Chapter 2.

4.6 Immobilization methods

As we saw in Chapters 2 and 3, the fluorescence emitted from single molecules can be collected by two distinct methods. In diffusion experiments the fluorescence emitted by a single molecule is collected as it passes through a small (~0.1 fl) detection volume. This is repeated many thousands of times and burst analyses (such as FRET) and correlation techniques (FCS) can be used to extract both equilibrium and kinetic parameters of interest. The advantage of this technique lies in its relative simplicity, however the time over which a *single molecule* can be observed is limited to the time

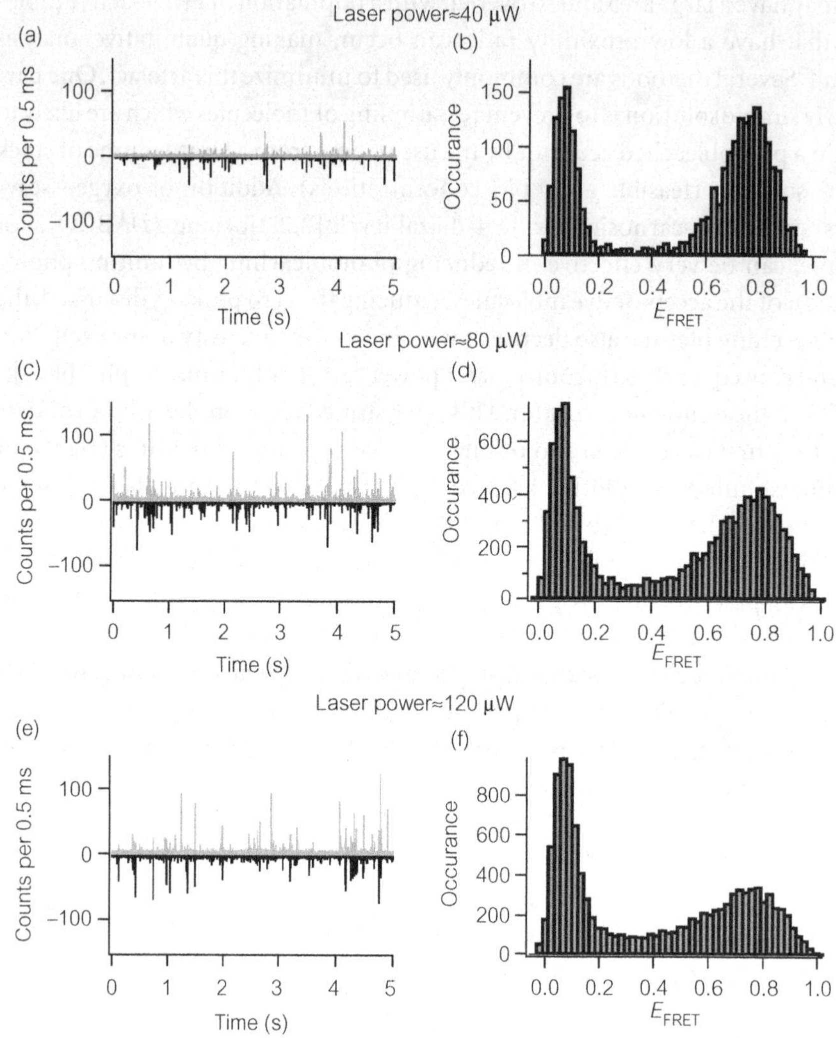

Figure 4.9 The effect of laser power upon the photobleaching of a freely diffusing spFRET labelled protein. Upon increasing the laser power from 40 to 120 μW (top to bottom, laser power measured before the objective, see Chapter 3), the raw data (left, donor channel grey, acceptor channel black) and the calculated proximity ratio histograms (right) both show the increase in donor only fluorescence (increase in donor channel in the raw data, increase in the relative magnitude of the zero peak) that is indicative of photobleaching.

taken for it to diffuse across the observation volume (which is typically less than one millisecond). In order to observe slower events, or to characterize the temporal behaviour of a *single* molecule or complex over an extended time, which is limited only by photobleaching, it is necessary to limit the movement of a single-labelled biomolecule so that it remains fixed in a volume smaller than the observation volume of the instrument. Many individual molecules can then be measured independently:

either in a scanning configuration, by moving the microscope stage until a new, unbleached, molecule is discovered, or simultaneously in an imaging system (both of these configurations are discussed in detail in Chapter 3). The difficulty of this approach lies in selecting an immobilization method that does not affect either the stability of the system or the process being investigated. Nucleic acids and proteins are immobilized onto passive substrates in many applications and, consequently, many solutions have been found which can be broadly divided into two groups: tethering methods and entrapment methods (for reviews see [55,56]). The principles of these methods are illustrated in Figure 4.10: (a) non-specific adsorption of the molecule onto a surface, (b) specific tethering via a linker from a specific location to a functionalized surface, (c) entrapment in the solvated pores of aqueous gels, (d) entrapment inside immobilized vesicles.

4.6.1 **Tethering onto a surface**

It is possible to immobilize biomolecules to surfaces non-specifically due to their innate affinity for substrates such as glass. This technique is far from ideal as the orientation of biomolecule on the surface is unknown and can, for example, also result in significant denaturation of a protein's native structure [57]. The non-specificity of simple adsorption can be surmounted by introduction of a 'tag' at a known location on the biomolecule, the tag then specifically binds to the substrate itself or one derivatized with a binding partner. A common method for nucleic acids is to utilize the interaction that occurs between the ligand biotin and the tetrameric protein streptavidin, each monomer of which is capable of binding a single biotin ligand tightly ($K_d \sim 10^{-13}$ M [58]). This provides a simple, highly effective solution as biotin can be introduced at either end, or at internal positions of oligonucleotides by using the same techniques described in Section 4.3.1. Tethering of complex nucleotide systems that possess significant secondary structure (such as the hairpin riobozyme [30,59–61]) has been achieved in this way and been shown not to perturb the activity or function of the system [62]. Unfortunately, specific biotinylation of a protein poses as many problems as that for the specific conjugation of dyes to proteins, especially as biotin is usually available derivatized only with NHS-esters. One solution may be to fuse the protein to be investigated to a protein that is biotinylated *in vivo* in *E.coli* as part of its function (Biotin Carboxy Carrier Protein, BCCP). This elegant solution was used to immobilize *E.coli* Rep helicase onto a surface ([63], see Chapter 6). Tags which are considerably smaller than BCCP (which is ~85 amino acids) can also be attached either at the DNA level for recombinant protein expression in bacteria or, for small proteins and peptides during solid phase synthesis. Addition of, typically, six histidine residues to either the N- or C-terminus of a protein allows

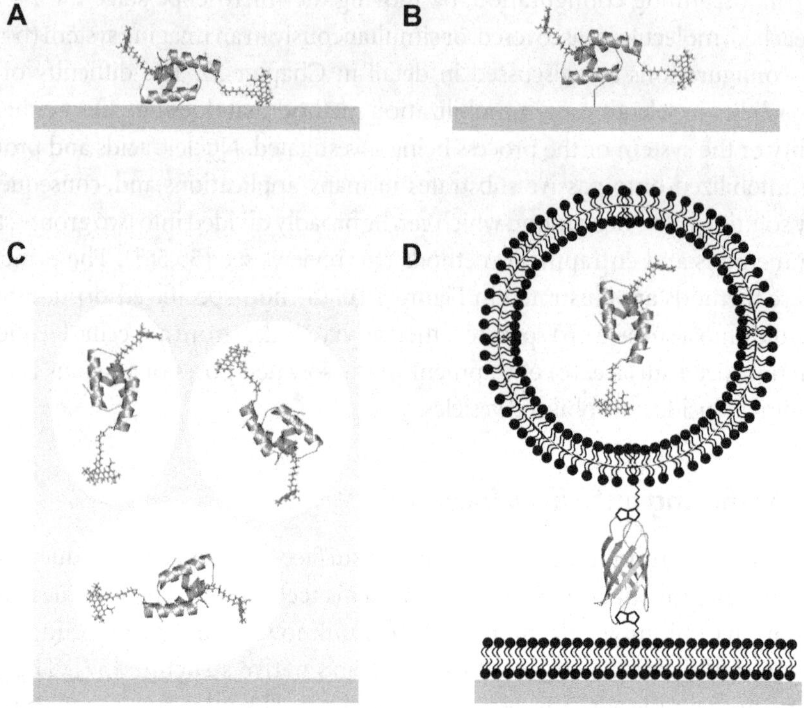

Figure 4.10 Commonly employed immobilization techniques. (a) Passive absorption, (b) tethering to a surface by a site specific tag, (c) encapsulation in a pore of a gel matrix, and (d) localization in a water-filled lipid vesicle. This figure is inspired by a similar figure in [86].

immobilization to a surface derivatized with nitriloacetic acetic acid *via* mutual chelation to a metal ion (usually Ni^{2+}). As this method is commonly used to purify proteins, many potentially useful products such as derivatized resins, His-tag antibodies, and reagents to derivatize various substrates are readily available. Talaga *et al.* [34] studied the dynamics and folding of yeast transcription factor GCN4 by tethering synthesized peptides, which terminated with four negatively charged glutamic acid residues, onto an amino-silanized (positively charged) glass surface. In order to minimize non-specific adsorption of biomolecules onto the substrate it is necessary to passivate the surface. For nucleic acids this is conveniently achieved by pre-treatment of the substrate with bovine serum albumin (BSA) especially when using biotin based systems as biotinylated BSA is commercially available. The system is thus assembled by creating a streptavidin 'sandwich' that links the biotinylated BSA surface layer with the dye-labelled nucleic acid to be studied. These surfaces have, however, been found to be problematic for some protein systems [55,64] resulting in either denaturation or high avidity. In these cases, the surface can be passivated using polyethylene glycol (PEG), a technique that has also been used with great success for nucleotide immobilization [65]. PEG can be purchased derivatized at either or both ends with a wide variety of functional groups. Pal *et al.* [66] solved both problems associated with tethering molecules at surfaces by employing many of the techniques described earlier by using a 'brick laying' approach. Glass coverslips are first passivated with an amine reactive PEG derivative in the presence of 0.25% heterobifunctional PEG (biotin and N hyroxysuccinamide). Streptavidin is then used to attach a biotinylated protein spacer to the surface. This spacer, protein L, consists of a series of highly homologous domains that bind tightly to antibodies without affecting their ability to bind antigens. This allows a His-tagged dye-labelled protein to be tethered away from the surface by using an anti$(His)_6$ antibody as a bridge between protein L and the target protein.

Tethering molecules to a surface allows the solvent conditions to be rapidly altered and its effects monitored, this is not the case when a biomolecule has been encapsulated in a gel pore or in a lipid vesicle. However, even if surface effects can be avoided, the tethering of one end of a molecule may have a profound effect on the property being studied and indeed it is essential this is monitored through some activity/structure/function test to ensure experiments are not biased. In particular with direct surface attachment one might have concerns that any rare heterogeneity in the data that is measured may simply be a result of the heterogeneity in the geometry of immobilization, indeed this is perhaps difficult to account for. In these cases, such as the study of protein folding dynamics, it may be more desirable to use encapsulation methods.

4.6.2 **Encapsulation**

Rather than physically attach the analyte to a surface, another methodology is to simply confine the molecule of interest to a small solvated volume in which the analyte is free to diffuse. Instrumentation for single molecule spectroscopy generally relies on the reduction of the sample volume in which fluorescence is monitored in order to reduce the contribution to the signal from scattering by the solvent and impurities in the solution. This arrangement, combined with a low analyte concentration, also ensures that only single molecules are interrogated at any instant. The two common instrumentation methods (far field confocal/multiphoton diffraction limited microscopy and total internal reflection fluorescence microscopy, TIRFM) all employ a highly non-homogenous excitation/detection volume that is effectively <1 μm in any dimension. As such, the requirements for analyte encapsulation are to restrict diffusion to a volume of smaller than this size but to allow rapid diffusion within the volume to ensure that the inhomogeneity in the excitation/detection profile does not contribute to the detected signal (these topics are covered more fully in Chapter 3).

Encapsulation is a technique that is instantly attractive as it clearly closely matches the conditions that might be found in cells. Indeed experiments on single molecules in live cells [67] and other systems that provide intrinsic immobilization, are also worthy of mention. These experiments essentially circumvent any concerns that the encapsulation method might perturb analyte function, as the spatial confinement is intrinsic. Mashanov and co-workers were able to monitor GFP-protein fusion constructs bound to the plasma membrane of live mouse myoblastic cells [33,68], this work is reviewed in Chapter 6. Ion-channel membrane proteins have also received significant attention and measurements of the fluorescence and spFRET from single membrane proteins, for example on gramicidin ion channels in lipid bilayers has been achieved [69,70]. In a novel study, the effect of a glassy sugar matrix (trelhose) on the fluorescence versus time trajectories of dye-labelled cytochrome was studied [71]. The motivation for this study being the observation that the survival of lower organisms in dehydrated environments is often concomitant with the presence of the sugar. The single molecule study suggested that the sugar matrix reduced the dye (and therefore proteins) exposure to oxygen (extended times to photobleaching were observed, suggesting a reduction in photo-oxidation). A number of studies to probe the local segmental dynamics in polymer films have also been performed [72,73]. These experiments rely on fluorophores that are very sensitive to the environment in which they are immobilized.

A number of strategies to create an artificial and passive environment for encapsulation have been suggested and we briefly review them here. Confinement in the solvated pores of gels [74–77], in particular agarose and acrylamide gels with high water content (see Figure 4.10 (c)), has been shown to be effective in immobilizing cholesterol oxidase [78] and GFP [11,74,79]. The drawback of such a method is that the polymerised gels, even at relatively high percentage contents, do not result in a completely isolated compartment and long-range diffusion (over several microns) is seen for small molecules. In addition, accurate control of the pore size is difficult and a broad distribution of pore size diameter is present (so the immobilization medium is not homogenous). Another concern is that the average pore size diameter in such systems can be low (<2 nm [76]) suggesting that the time spent interacting with the cross-linked gel matrix may be large. Allen and co-workers [80] noted hindered diffusion (rather than complete immobilization to a volume similar to that of the probe) in their work on spFRET labelled calmodulin (CaM) [80] and so instead conducted studies on a fusion of the labelled CaM with the maltose binding protein (MBP). This significantly larger protein chimera showed no translational motion in the same agarose gels and calcium-binding assays of CaM activity confirmed that the coupling did not perturb the structure of CaM significantly. A further concern with gel encapsulation is the method by which the analyte is incorporated into the gel matrix. This is usually achieved by adding the analyte to the gel solution before polymerization either by keeping the gel at a high temperature (agarose) or by adding the cross-linking agents afterwards. The formation of the matrix or the heat (either generated by the exothermic cross-linking reaction or due to the relatively high melting temperature) may be a concern in terms of maintaining analyte function. Silica gels have also been suggested as ideal hosts for molecules of biological interest. Chirico *et al.*[81] showed that GFP photobleaching lifetimes were significantly extended when the protein was encapsulated in a silica gel compared to bare substrates where significant interaction with the surface was apparent.

Another elegant method following a principle suggested by Chiu [82] and realized by Haran and co-workers [83–85] is encapsulation of single protein molecules in small unilamellar lipid vesicles (see Figure 4.10 (d)). In this method small (~100 nm diameter) liposomes are tethered to a lipid bilayer on glass via the avidin–biotin interaction. This provides a simple method by which a protein is encapsulated but allowed to diffuse. The authors were also able to demonstrate that the protein molecules studied were freely diffusing within the compartment, and spent little time interacting with the lipid walls by comparing the polarization distribution for liposome encapsulated molecules with that for molecules directly adsorbed onto a glass substrate (which show a broad polarization distribution indicating significantly hindered rotation). Aspects of this technique are discussed more fully in Chapter 6, in which a paper from this group appliying this methodology is reviewed [85]. This technique was used more recently for studies of encapsulated ribozymes [62]. Okumus and co-workers were also able to show that this immobilization methodology was useful for TIRF geometries where a strongly non-uniform illumination profile exists (see Chapter 3). The resultant data showed no artefacts that might be expected from transient adsorption onto the vesicle walls, again supporting the notion that the molecules of interest are diffusing rapidly in the compartment (as variation in the signal intensity caused by diffusion through the highly non-uniform excitation volume was not observed, suggesting that it is averaged out on the timescale on the measurement). This method is perhaps more desirable than the gel method as much more control is available on the size and uniformity of the liposomes. Furthermore, the pore size is much larger than can be achieved with gels but also small enough to ensure immobilization within the instrument's detection volume. This method of analyte encapsulation is also somewhat less likely to disrupt the analyte structure or function as it is simply added, at a low concentration, to a solution of large liposomes that are then extruded to form smaller liposomes, some of which will incorporate the analyte.

References

[1] **Lakowicz, JR**, *Principles of Fluorescence Spectroscopy*, Plenum Press, New York, 1983.

[2] **Wilson, J** and **Hawkes, JFB**, *Optoelectronics: An Introduction*, Prentice Hall Int., Cambridge, 1989.

[3] **Berlier, JE, Rothe, A, Buller, G, Bradford, J, Gray, DR, Filanoski, BJ**, *et al.*, Quantitative comparison of long-wavelength Alexa Fluor dyes to CY dyes: Fluorescence of the dyes and their bioconjugates. *Journal of Histochemistry* and *Cytochemistry* **51** (2003) 1699–1712.

[4] **van der Meer, BW, Coker III, G**, and **Chen, S-Y**, *Resonance Energy Transfer: Theory and Data*, Wiley-VCH, Chichester, New York 1994.

[5] **Kapanidis, AN** and **Weiss, S**, Fluorescent probes and bioconjugation chemistries for single-molecule fluorescence analysis of biomolecules. *Journal of Chemical Physics* **117** (2002) 10953–10964.

[6] **Wilkinson, F, Mcgarvey, DJ**, and **Olea, AF**, Excited triplet-state interactions with molecular-oxygen—influence of charge-transfer on the bimolecular quenching rate constants and the yields of singlet oxygen (O-2(asterisk),(1)delta(G)) for substituted naphthalenes in various solvents. *Journal of Physical Chemistry* **98** (1994) 3762–3769.

[7] **Kapanidis, AN, Lee, NK, Laurence, TA, Doose, S, Margeat, E**, and **Weiss, S**, Fluorescence-aided molecule sorting: Analysis of structure and interactions by alternating-laser excitation of single molecules. *Proceedings of the National Academy of Sciences of the United States Of America* **101** (2004) 8936–8941.

[8] **English, DS, Furube, A**, and **Barbara, PF**, Single-molecule spectroscopy in oxygen-depleted polymer films. *Chemical Physics Letters* **324** (2000) 15–19.

[9] **Bates, M, Blosser, TR**, and **Zhuang, XW**, Short-range spectroscopic ruler based on a single-molecule optical switch. *Biophysical Journal* **88** (2005) 363A–364A.

[10] **Moerner, WE**, Single-molecule optical spectroscopy of autofluorescent proteins. *Journal of Chemical Physics* **117** (2002) 10925–10937.

[11] **Dickson, RM, Cubitt, AB, Tsien, RY**, and **Moerner, WE**, On/off blinking and switching behaviour of single molecules of green fluorescent protein. *Nature* **388** (1997) 355–358.

[12] **Haupts, U, Maiti, S, Schwille, P**, and **Webb, WW**, Dynamics of fluorescence fluctuations in green fluorescent protein observed by fluorescence correlation spectroscopy. *Proceedings of the National Academy of Sciences of the United States of America* **95** (1998) 13573–13578.

[13] **Peterman, EJG, Brasselet, S**, and **Moerner, WE**, The fluorescence dynamics of single molecules of green fluorescent protein. *Journal of Physical Chemistry A* **103** (1999) 10553–10560.

[14] **Malvezzi-Campeggi, F, Jahnz, M, Heinze, KG, Dittrich, P**, and **Schwille, P**, Light-induced flickering of DsRed provides evidence for distinct and interconvertible fluorescent states. *Biophysical Journal* **81** (2001) 1776–1785.

[15] **Pierce, DW, HomBooher, N**, and **Vale, RD**, Imaging individual green fluorescent proteins. *Nature* **388** (1997) 338–338.

[16] **McAnaney, TB, Zeng, W, Doe, CFE, Bhanji, N, Wakelin, S, Pearson, DS**, *et al.*, Protonation, photobleaching, and photoactivation of yellow fluorescent protein (YFP 10C): A unifying mechanism. *Biochemistry* **44** (2005) 5510–5524.

[17] **Hou, YW, Bardo, AM, Martinez, C**, and **Higgins, DA**, Characterization of molecular scale environments in polymer films by single molecule spectroscopy. *Journal of Physical Chemistry B* **104** (2000) 212–219.

[18] **Förster, T**, Transfer Mechanisms of electronic excitation. *Discussions of the Faraday Society* **27** (1959) 7–17.

[19] http://www.shep.net/resources/curricular/physics/java/physengl/cpendula.htm

[20] **Verheyden, S, Sillen, A, Gils, A, Declerck, PJ**, and **Engelborghs, Y**, Tryptophan properties in fluorescence and functional stability of plasminogen activator inhibitor 1. *Biophysical Journal* **85** (2003) 501–510.

[21] **Callis, PR** and **Vivian, JT**, Understanding the variable fluorescence quantum yield of tryptophan in proteins using QM-MM simulations. Quenching by charge transfer to the peptide backbone. *Chemical Physics Letters* **369** (2003) 409–414.

[22] **Lippitz, M**, **Erker, W**, **Decker, H**, **van Holde, KE**, and **Basche, T**, Two-photon excitation microscopy of tryptophan-containing proteins. *Proceedings of the National Academy of Sciences of the United States of America* **99** (2002) 2772–2777.

[23] **Diehl, H** and **Markuszewski, R**, Studies on fluorescein 7: The fluorescence of fluorescein as a function of pH. *Talanta* **36** (1989) 416–418.

[24] **Kapanidis, AN**, **Ebright, YW**, and **Ebright, RH**, Site-specific incorporation of fluorescent probes into protein: Hexahistidine-tag-mediated fluorescent labeling with (Ni2+: nitrilotriacetic acid)(N)-fluorochrome conjugates. *Journal of the American Chemical Society* **123** (2001) 12123–12125.

[25] **Ratner, V**, **Kahana, E**, **Eichler, M**, and **Haas, E**, A general strategy for site-specific double labeling of globular proteins for kinetic FRET studies. *Bioconjugate Chemistry* **13** (2002) 1163–1170.

[26] **Mullis, KB** and **Faloona, FA**, Specific synthesis of DNA invitro via a polymerase-catalyzed chain-reaction. *Methods in Enzymology* **155** (1987) 335–350.

[27] **Tomschik, M**, **Zheng, HC**, **van Holde, K**, **Zlatanova, J**, and **Leuba, SH**, Fast, long-range, reversible conformational fluctuations in nucleosomes revealed by single-pair fluorescence resonance energy transfer. *Proceedings of the National Academy of Sciences of the United States of America* **102** (2005) 3278–3283.

[28] **Deniz, AA**, **Dahan, M**, **Grunwell, JR**, **Ha, T**, **Faulhaber, AE**, **Chemla, DS**, *et al.*, Single-pair fluorescence resonance energy transfer on freely diffusing molecules: Observation of Förster distance dependence and subpopulations. *Proceedings of the National Academy of Sciences of the United States of America* **96** (1999) 3670–3675.

[29] **Sabanayagam, CR**, **Eid, JS**, and **Meller, A**, Using fluorescence resonance energy transfer to measure distances along individual DNA molecules: Corrections due to nonideal transfer. *Journal of Chemical Physics* **122** (2005) art. no.-061103.

[30] **Bartley, LE**, **Zhuang, XW**, **Das, R**, **Chu, S**, and **Herschlag, D**, Exploration of the transition state for tertiary structure formation between an RNA helix and a large structured RNA. *Journal of Molecular Biology* **328** (2003) 1011–1026.

[31] **Zhang, ZW**, **Smith, BAC**, **Wang, L**, **Brock, A**, **Cho, C**, and **Schultz, PG**, A new strategy for the site-specific modification of proteins in vivo. *Biochemistry* **42** (2003) 6735–6746.

[32] **Philipps, B**, **Hennecke, J**, and **Glockshuber, R**, FRET-based in vivo screening for protein folding and increased protein stability. *Journal of Molecular Biology* **327** (2003) 239–249.

[33] **Mashanov, GI**, **Tacon, D**, **Peckham, M**, and **Molloy, JE**, The spatial and temporal dynamics of pleckstrin homology domain binding at the plasma membrane measured by imaging single molecules in live mouse myoblasts. *Journal of Biological Chemistry* **279** (2004) 15274–15280.

[34] **Talaga, DS**, **Lau, WL**, **Roder, H**, **Tang, JY**, **Jia, YW**, **Degrado, WF**, et al., Dynamics and folding of single two-stranded coiled-coil peptides studied by fluorescent energy transfer confocal microscopy. *Proceedings of the National Academy of Sciences of the United States of America* **97** (2000) 13021–13026.

[35] **Dawson, PE** and **Kent, SBH**, Synthesis of native proteins by chemical ligation. *Annual Review of Biochemistry* **69** (2000) 923–960.

[36] **Beligere, GS** and **Dawson, PE**, Conformationally assisted protein ligation using C-terminal thioester peptides. *Journal of the American Chemical Society* **121** (1999) 6332–6333.

[37] **Deniz, AA**, **Laurence, TA**, **Beligere, GS**, **Dahan, M**, **Martin, AB**, **Chemla, DS**, *et al.*, Single-molecule protein folding: Diffusion fluorescence resonance energy transfer studies of the denaturation of chymotrypsin inhibitor 2. *Proceedings of the National Academy of Sciences of the United States of America* **97** (2000) 5179–5184.

[38] **Offer, J, Boddy, CNC,** and **Dawson, PE,** Extending synthetic access to proteins with a removable acyl transfer auxiliary. *Journal of the American Chemical Society* **124** (2002) 4642–4646.

[39] **Mujumdar, RB, Ernst, LA, Mujumdar, SR, Lewis, CJ,** and **Waggoner, AS,** Cyanine dye labeling reagents—sulfoindocyanine succinimidyl esters. *Bioconjugate Chemistry* **4** (1993) 105–111.

[40] **Notling, B,** *Protein Folding Kinetics: Biophysical Methods,* Springer, Berlin, 1999.

[41] **Brinkley, M,** A brief survey of methods for preparing protein conjugates with dyes, haptens, and cross-linking reagents. *Bioconjugate Chemisty* **3** (1992) 2–13.

[42] **Kabsch, W** and **Sander, C,** Dictionary of protein secondary structure—pattern-recognition of hydrogen-bonded and geometrical features. *Biopolymers* **22** (1983) 2577–2637.

[43] http://www.rcsb.org/pdb/

[44] **Schuler, B, Lipman, EA,** and **Eaton, WA,** Probing the free-energy surface for protein folding with single-molecule fluorescence spectroscopy. *Nature* **419** (2002) 743–747.

[45] **Capaldi, AP, Kleanthous, C,** and **Radford, SE,** Im7 folding mechanism: Misfolding on a path to the native state. *Nature Structural Biology* **9** (2002) 209–216.

[46] **Friel, CT, Beddard, GS,** and **Radford, SE,** Switching two-state to three-state kinetics in the helical protein Im9 via the optimisation of stabilising non-native interactions by design. *Journal of Molecular Biology* **342** (2004) 261–273.

[47] **Spence, GR, Capaldi, AP,** and **Radford, SE,** Trapping the on-pathway folding intermediate of Im7 at equilibrium. *Journal of Molecular Biology* **341** (2004) 215–226.

[48] **Ferguson, N, Capaldi, AP, James, R, Kleanthous, C,** and **Radford, SE,** Rapid folding with and without populated intermediates in the homologous four-helix proteins Im7 and Im9. *Journal of Molecular Biology* **286** (1999) 1597–1608.

[49] **Ferguson, N, Li, W, Capaldi, AP, Kleanthous, C,** and **Radford, SE,** Using chimeric immunity proteins to explore the energy landscape for α-Helical protein folding. *Journal of Molecular Biology* **307** (2001) 393–405.

[50] **Gorski, SA, Capaldi, AP, Kleanthous, C,** and **Radford, SE,** Acidic conditions stabilise imtermediates populated during the folding of Im7 and Im9. *Journal of Molecular Biology* **312** (2001) 849–863.

[51] **Cranz-Mileva, S, Friel, CT,** and **Radford, SE,** Helix stability and hydrophobicity in the folding mechanism of the bacterial immunity protein Im9. *Protein engineering design or selection*: **18** (2005) 41–50.

[52] **Friel, CT, Capaldi, AP,** and **Radford, SE,** Structural analysis of the rate-limiting transition states in the folding of Im7 and Im9: Similarities and differences in the folding of homologous proteins. *Journal of Molecular Biology* **326** (2003) 293–305.

[53] **Kapanidis, AN, Laurence, TA, Lee, NK, Margeat, E, Kong, XX,** and **Weiss, S,** Alternating-laser excitation of single molecules. *Accounts of Chemical Research* **38** (2005) 523–533.

[54] **Lee, NK, Kapanidis, AN, Wang, Y, Michalet, X, Mukhopadhyay, J, Ebright, RH,** et al., Accurate FRET measurements within single diffusing biomolecules using alternating-laser excitation. *Biophysical Journal* **88** (2005) 2939–2953.

[55] **Rasnik, I, McKinney, SA,** and **Ha, T,** Surfaces and orientations: Much to FRET about? *Accounts of Chemical Research* **38** (2005) 542–548.

[56] **Heyes, CD, Kobitski, AY, Amirgoulova, EV,** and **Nienhaus, GU,** Biocompatible surfaces for specific tethering of individual protein molecules. *Journal of Physical Chemistry B* **108** (2004) 13387–13394.

[57] **Zhuang, XW, Ha, T, Kim, HD, Centner, T, Labeit, S,** and **Chu, S,** Fluorescence quenching: A tool for single-molecule protein- folding study. *Proceedings of the National Academy of Sciences of the United States of America* **97** (2000) 14241–14244.

[58] **Le Trong, I, Freitag, S, Klumb, LA, Chu, V, Stayton, PS,** and **Stenkamp, RE,** Structural studies of hydrogen bonds in the high-affinity streptavidin-biotin complex: Mutations of amino acids interacting with the ureido oxygen of biotin. *Acta Crystallographica Section D-Biological Crystallography* **59** (2003) 1567–1573.

[59] **Zhuang, XW, Bartley, LE, Babcock, HP, Russell, R,** Ha, TJ, **Herschlag, D,** *et al.*, A single-molecule study of RNA catalysis and folding. *Science* **288** (2000) 2048–2051.

[60] **Zhuang, XW, Kim, H, Pereira, MJB, Babcock, HP, Walter, NG,** and **Chu, S,** Correlating structural dynamics and function in single ribozyme molecules. *Science* **296** (2002) 1473–1476.

[61] **Zhuang, XW** and **Rief, M,** Single-molecule folding. *Current Opinion in Structural Biology* **13** (2003) 88–97.

[62] **Okumus, B, Wilson, TJ, Lilley, DMJ,** and **Ha, T,** Vesicle encapsulation studies reveal that single molecule ribozyme heterogeneities are intrinsic. *Biophysical Journal* **87** (2004) 2798–2806.

[63] **Ha, T, Rasnik, I, Cheng, W, Babcock, HP, Gauss, GH, Lohman, TM,** *et al.*, Initiation and reinitiation of DNA unwinding by the Escherichia coli Rep helicase. *Nature* **419** (2002) 638–641.

[64] **Ha, T,** Single-molecule fluorescence resonance energy transfer. *Methods* **25** (2001) 78–86.

[65] **Rasnik, I, Myong, S, Cheng, W, Lohman, TM,** and **Ha, T,** DNA-binding orientation and domain conformation of the E-coli Rep helicase monomer bound to a partial duplex junction: Single-molecule studies of fluorescently labeled enzymes. *Journal of Molecular Biology* **336** (2004) 395–408.

[66] **Pal, P, Lesoine, JF, Lieb, MA, Novotny, L,** and **Knauf, PA,** A novel immobilization method for single protein spFRET studies. *Biophysical Journal* **89** (2005) L11–L13.

[67] **Moerner, WE,** Optical measurements of single molecules in cells. *Trends in Analytical Chemistry: TRAC* **22** (2003) 544–548.

[68] **Mashanov, GI, Tacon, D, Knight, AE, Peckham, M,** and **Molloy, JE,** Visualizing single molecules inside living cells using total internal reflection fluorescence microscopy. *Methods* **29** (2003) 142–152.

[69] **Borisenko, V, Lougheed, T, Hesse, J, Fureder-Kitzmuller, E, Fertig, N, Behrends, JC,** *et al.*, Simultaneous optical and electrical recording of single gramicidin channels. *Biophysical Journal* **84** (2003) 612–622.

[70] **Harms, G, Orr, G,** and **Lu, HP,** Probing ion channel conformational dynamics using simultaneous single-molecule ultrafast spectroscopy and patch-clamp electric recording. *Applied Physics Letters* **84** (2004) 1792–1794.

[71] **Mei, E, Tang, JY, Vanderkooi, JM,** and **Hochstrasser, RM,** Motions of single molecules and proteins in trehalose glass. *Journal of the American Chemical Society* **125** (2003) 2730–2735.

[72] **Willets, KA, Callis, PR,** and **Moerner, WE,** Experimental and theoretical investigations of environmentally sensitive single-molecule fluorophores. *Journal of Physical Chemistry B* **108** (2004) 10465–10473.

[73] **Vallee, RAL, Tomczak, N, Kuipers, L, Vancso, GJ,** and **van Hulst, NF,** Single molecule lifetime fluctuations reveal segmental dynamics in polymers. *Physical Review Letters* **91** (2003) art. no.-038301.

[74] **Dickson, RM, Norris, DJ, Tzeng, YL,** and **Moerner, WE,** Three-dimensional imaging of single molecules solvated in pores of poly(acrylamide) gels. *Science* **274** (1996) 966–969.

[75] **Dickson, RM, Norris, DJ, Tzeng, YL, Sakowicz, R, Goldstein, LSB,** and **Moerner, WE,** Single molecules solvated in pores of polyacrylamide gels. *Molecular Crystals and Liquid Crystals Science and Technology Section A-Molecular Crystals and Liquid Crystals* **291** (1996) 31–39.

[76] **Kummer, SD, Dickson, RM, Moerner, WE,** Probing single molecules in polyacrylamide gels. *Proceedings of the SPIE* 3273 (1998) 165–173.

[77] **Fatin-Rouge, N**, **Starchev, K**, and **Buffle, J**, Size effects on diffusion processes within agarose gels. *Biophysical Journal* **86** (2004) 2710–2719.

[78] **Lu, HP**, **Xun, L**, and **Xie, XS**, Single-molecule enzymatic dynamics. *Science* **282** (1998) 1877.

[79] **Peterman, EJG**, **Brasselet, S**, and **Moerner, WE**, The fluorescence dynamics of single molecules of green fluorescent protein. *Journal of Physical Chemistry A* **103** (1999) 10553–10560.

[80] **Allen, MW**, **Urbauer, RJB**, **Zaidi, A**, **Williams, TD**, **Urbauer, JL**, and **Johnson, CK**, Fluorescence labeling, purification, and immobilization of a double cysteine mutant calmodulin fusion protein for single-molecule experiments. *Analytical Biochemistry* **325** (2004) 273–284.

[81] **Chirico, G**, **Cannone, F**, **Beretta, S**, **Diaspro, A**, **Campanini, B**, **Bettati, S**, *et al.*, Dynamics of green fluorescent protein mutant2 in solution, on spin-coated glasses, and encapsulated in wet silica gels. *Protein Science* **11** (2002) 1152–1161.

[82] **Chiu, DT**, **Wilson, CF**, **Ryttsen, F**, **Stromberg, A**, **Farre, C**, **Karlsson, A**, *et al.*, Chemical transformations in individual ultrasmall biomimetic containers. *Science* **283** (1999) 1892–1895.

[83] **Boukobza, E**, **Sonnenfeld, A**, and **Haran, G**, Immobilization in surface-tethered lipid vesicles as a new tool for single biomolecule spectroscopy. *Journal of Physical Chemistry B* **105** (2001) 12165–12170.

[84] **Rhoades, E**, **Cohen, M**, **Schuler, B**, and **Haran, G**, Two-state folding observed in individual protein molecules. *Journal of the American Chemical Society* **126** (2004) 14686–14687.

[85] **Rhoades, E**, **Gussakovsky, E**, and **Haran, G**, Watching proteins fold one molecule at a time. *Proceedings of the National Academy of Sciences of the United States of America* **100** (2003) 3197–3202.

[86] **Haran, G**, Single-molecule fluorescence spectroscopy of biomolecular folding. *Journal of Physics-Condensed Matter* **15** (2003) R1291–R1317.

[87] **Bestvater, F**, **Spiess, E**, **Stobrawa, G**, **Hacker, M**, **Feurer, T**, **Porwol, T**, *et al.*, Two-photon fluorescence absorption and emission spectra of dyes relevant for cell imaging. *Journal of Microscopy-Oxford* **208** (2002) 108–115.

[88] http://www.synthegen.com/Action.lasso?-Response = /products/fluorescent/table.lasso&-Token.SortColumn = sort_order&-Nothing.

[89] http://www.biophys.leidenuniv.nl/research/fvl/TSLesHouches2001_2A.pdf

[90] **Schmidt, T**, **Kubitscheck, U**, **Rohler, D**, and **Nienhaus, U**, Photostability data for fluorescent dyes: An update. *Single Molecules* **3** (2002) 327.

[91] **Soria, S**, **Katchalski, T**, **Teitelbaum, E**, **Friesem, AA**, and **Marowsky, G**, Enhanced two-photon fluorescence excitation by resonant grating waveguide structures. *Optics Letters* **29** (2004) 1989–1991.

[92] **Brackmann, U**, Lambdachrome Laser Dyes, Lambda Physik AG, Goettingen, 2000.

[93] http://pingu.salk.edu/flow/fluo.html

[94] http://www.clontech.com/clontech/archive/OCT99UPD/RFP.shtml

[95] http://probes.invitrogen.com/servlets/datatable?item = 10168&id = 38089

[96] http://www.atto-tec.com/ATTO-TEC.com/Products/data_table.htm

[97] **Margittai, M**, **Widengren, J**, **Schweinberger, E**, **Schroder, GF**, **Felekyan, S**, **Haustein, E**, *et al.*, Single-molecule fluorescence resonance energy transfer reveals a dynamic equilibrium between closed and open conformations of syntaxin 1. *Proceedings of the National Academy of Sciences of the United States of America* **100** (2003) 15516–15521.

[98] **Schuler, B**, **Lipman, EA**, **Steinbach, PJ**, **Kumke, M**, and **Eaton, WA**, Polyproline and the 'spectroscopic ruler' revisited with single-molecule fluorescence. *Proceedings of the National Academy of Sciences of the United States Of America* **102** (2005) 2754–2759.

[99] **Xie, Z**, **Srividya, N**, **Sosnick, TR**, **Pan, T**, and **Scherer, NF**, Single-molecule studies highlight conformational heterogeneity in the early folding steps of a large ribozyme. *Proceedings of the National Academy of Sciences of the United States of America* **101** (2004) 534–539.

FIVE

Fluorescence spectroscopy of freely diffusing single molecules: examples

5.1 Introduction

Perhaps the greatest advantage of studying single molecules that are freely diffusing in solution is the simplicity of the approach. Immobilization of molecules and complexes whilst maintaining their integrity and function is not straightforward and a higher level of rigour must be adopted in such experiments to ensure that artefacts due to the immobilization protocol or interactions with the surface are not encountered. In experiments in which the molecules are freely diffusing, burst analysis or correlation techniques can be used to extract both equilibrium and kinetic parameters of interest. It is vital however that the interplay between the kinetic rate constants of the system, the transit time through the observation volume, and experimental parameters such as the integration time are thoroughly understood to avoid erroneous conclusions about molecular dynamics and heterogeneity. Here we discuss several studies of freely diffusing nucleic acids and proteins that illustrate the potential of these experiments and help to highlight the limitations which must be borne in mind when designing single molecule experiments.

5.2 Single molecule studies of freely diffusing molecules

5.2.1 DNA hairpin loop dynamics

Oligonucleotide structures such as the hairpin loop of a single-stranded DNA, shown in Figure 5.1, are not static. These structures fluctuate between fully closed

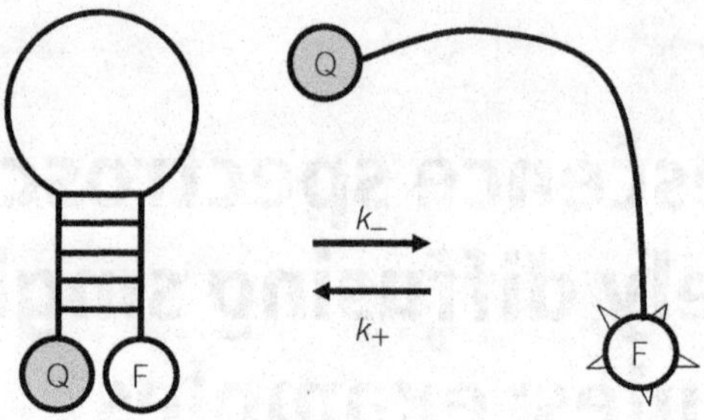

Figure 5.1 Sketch of the DNA molecular beacon. The five bases at the two ends of the beacon are complementary to each other. The size of the loop and its content are varied. The beacon flips between open and closed states with the characteristic rates k_- and k_+. The fluorophore (F) and the quencher (Q) are covalently linked to the two arms of the beacon. In the open state the beacon fluoresces, in the closed state the fluorescence is quenched. Reprinted with permission from Bonnet *et al.*, Kinetics of conformational fluctuations in DNA hairpin-loops. *Proceedings of the National Academy of Sciences of the United States of America* 95 (1998) 8602–8606. Copyright 1998 National Academy of Sciences, USA.

states, open random coils, and possibly also between intermediate partially folded states. The dynamic behaviour of these oligonucleotide structures probably plays an important role in their function since recognition events, such as those involved in regulation of gene expression, are likely to be affected by the kinetics of conformational changes.

The conformational fluctuations of DNA hairpin loops can be addressed using single molecule techniques. The autocorrelation of a FRET or fluorescence signal that is affected by the fluctuations can be used to determine the forward and reverse kinetic rate constants for this (or any other) two-state system (see Chapter 2). Bonnet and colleagues used this approach to study the effects of sequence, size, and experimental conditions on a small DNA hairpin loop [1] (Figure 5.1). The DNA hairpin loop that they chose to study was formed from the sequence 5′-CCCAA-$(N)_n$-TTGGG-3′. Within this scaffold the sequence $(N)_n$ was varied to explore changes in loop length $(T)_{12\text{–}30}$ and the composition $(A)_{21}$. The 5′ and 3′ ends of each sequence were modified with a fluorophore (6-carboxyrhodamine, or 6G) and a quencher of the fluorophore (4-{[4(dimethylamino)phenyl]azo} benzoic acid, or Dabcyl), respectively. Dabcyl is an efficient quencher of 6G fluorescence and therefore when the hairpin loop is closed the intensity of fluorescence is considerably lower (by a factor of 50) than when the DNA is in the open conformation and the quencher and fluorophore are, on average, much further apart. In an ensemble fluorescence experiment the 6G emission was monitored as a function of temperature (10–80°C) and the equilibrium constant K(T) obtained from fitting the fraction of folded hairpins as a function of temperature $p(T)$ by using a simple thermodynamic two-state model ($K(T) = p(T)/[1-p(T)]$). K(T) thereby provides the ratio of the opening to closing kinetic rate constants

(K(T) = k_{-}/k_{+}). As is typical in such studies of assumed two-state folding, this equilibrium measurement can be complemented by a measurement of the observed kinetic rate constant k_{obs} which is equal to the sum of the forward and reverse rate constants, k_{-} and k_{+} in this case. These two measurements therefore provide sufficient information to determine the values of the kinetic microscopic rate constants and therefore the relative size of the activation barrier $\Delta G_i^{\dagger}$ for forward and reverse transitions since $k_i = A\exp(-\Delta G_i^{\dagger}/k_B T)$ where subscript i represents the opening or closing process, A is a pre-exponential factor [2] whose value depends on the system in question and k_B is Boltzmann's constant.

Conventional ensemble methods, such as stopped flow [3], provide one means of probing the kinetic behaviour of the system and extracting the observed rate constant. Ensemble methods such as these work by introducing a perturbation to the ensemble (such as the rapid dilution out of a chemical denaturant) and monitoring the ensemble as it relaxes to the new equilibrium. These methods then suffer the disadvantage of a large dead time due to the time taken to induce the perturbation, and as such would be inappropriate for the fast rates expected for simple nucleotide structures such as these (and many, more complex biomolecules, such as proteins). Ensemble methods such as continuous flow rapid mixing [4] or temperature jump [2] can provide access to the time window necessary to monitor these fast kinetics but single molecule resolution provides a novel approach: perturbation is necessary in ensemble studies due to the lack of synchronization of molecules within the ensemble, despite the fact that in favourable conditions all members of the ensemble are in fact in a dynamic equilibrium between available conformational states. Thus if one is able to monitor the timescale of these intrinsic fluctuations at the resolution of single, or a few, molecules then no perturbation is necessary and one is only limited by the time resolution with which the experiments can be performed. To measure the observed kinetic rate constant, Bonnet and co-workers calculate the autocorrelation function of the 6G fluorescence of freely diffusing DNA molecules in solution at a concentration of 10 nM. A confocal arrangement was used (a water immersion Olympus 60x objective with a numerical aperture of 1.2) to both illuminate the solution and collect fluorescence. The excitation light was rejected with a high pass dichroic mirror and notch filter, then the fluorescence was focused through a 25 μm pinhole then split using a beamsplitting cube onto two avalanche photodiode detectors. The correlation of the two signals from the detectors was derived in real time using a hardware correlator. This correlation function reveals information about fluctuations (see Chapter 2), which in this case arise from the diffusion of molecules in and out of the focal volume and from the opening and closing dynamics of the molecule (the quencher modulated 6G fluorescence). In order to extract information about the folding dynamics from the measurement, a control sample has to be measured in which no folding dynamics occur (or are not measured) and therefore the fluctuations arising solely from the diffusion can be determined. This diffusional contribution to the correlation function of the DNA hairpin loop can thereby be removed and the folding dynamics specifically extracted (see Chapter 2 for a detailed discussion of this approach).

In this manner, the kinetics of the two-state system were determined and are shown in Figure 5.2. Figure 5.2(a) shows the autocorrelation of the fluorescence for the 6G labelled DNA hairpin loop and the control sample, which in this case was a DNA hairpin loop with a 6G label and no quencher. The 6G fluorescence is assumed to be unaffected by the dynamics of the DNA in the control sample. Similarly, the difference in diffusion properties for this moderately lower mass molecule is assumed negligible in the control. The ratio of the two curves in

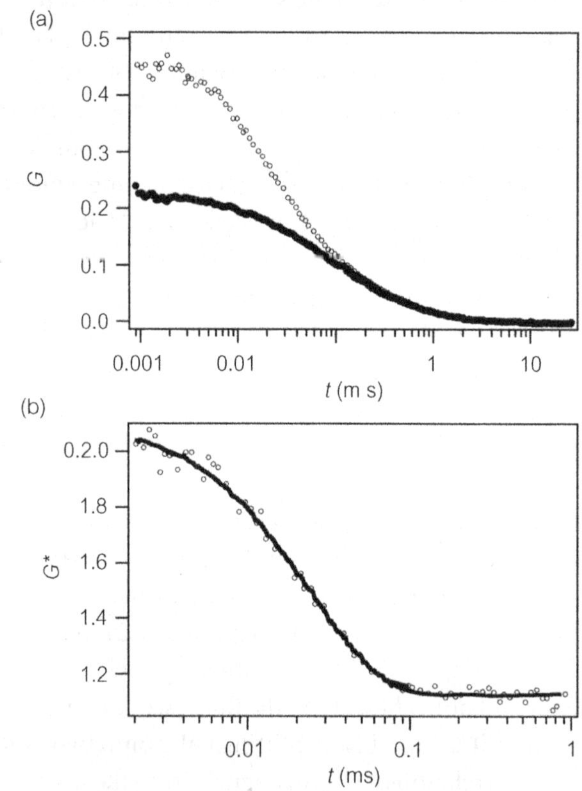

Figure 5.2 Example of the experimental procedure. (a) Autocorrelation curves for beacon (○) and control (●) at $T = 45°C$, 0.2 M NaCl. Both beacon and control have loops of 21 T residues. (b) Ratio of the two curves shown in (a). The line is a three-parameter exponential fit to the data giving $\tau_{reaction} = 24.2 \pm 0.6$ s. Reprinted with permission from Bonnet *et al.*, Kinetics of conformational fluctuations in DNA hairpin-loops. *Proceedings of the National Academy of Sciences of the United States of America* 95 (1998): 8602–8606. Copyright 1998 National Academy of Sciences, USA.

Figure 5.2(a) is the correlation function of the fluctuations of the fluorescence signal arising from the folding dynamics alone ($G^*(t)$) and is shown in Figure 5.2(b). Bonnet *et al.* found that this correlation function could be fitted very well with a single exponential function $G^*(t) = B + C\exp(-k_{obs}t)$ which is to be expected for a two-state system. By contrast, if the system is not two-state, that is, there are folding intermediate states that are populated on the timescale of the measurement, then the autocorrelation function would not be expected to be single exponential and might be fitted better with a stretched exponential for example [5]. By combining the ensemble equilibrium and single moelcule kinetic measurements Bonnet *et al.* obtained the opening and closing rate constants for a range of DNA hairpin loops as a function of temperature (referred to as an Arrhenius plot, see Figure 5.3). Figure 5.3 shows the dependence of the rate constants on the length of the loop (T_{12}, T_{16}, T_{21}, and T_{30}). The rate constants for opening depend quite strongly on temperature but not on the loop length and cover a range 10^2–10^4 s^{-1}. However, the closing rates have a much smaller dependence on temperature (less than an order of magnitude from 10–50°C) but a more

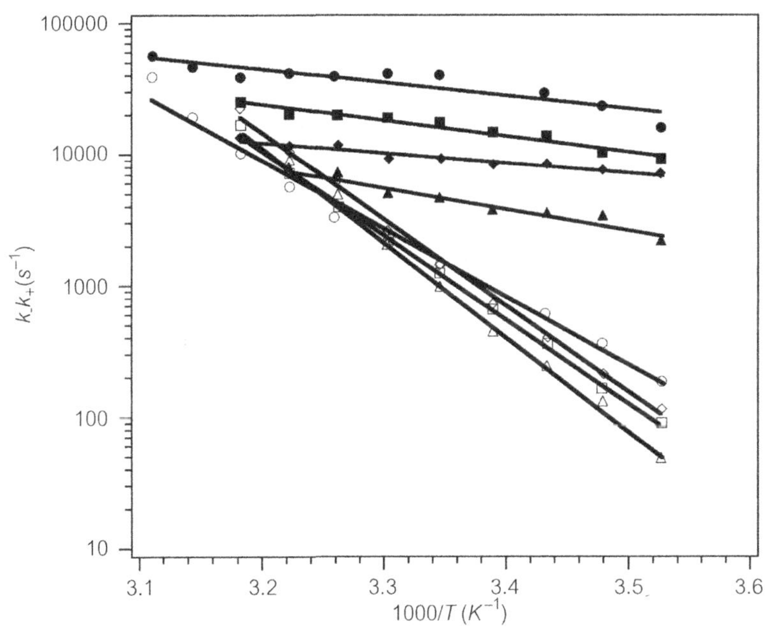

Figure 5.3 Arrhenius plots of the opening rates (open symbols) and the closing rate constants (filled symbols) of beacons with different loop lengths: $(T)_{12}$ (circles), $(T)_{16}$ (squares), $(T)_{21}$ (diamonds), and $(T)_{30}$ (triangles). The lines are exponential fits to the data. The lines corresponding to opening and closing rate constants intersect at the melting temperature. The buffer contained 0.1 M NaCl for all the data. Reprinted with permission from Bonnet *et al.*, Kinetics of conformational fluctuations in DNA hairpin-loops. *Proceedings of the National Academy of Sciences of the United States of America* 95 (1998) 8602–8606. Copyright 1998 National Academy of Sciences, USA.

significant dependence on the loop length. Further experiments to test the effect of changing the loop sequence from all *T* to all *A* showed that the opening rate was largely unaffected by this variation but the closing rate was dramatically reduced with increasing loop length. The effects of varying the ionic strength of the solution through increasing the Na^+ concentration were also studied and the closing rates once again were the most affected increasing with increasing salt.

These experiments showed that DNA hairpin loop opening, which involves an unzipping of all the base pairs in the stem, is largely independent of loop length and its sequence, as might be expected. The number and type of base pairs in the stem would be expected to affect the opening rate but this was not studied. The closing rates were shown to decrease with an increase in loop length and also in the case of the poly-A loop which is more rigid. The addition of salt has the effect of screening the electrostatic charges associated with the bases which decreases the opening rates and increases the closing rates. There is a hint of a much faster kinetic process on a sub-microsecond timescale in the data that the authors suggest is due to the unzipping of the individual base pair interactions in the loop. In principle, autocorrelation ought to be able to achieve a time resolution of better than 1 μs and a range of interesting dynamical studies of oligonucleotides and proteins in this time regime ought to be feasible in well-designed systems. These experiments are an elegant demonstration of the power and resolution of both single molecule resolution experiments and autocorrelation, both of these areas being greatly expanded on in other chapters of this text. In particular, in Chapter 2, we provide a more complete description of this method of extracting kinetic information by normalization of the autocorrelation function. We also discuss alternative methodologies [5] that are conceptually similar but remove the need to normalization with a control sample, indeed we note that such a control can be difficult to obtain for more complex molecules as often the label fluorescence is differentially quenched in the accessible conformations.

5.2.2 **Observation of subpopulations in freely diffusing DNA molecules**

When dye-labelled molecules diffuse through a focused laser beam, fluorescence photon bursts are generated. Although the observation period is short (~1 ms), which precludes studies of longer timescale processes, this data can be analysed to provide information about the equilibrium distribution of the molecular properties of the system such as hydrodynamic radius, diffusion coefficient, identity, complex formation, and concentration of the analyte. Deniz and co-workers [6] exploited this aspect of single molecule fluorescence measurements of freely diffusing molecules to study the structural heterogeneity of solutions of DNA molecules. They did this by analysing the FRET efficiency of each molecule in a ratiometric manner, which has the advantage of minimizing the effect of the inherent intensity fluctuations in single molecule experiments.

To study the Förster transfer process between donor and acceptor dyes [7], labelled double-stranded DNA (dsDNA) of varying lengths were used. A 40 base pair single-stranded DNA was labelled at the 5′ end with tetramethylrhodamine (TMR) and its complimentary strand was labelled with Cy5 at various positions along its length (7, 12, 14, 19, 24, 27 base pairs) using an C6 amino modified thymine (C6 dT). This dye pair has a Förster distance (R_0) of about 53 Å [7] and the range of possible inter-dye distances provided by these dsDNA molecules, which are assumed quite rigid at this length, is 35–100 Å. Since the overall length of all the DNA constructs is the same (40 bp), there are no additional effects of changes in the diffusion coefficient to be taken into account when comparing the behaviour of the different systems. The FRET labelled dsDNA samples were then made by mixing together the two complimentary strands in a 1 : 1.5 ratio. The single pair FRET measurements were made using a confocal microscope (Zeiss Axiovert SV 100) equipped with two avalanche photodiode detectors (EG&G SPCM AQ-141). The sample concentration was 30 pM which ensured that there was on average 0.01 molecules in the focal volume.

When a FRET labelled DNA molecule diffuses through the focal volume a burst of photons is detected by the two detectors. A thresholding criterion was applied to this data as is common in this type of experiment in order to reject the background signal with care taken to avoid biasing the accepted data. Deniz chose to use a simple criterion which required the sum of the signals in the two detector channels to be greater than a certain threshold value $I_A + I_D > T$. In this experiment a threshold value of around 20 was chosen. The FRET efficiency E of each accepted event was then calculated according to $E = I_A/(I_A + \gamma I_D)$ where γ is a factor which takes into account the ratio of the quantum efficiencies of the two dyes and the efficiency of the two detector channels. In fact γ was approximated to unity in this experiment. The FRET efficiency is calculated for each photon burst and plotted as a histogram, as shown in Figure 5.4. Figure 5.4(a) shows three representative FRET histograms for oligonucleotides with 7, 12, and 19 base pairs separation between the two dyes. The so-called zero peak in each of these histograms is suggested to arise due to photobleaching of acceptors leaving donor-only labelled DNA which yields an observed FRET efficiency near to zero (see Chapter 2 for a discussion of the 'zero' peak). The second peak in the histogram, which indicates the FRET efficiency between the donor and acceptor fluorophores of the labelled dsDNA, clearly shifts to a lower mean value as the distance between the fluorophores increases. This would be expected for the Förster transfer process in which the efficiency has a strong distance dependence (see

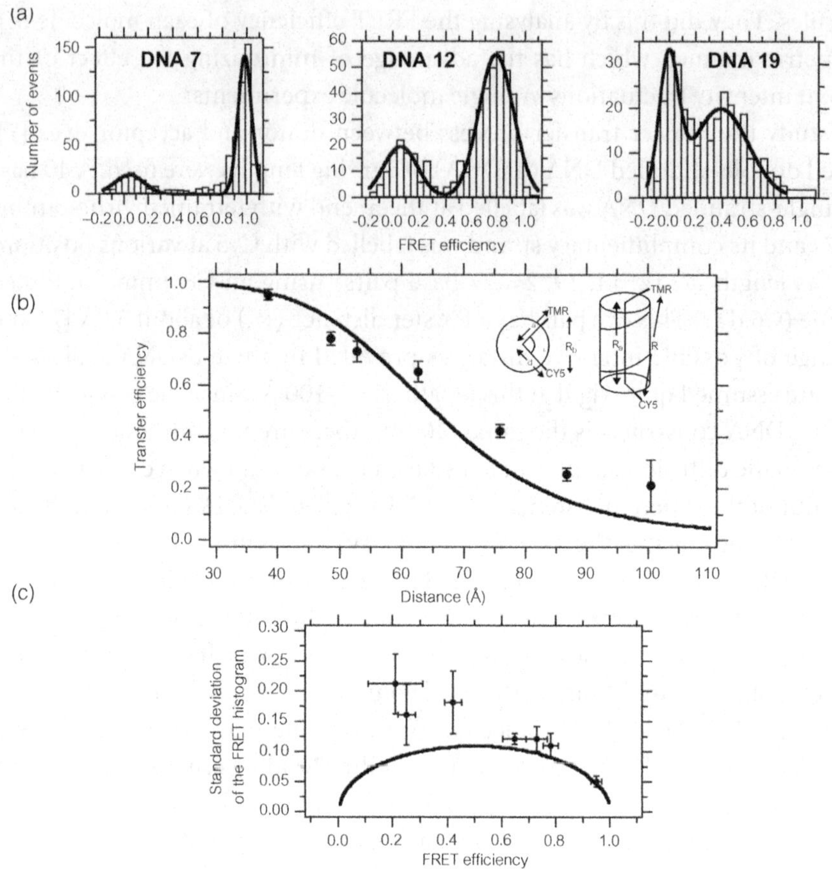

Figure 5.4 (a) FRET histograms extracted from time traces for DNA 7, 12, and 19 using a threshold of 20. Double Gaussian fits extract numbers for the mean (and width) of the higher efficiency peak of 0.95 (0.05), 0.75 (0.13), and 0.38 (0.21) respectively. (b) Mean FRET efficiencies extracted from FRET histograms plotted as a function of distance for the seven DNA constructs, DNA 7, 12, 14, 16, 19, 24, and 27. The error bars represent two SD ($\pm 1\sigma$) from multiple measurements and increase with distance. The solid line is the theoretical curve with $R_0 = 65$ Å for comparison. (c) Mean widths extracted from the histograms are plotted as a function of the mean FRET efficiencies. The x-axis error bars are the same as in (b). The y-axis error bars represent two SD ($\pm 1\sigma$) from multiple measurements. The solid line shows widths calculated by using a simple model for the effect of shot noise. Reprinted with permission from Deniz *et al.*, Single-pair fluorescence resonance energy transfer on freely diffusing molecules: observation of Förster distance dependence and subpopulations. *Proceedings of the National Academy of Sciences of the United States of America* 96 (1999) 3670–3675. Copyright 1999 National Academy of Sciences, USA.

Chapter 2). Systems with dye separation greater than 19 base pairs resulted in a FRET efficiency that could not be clearly resolved from the zero peak. The peaks in Figure 5.4 were fitted using Gaussian functions to extract the mean and the width of the distributions. Figure 5.4(b) shows a plot of mean FRET efficiency as a function of dye separation and the solid line represents the prediction of Förster theory: $E = 1/(1+ (R/R_0)^6)$. In this plot a value of R_0 of 65 Å had to be used to achieve close agreement with the measured single molecule data, rather than 53 Å which is determined from ensemble spectroscopic data. This discrepancy is attributed to the assumption made that the dyes are completely free to rotate (in both the system used to determine the R_0 value for this dye pair and the experimental system used by Deniz *et al.*) and therefore there is no bias towards a fixed angle between the emission dipole moment of the donor and the absorption dipole moment of the acceptor. As well as their position, the width of each peak in a FRET efficiency histogram can also be analysed. Distributions that show significant broadening may indicate the existence of an equilibrium ensemble of similar states or conformational dynamics between states during the time the molecule takes to diffuse through the laser spot. However, in this particular experiment the dsDNA molecules are quite rigid with no folding taking place and so the width of the peaks arises mainly from the presence of shot noise, which places a limit on the separation resolution of this approach. Figure 5.4(c) shows the mean widths of the distributions observed in these systems as a function of the FRET efficiency along with the result of a simple model which estimates the shot noise limit (solid line). This reveals that the width of the distributions strongly depends on FRET efficiency and the widths at low FRET efficiency are considerably greater than this theoretical minimum width despite the absence of any conformational dynamics. In this case the excess width is suggested to be partly due to DNA-dye interactions (which causes the assumption that the orientational factor κ averages to a value of 2/3 to break down). Distance fluctuations due to DNA conformational changes or movement of the dyes on the linkers are expected to be on a faster timescale than the ms time resolution of the measurement and are therefore averaged out, although it was found that increasing the integration time reduced the widths of the peaks. Thus there is a trade-off between time resolution and the ability to resolve sub-populations in a heterogeneous mixture.

The ability of the technique to resolve subpopulations in a mixture of conformers was demonstrated by mixing, in equal quantity, two of the oligonucleotides with 7 and 17 base pair separation between the dyes. The FRET histograms in Figure 5.5(a) clearly show that these two subpopulations in the mixture can be resolved. The DNA with the greater distance between the dyes also had a *Eco*RI endonuclease restriction site placed between the dyes. On addition of *Eco*RI the longer DNA is cleaved with a resulting loss in the amplitude of the lower FRET efficiency peak (Figure 5.5(b)). This cleavage also results in an increase in the size of the 'zero' peak in the FRET histogram which arises from the excitation of the population of donor fluorophores which are now conjugated to a short oligonucleotide that has no acceptor attached.

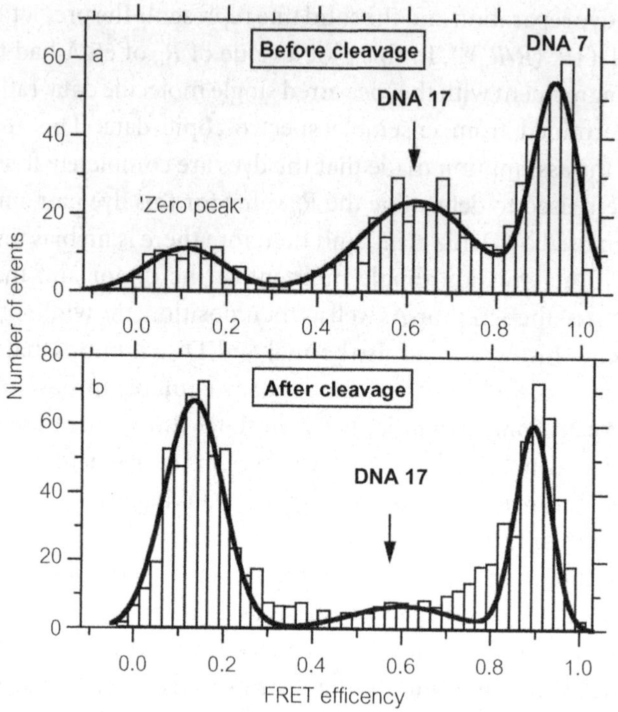

Figure 5.5 Restriction endonuclease cleavage of DNA. Histograms of a mixture of DNA 7 (no *Eco*RI site) and DNA 17 (with *Eco*RI site) before (a) and after (b) the cleavage reaction. The FRET peak corresponding to DNA 17 virtually disappears after the cleavage reaction, and there is a simultaneous increase in the 'zero peak'. Reprinted with permission from Deniz *et al.*, Single-pair fluorescence resonance energy transfer on freely diffusing molecules: observation of Förster distance dependence and subpopulations. *Proceedings of the National Academy of Sciences of the United States of America* 96 (1998) 3670–3675. Copyright 1999 National Academy of Sciences, USA.

This work shows nicely that single molecule measurements on freely diffusing FRET labelled molecules can be used to provide information similar to that which an ensemble experiment would provide, but in addition allows subpopulations in a complex system to be resolved. Improvement in fluorescence probes to reduce photobleaching and conformational flexibility through the use of short, rigid linkers will be necessary to get the best results from this approach.

5.2.3 **Studies of protein folding with single molecule sensitivity**

The conformational transformation from the unfolded to the native state of a protein is highly likely to be a heterogeneous process in which many different pathways (i.e. sequences of intermediate conformations) can be taken.

Understanding the kinetic search across a complex free energy landscape is a major challenge that is ideally addressed by single molecule studies in which the inherent heterogeneity can potentially be unravelled.

Schuler and colleagues [8, 9] have reported some of the most advanced studies in this area. In one study they labelled the termini of the cold shock protein from the hyperthermophilic bacterium *Thermotoga Maritima* (CspTm) with a FRET dye pair and studied the system in free diffusion under varying concentrations of chemical denaturant. When folded in the native state, the donor and acceptor dyes are about 1 nm apart resulting in a high FRET efficiency, whilst under denaturing conditions the average displacement of the dyes is much greater and the FRET efficiency falls. In order to eliminate other effects of the chemical denaturant two control samples were also studied. These controls were polyproline rods of six and twenty amino acids in length, $(Pro)_6$ and $(Pro)_{20}$, each labelled at their termini with the same FRET pair as CspTm.

The diffusion single pair FRET data were acquired using a confocal fluorescence microscope with a 1.4 NA x100 objective (Nikon CFN Plan Apo 85025). The average background counts in a 1 ms integration time were subtracted from the data and a sum threshold criterion of 25 counts was used to select significant photon bursts. The FRET efficiency was calculated as described in Chapter 2. Figure 5.6(a) and (b) show the single pair FRET data for the two controls. The shorter polyproline rod exhibits much higher counts in the acceptor channel as would be expected and this is reflected in the histograms of FRET efficiency shown in Figure 5.6(c) and (d). The additional peak near zero FRET efficiency is attributed to $(Pro)_{20}$ molecules in which the acceptor dye has been photochemically altered or in which acceptor labelling did not occur. This peak is also seen for the shorter polyproline rod but the different scales in Figure 5.6(c) and (d) make it difficult to resolve. The addition of chemical denaturant, in this case guanidinium hydrochloride (GdnHCl), has the effect of shifting the peaks to slightly lower apparent mean FRET efficiencies (there is a dependence on the R_0 value for a particular dye pair with solvent refractive index, see Chapter 2). The change is however modest and can easily be corrected for (as in Figure 5.7(a)). In the case of the protein, the FRET efficiency histogram is much more strongly dependent on denaturant. Since CspTm shows ensemble two-state unfolding behaviour then it would be expected, at some denaturant concentrations, to resolve two peaks corresponding to the compact native state and slightly expanded unfolded state. Further, the relative areas under the high and low FRET efficiency peaks in the histogram should follow the relative populations of the native and unfolded states described by the ensemble equilibrium denaturation curve. This behaviour is indeed observed in the histograms for CspTM as a function of denaturant as is shown in Figure 5.6(e). Several additional interesting observations are made in relation to the mean and width of the folded and unfolded FRET peaks in the data in Figure 5.6(e). The mean FRET efficiency of the unfolded state shows a strong dependency on the denaturant concentration between 0 and 3 M GdnCl (Figure 5.7(b)). Since the polyproline controls show no such dependency it is

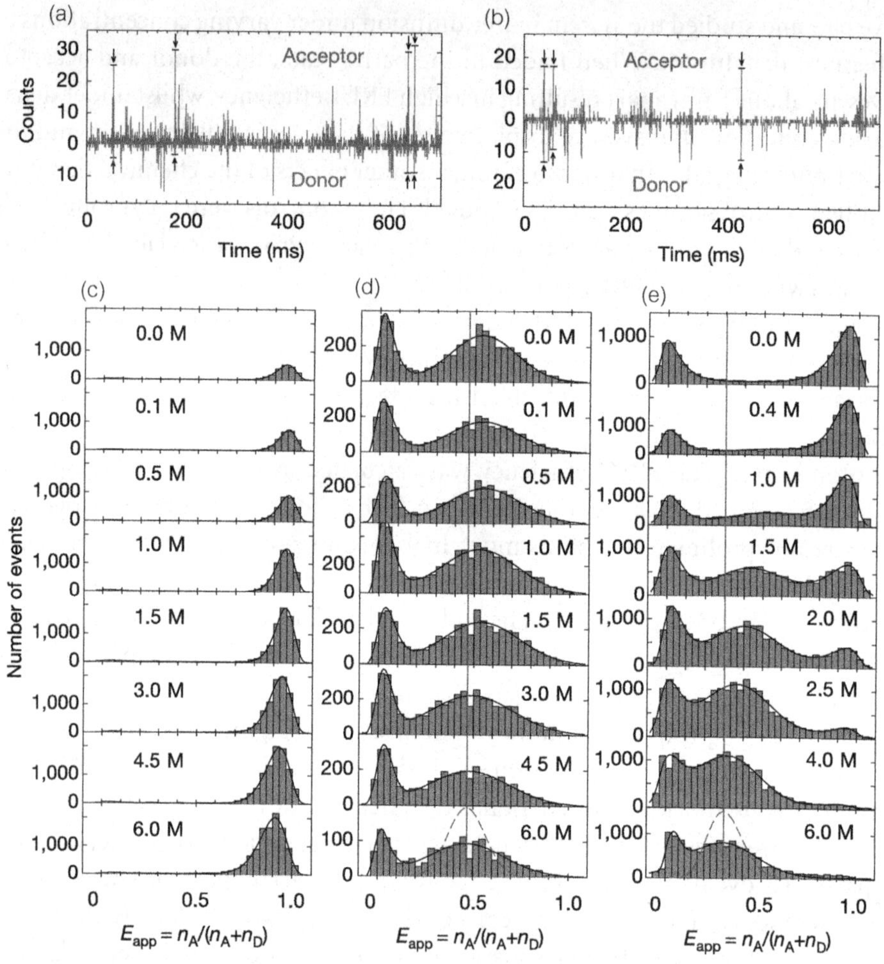

Figure 5.6 FRET trajectories and histograms. (a, b) Donor and acceptor channel time traces using 1-ms bins for labelled $(Pro)_6$ (a) and $(Pro)_{20}$ (b). Arrows indicate photon bursts for which the sum of the counts in the two channels is greater than 25. (c–e) Histograms of measured FRET efficiencies (E_{app}) at various GdnHCl concentrations for labelled $(Pro)_6$ (c), $(Pro)_{20}$ (d), and CspTm (e). The solid curves are the best fits to the data using lognormal and/or Gaussian functions. The dashed curves were calculated from the β-distribution, $P(E_{app}) = E_{app}^{\langle n_A \rangle}(1-E_{app})^{\langle n_A \rangle}$, where $\langle n_D \rangle$ and $\langle n_D \rangle$ are the average number of detected acceptor and donor photons in the significant bursts. A colour version of this figure may be found in the authors original publication [8]. Reproduced from Schuler, *et al.*, Probing the free-energy surface for protein folding with single-molecule fluorescence spectroscopy. *Nature* 419 (2002) 743–747 with permission from Nature Publishing Group (Copyright 2002).

concluded that the denatured state must be collapsing to a more compact form at low denaturant concentration. Interestingly, theoretical models predict this, but the same theoretical models suggest that there should be a continuous expansion of the unfolded state above 3 M GdmCl, which is not observed in the data in Figure 5.7(b). Further interesting information can be gleaned about the dynamics of the unfolded polypeptide by consideration of the widths of the distribution. If the motion of the polypeptide chain of the unfolded protein was very slow compared with the transit time through the laser beam then each protein measured in

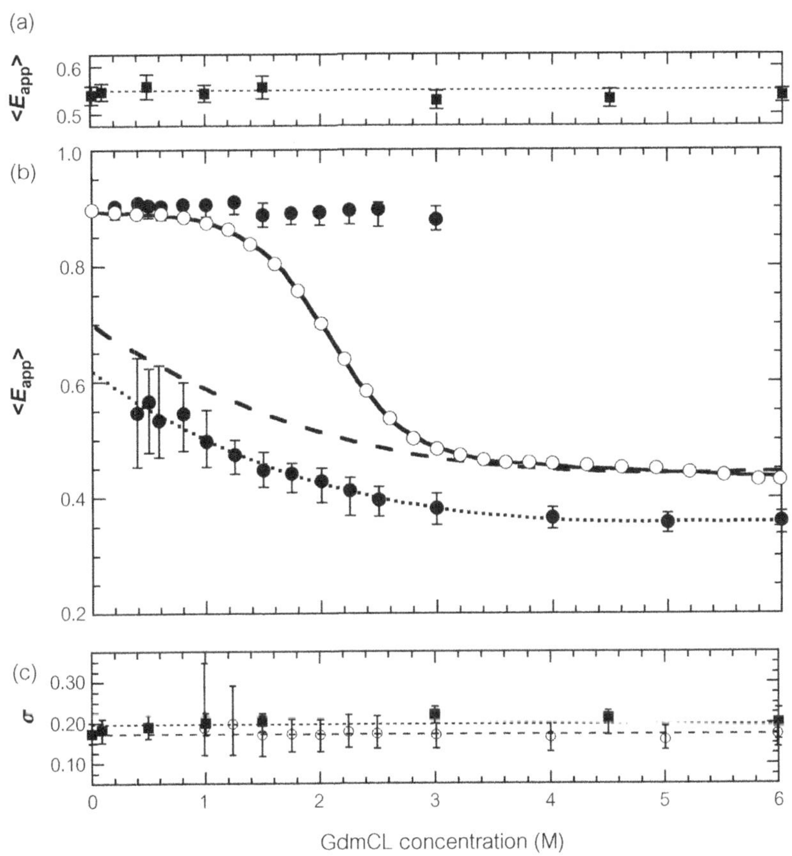

Figure 5.7 Dependence of the means and widths of the measured FRET efficiency (E_{app}) on the concentration of GdnHCl. (a) $<E_{app}>$ for $(Pro)_{20}$. (b) Single molecule mean values (filled circles), ensemble FRET efficiencies (open circles), and associated two-state fit (unbroken curve) for CspTm. The dotted curve is a third-order polynomial fit to the unfolded protein data that was matched (dashed curve) to the ensemble data between 4 and 6 M GdnHCl. (c) Standard deviations (σ) taken from the gaussian fits to the $(Pro)_{20}$ data (squares) and unfolded CspTm data (circles) in Figure 5.6. Error bars indicate uncertainty in the fits. Reproduced from Schuler *et al.*, Probing the free-energy surface for protein folding with single-molecule fluorescence spectroscopy. *Nature* 419 (2002) 743–747 with permission from Nature Publishing Group (Copyright 2002).

the experiment would exhibit a different unfolded FRET efficiency resulting in a broad measured FRET efficiency distribution. However, if the conformational changes of the unfolded state were rapid such that the molecule could explore all of its possible conformational space during its passage through the laser beam then the same FRET efficiency would be observed for all molecules and a more narrow peak would be measured. The width of the unfolded FRET peak in the histogram for CspTm is no greater than for the $(Pro)_{20}$ control which is thought to have a relatively narrow range of end-to-end distances. Thus the range of transfer efficiencies for the unfolded protein does not appear to be affected by the large range of possible end-to-end distances. The authors suggest therefore that the protein must be reconfiguring rapidly with respect to the integration time used to collect the data (1 ms). Using the relationship between the autocorrelation of the FRET efficiency and the transit time, the authors place an upper limit on the polypeptide reconfiguration time (i.e. the pre-exponential in the Kramers description of barrier crossing [2]) of 0.2 ms (note that the figure in the main text of the report is incorrect and is modified in an *erratum* [9]). This number is very important since a knowledge of its magnitude is essential if absolute activation energies are to be calculated [2]. A reconfiguration time of 0.2 ms places a lower bound of 2 $k_B T$ on the activation energy for folding. This is a rather small value that implies that the folding of this protein is close to down hill, which does not agree with the body of data that indicate that it is clearly a two-state system. However, this value is only a lower bound and an upper bound can be estimated from Gaussian chain theory, which the authors estimate to be 11 $k_B T$.

The authors extended this work by combining single molecule FRET measurements with a laminar flow mixing device [10]. The mixing device causes an abrupt change in denaturant concentration as two solutions, one containing the protein in denaturant and the other a buffer, mix. This approach allows proteins to be observed under conditions far from equilibrium. Furthermore, measurement of FRET efficiency between the donor and acceptor at increasing distance from the point of mixing allows different points (times) along the folding trajectory to be measured. The refolding of CspTm in ~0.5 M GdnHCl (after mixing) was studied. The data showed that the protein folded in a two-state fashion with a shift in the population from unfolded to folded along the mixer channel.

A sophisticated experimental setup capable of simultaneous measurement of intensity, lifetime and anisotropy of both the donor and acceptor dyes has been reported by Margittai *et al.* [11]. The advantage of this demanding approach which they have termed single molecule multiparameter fluorescence detection (smMFD) is that it overcomes the problems associated with simpler FRET schemes such as the bleaching of donor or acceptor (see also the ALEX technique Chapter 2 and [12,13]) and assumptions concerning the quantum yields and changes in the relative orientation of the two dyes. They studied the conformational dynamics of syntaxin 1: a soluble N-ethlymaleimide-sensitive factor attachment protein receptor or SNARE. Such proteins are thought to play an important role in mediating membrane fusion and have, what is referred to as, open and closed conformations, in which complexes may be formed or are blocked, respectively. The group studied the conformational heterogeneity of syntaxin 1 in the presence and absence of munc-18, a regulatory protein that arrests syntaxin 1 in its closed conformation, and also syntaxin 1 as part of a ternary SNARE complex with SNAP-25 and synaptobrevin. The crystal structure of syntaxin and schematics of these complexes are shown in Figure 5.8(a). Fifteen double cysteine mutants were prepared and labelled randomly with donor acceptor (Alexa 488 and Alexa 594). This results in all four possible combinations of protein modification at the two cysteines (i.e. DA, AD, AA, and DD) but the technique allows the homogeneously labelled molecules to be discarded in the data sets. Freely diffusing molecules were detected in a confocal microscope system with pulsed laser excitation at 477 nm. The data selection criteria were:(1) accepted photon bursts must contain more than 160 photons, and (2) mean interphoton time should be $< 42.1\ \mu s$.

The closed conformation of syntaxin when complexed with munc-18 was characterized by measurement of the lifetime of the donor with the acceptor present τ_{DA} and the inter-dye distance R_{DA} (see Figure 5.8(c)). The tight clustering of R_{DA}, around a single value with a distribution no greater than the shot noise limit for all the mutants with different dye positions suggests a highly homogeneous population of molecules. The value of R_{DA} indicates that this population is dominated by the closed conformation as expected. In order to obtain accurate values for R_{DA} the rotational freedom of the dyes (i.e. the anisotropy in Figure 5.8) and local quenching must be fully characterized. These parameters are usually given assumed values. These authors use the extra data that is obtained from the smMFD to check these assumptions. They calculate a theoretical relationship between R_{DA} and τ_{DA} which is shown by the line in Figures 5.8(c–e) upon which the data should fall. They also calculate a theoretical relationship between the anisotropy r_D and lifetime τ_{DA} in the case of the dyes freely rotating upon which the data in the lower plots in Figure 5.8 should fall for maximum confidence.

In contrast to the closed conformation, free syntaxin yielded a distribution in R_{DA} much broader than the shot noise limit, which was fitted by two overlapping peaks which represent different conformations. The maximum of one of these peaks coincides with that observed for the closed conformation formed in the presence of munc-18 (see Figure 5.8(d)). The authors concluded that a subpopulation of 15–30% of free syntaxin was in the closed conformation whilst most molecules adopt a conformation with a much larger mean R_{DA} commensurate

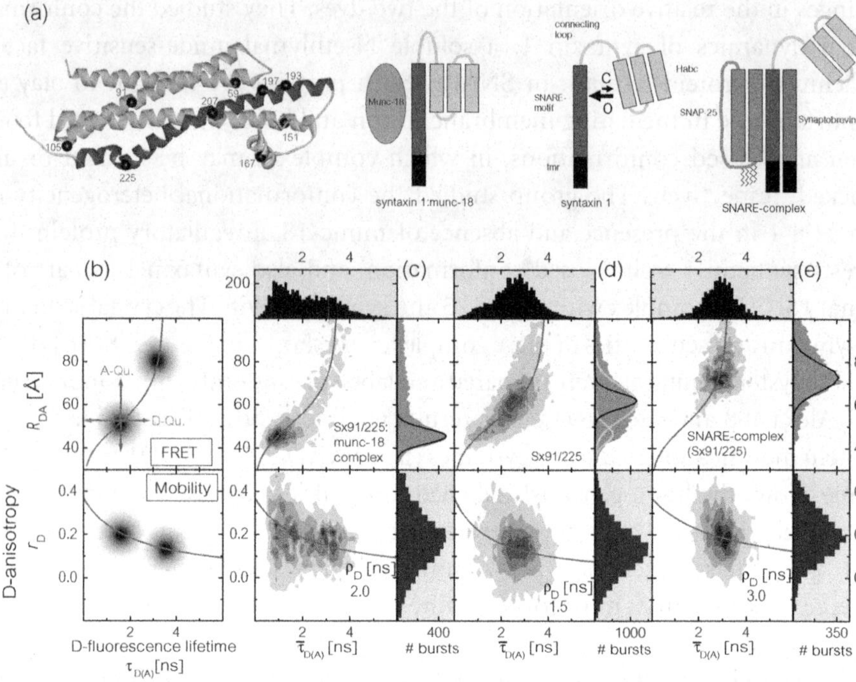

Figure 5.8 ((a), left-hand structure) Crystal structure of the cytosolic part of syntaxin in complex with munc-18. Munc-18 and the transmembrane region are not shown. Habc (light grey) domain and linker region; SNARE motif (H3 domain, dark grey). Black spheres indicate residues (numbers) substituted by cysteines for dye labelling. (b) The combination of different observables has advantages in precision and accuracy allowing confident measurement of accurate absolute interdye distances, R_{DA}. The two main sources of errors in distance calculations are individual local quenching effects (Upper panel in (b)) and restricted mobility of the reporter dyes (Lower panel in (b)), can be excluded by SmMFD. First, in $\overline{\tau}_{D(A)}$-R_{DA} plots ((b), upper panel), local quenching of (d) and (a), respectively, is distinguished from distance-dependent FRET effects (measurements should lie on the calculated sigmoid curve). Second, two-dimensional plots of $\overline{\tau}_{D(A)}$. versus the donor anisotropy r_D ((b), lower panel) allows one to analyse the dye motilities (points should be distributed around the overlaid solid curve). (c–e) SmMFD analysis of burst-averaged observables (number of bursts increases from white to black) for different FRET experiments with associated SNARE protein's (cartoons above). Histograms were obtained for Sx91/255 in complex with munc-18 (c), alone (d), and as part of the ternary SNARE complex (e). Also shown are projections yielding one-dimensional distributions of the parameters. A colour version of this figure with extended figure caption may be found in the authors' original publication [11]. Reprinted with permission from Margittai *et al.*, Single-molecule fluorescence resonance energy transfer reveals a dynamic equilibrium between closed and open conformations of syntaxin 1. *Proceedings of the National Academy of Sciences of the United States of America* 100 15516–15521. Copyright 2003 National Academy of Sciences, USA.

with an open structure. In the case where the ternary complex with SNAP-25 was studied a single peak was observed regardless of which pair of cysteines was labelled with the dyes, reflecting the fact that in this complexed form the syntaxin is unable to form a closed conformation.

The data discussed so far indicate that free syntaxin exists in at least two conformations one of which is similar to the closed conformation when complexed with munc-18. This could reflect a static distribution between states or structural dynamics where individual molecules are able to interconvert in a dynamic equilibrium. The real-time data (Figure 5.9(a)) show anticorrelated red and green signals, which suggests that there are sub-millisecond fluctuations in the value of R_{DA}. By reducing the length of the time windows over which these data are averaged fluctuations between these states were observed, confirming the dynamic nature of the system. Figure 5.9(b) gives an example for one of the double mutants studied referred to as Sx91/225. When averaging over a window as short as 0.5 ms two subpopulations are observed. The distribution with the smaller mean R_{DA} has a width similar to that expected from the shot noise indicating a well-defined structure whereas the larger R_{DA} distribution is much broader than the shot noise limit suggesting that the open state is structurally much more heterogeneous. Further improvements in the level of kinetic analysis can be obtained by calculating the autocorrelation function for each fluorophore and the cross-correlation function of the real-time data for the FRET labelled mutants. The autocorrelation function shows a response typical of the translational diffusion and photochemistry of dyes in an FCS experiment (Figure 5.9(c)). In mutants where there is little conformational dynamics (i.e. those with narrow R_{DA} peaks such as Sx59/105 in Figure 5.9(c)) the auto- and cross-correlation curves are almost identical on the millisecond timescale since the cross-correlation in this case also only reflects the kinetics of diffusion and photochemistry. By contrast, in those mutants in which conformational dynamics are observed, the two correlation functions differ significantly (Figure 5.9(d)). Analysis of these data indicate a relaxation time for the intramolecular motion of ~0.8 ms. This very elegant and detailed single molecule fluctuation spectroscopy study explains why regulatory proteins are needed to maintain the protein in one state because of the relatively rapid interconversion between the open and the inactive closed state of free syntaxin.

5.2.4 Single molecule fluctuation spectroscopy as a high throughput screening tool

High throughput screening (HTS), the process of measuring many biological interactions in a short period of time, is a key tool in drug discovery and has enormous potential for clinical diagnostics and personalized medicine [14]. Typically, a large pharmaceutical company might have a library of 100,000 compounds, which could have therapeutic potential, and ideally one would like to

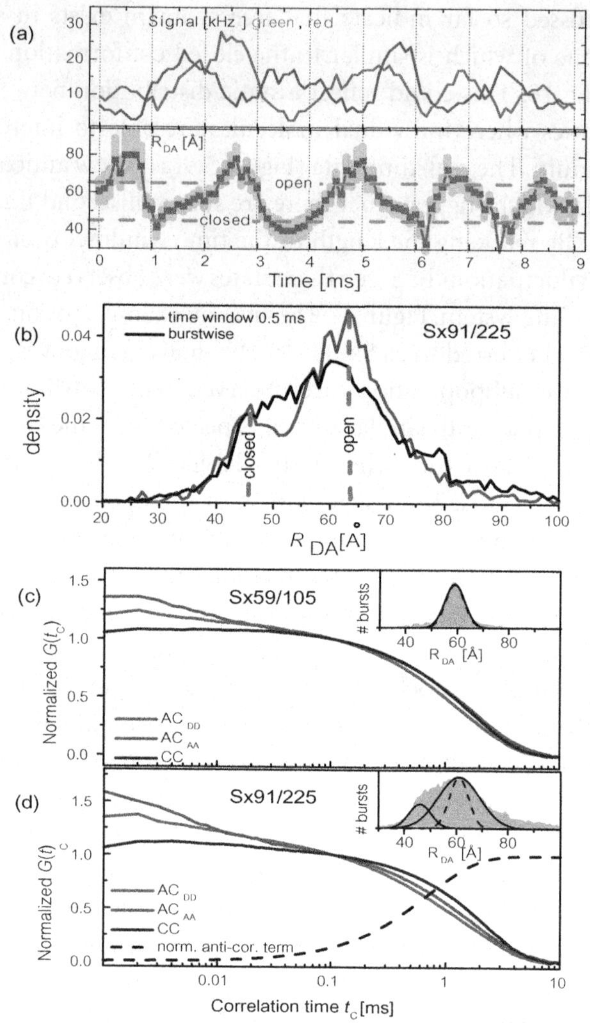

Figure 5.9 (a) Single burst of a Sx91/255 molecule with a passage time of 9 ms. The smoothed (sliding window 0.5 ms) trajectories of the green (donor) and red (acceptor) signal fluctuate in an anticorrelated manner (Upper). The calculated distance trajectory (error bars in light grey) shows the switching of syntaxin between the closed and open state. The dashed lines are taken from the time-window analysis in (b). (b) By analyzing many R_{DA} trajectories, two distance populations of Sx91/255 can be seen by time-window analysis at 0.5 ms. For comparison, the result of the burstwise analysis is also given. Two visible populations merge into one with decreasing time resolution that reflects the averaging of R_{DA}. (c and d) Autocorrelation and cross-correlation curves for two mutants Sx59/105 (c) and Sx91/255 (d) with shot-noise-limited and broad R_{DA} peaks, respectively. In the millisecond time range, the major differences between the autocorrelated (grey) and cross-correlated curves (black) for Sx91/255 are indicative of FRET dynamics. A colour version of this figure may be found in the authors' original publication [11]. Reprinted with permission from Margittai *et al.*, Single-molecule fluorescence resonance energy transfer reveals a dynamic equilibrium between closed and open conformations of syntaxin 1. *Proceedings of the National Academy of Sciences of the United States of America* 100 (2003): 15516–15521. Copyright 2003 National Academy of Sciences, USA.

test each of these molecules against one or more targets. It becomes obvious very quickly that if all of these compounds are to be tested against even a single target, let alone hundreds or thousands of targets, very short measurement times and small volumes are required to prevent the screening process becoming impractically long and expensive. Not only must large numbers of interactions be characterized in a short space of time but this must also be done quantitatively so that binding constants and other comparators (such as the IC50 which is the concentration at which half the targets are inhibited) can be determined. Typically such screens are carried out using a two-dimensional array on a surface, so-called high throughput screening chip technology. These assays are very widely used but there are issues with accurate quantitation and speed since often a number of washing steps are required.

Schaertl and co-workers [15] at EVOTEC Biosystems have developed a simple assay based on fluorescence intensity distribution analysis (FIDA), which is a form of photon counting histogram (PCH) analysis (see Chapter 2). In this analysis carrier particles, which may either be nanoscale (50–500 nm) colloidal particles or bacterial cells, have the target molecules immobilized on their surface. Two assay forms have been demonstrated—a competition assay and a sandwich assay (see Figure 5.10). In the competition assay, a fluorescently labelled analyte is pre-bound to the target on the carrier particle and displaced by the addition of the unlabelled analyte, which is to be detected. In the sandwich assay a fluorescently labelled antibody for the analyte is synthesized which will bind to analyte that is bound to the target rendering it detectable. In both cases it is the change in fluorescence intensity of the carrier particles, or overall 'molecular brightness', that is detected and analysed to provide a quantitative measure of binding.

In this report three targets are used to demonstrate the novel nanoparticle immunoassay system (NPIA). We will discuss two of the systems studied, one to demonstrate the principles of single molecule competition and the other describes sandwich assays. Estradiol plays a key role in the human menstrual cycle and its concentration in blood serum is indicative of normal function, pregnancy and some pathologies. Rabbit anti-estradiol was used as a target and a tetramethylrhodamine labelled estradiol compound was synthesized for a competition assay. Human chronic gonadotropin (hCG) was used in a sandwich assay with a fluorescently labelled anti-β-hCG antibody. The hCG stimulates the release of pregnancy sustaining steroids and again the relevance of the assay is that hCG concentration indicates certain pathologies.

The single molecule fluorescence experiments were carried out using a Confocor confocal microscope (Carl Zeiss, Germany). Nanocarriers (at 0.18 nM) labelled with the anti-estradiol antibody are loaded with fluorescently labelled

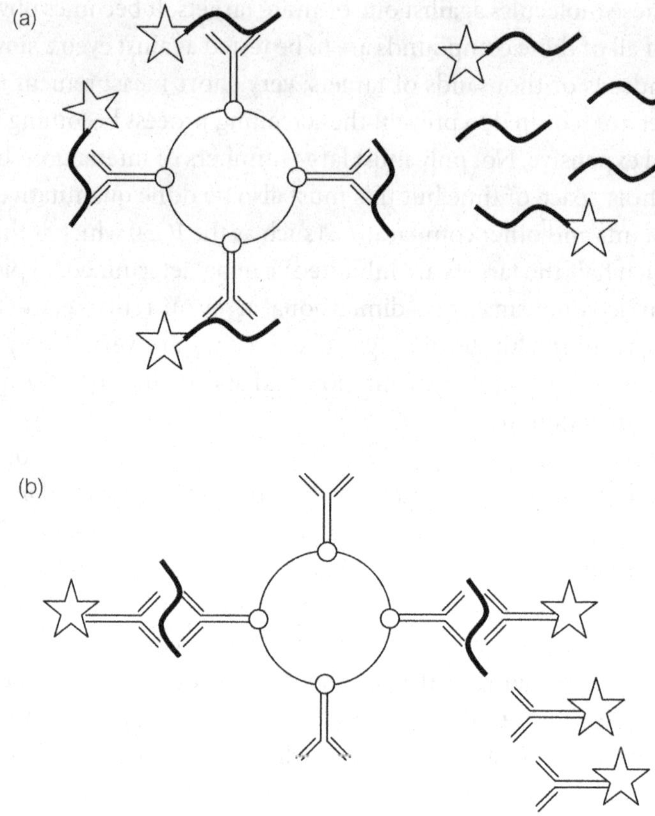

Figure 5.10 Schematic drawing of NPIA principle. (a) Competition assay (example: estradiol, theophylline). (b) ELISA-like assay (example: hCG). Big circles, bacterium or artificial nanoparticle: small circles, linker (protein A, streptavidin, secondary antibody, or covalent direct link); two-tailed double line, antibody: wavy black lines, analyte (protein or peptide); stars, fluorescent label. Reprinted with permission from Schaertl *et al.*, A novel and robust homogeneous fluorescence-based assay using nano particles for pharmaceutical screening and diagnostics. *Journal of Biomolecular Screening* 5 (2000) 227–237. Copyright 2000 Sage Publications, Inc.

estradiol by incubation with a 5 nM solution. Unlabelled estradiol competes with the bound labelled estradiol reducing the brightness of the nanocarriers (q_2) but not their concentration (c_2) and increases the concentration of free fluorescently labelled estradiol molecules (c_1) but not altering their brightness (q_1). An important aspect of the fitting procedure is that the concentration of nanocarriers c_2 and the brightness of the free molecules q_1 are assumed to be constant leaving only two free parameters thereby increasing the quantitative reliability of the results. Figure 5.11 shows the photon counting histograms in the estradiol competition assay in the presence and absence of 1 μm competitor. In the absence of competitor (Figure 11(b)) there are more occurrences of high numbers of fluorescence

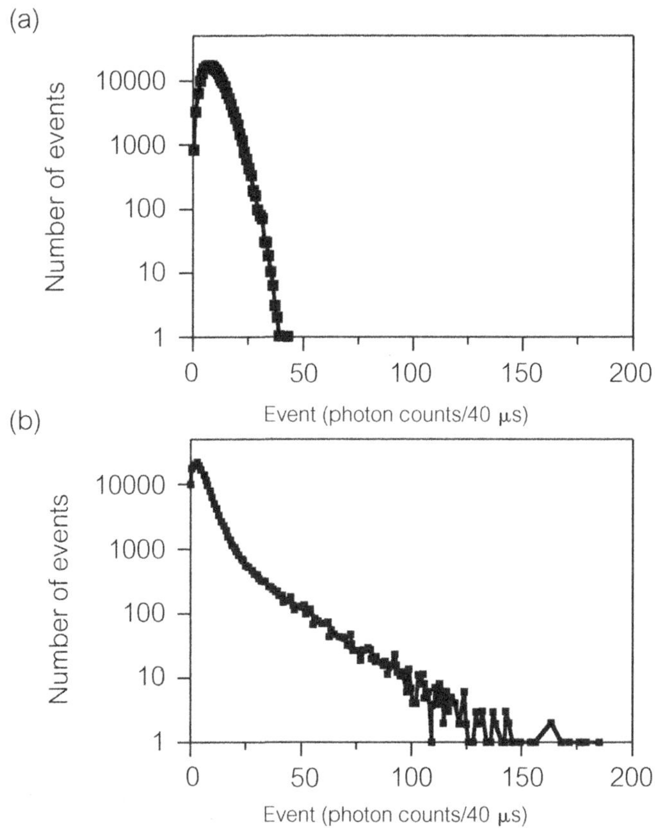

Figure 5.11 Photon Count histogram of 5 nM unbound conjugate (estradiol-Tamra) (a) in the presence of 1 μM competitor (estradiol) or (b) in the absence of competitor at full complex formation with nanoparticles. Total number of photons counted within the measurement time was 1,848,127 (a) and 1,504,390 (b). Reprinted with permission from Schaertl *et al.*, A novel and robust homogeneous florescence-based assay using nano particles from pharmaceutical screening and diagnostics. *Journal of Biomolecular Screening* 5 (2000) 227–237. Copyright 2000 Sage Publications, Inc.

photons in each 40 μs time bin because the nanocarriers are fully loaded with fluorescent estradiol. The displacement of these molecules when unlabelled estradiol is added is reflected in the data in Figure 5.12(a) and (b) which shows the variation in total 'intensity' $I_x = c_x \times q_x$ of the carrier (I_2, antibody coated nanoparticles in this case) and the fluorescent estradiol (I_1) as the unlabelled competitor is titrated into the assay. A further reduced parameter $I_3 = I_2/(I_1 + I_2)$ can also be calculated and a fit to any of these intensity titration results provides the IC50 parameter.

The sandwich assay for hCG binding to its antibody was carried out by attaching the antibody to the nanocarrier through a biotin-streptavidin linkage. The

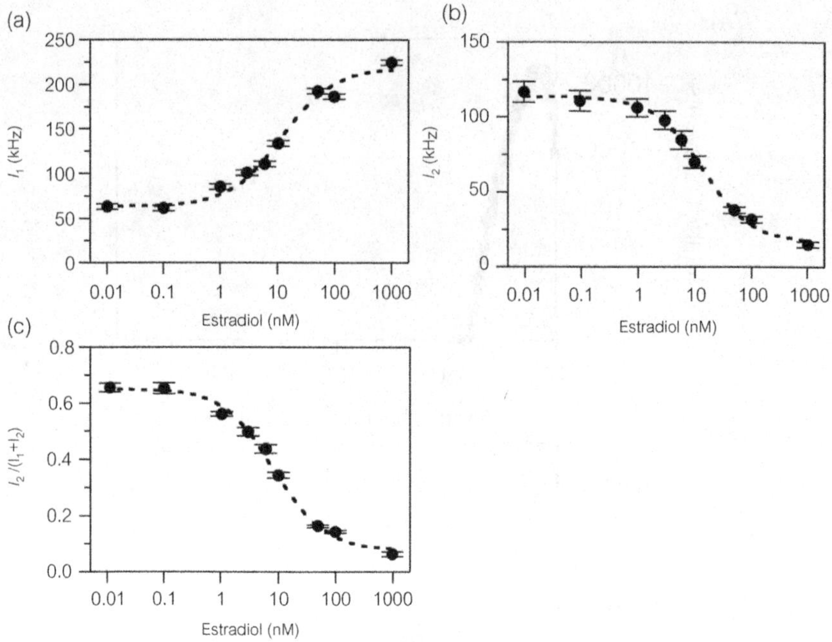

Figure 5.12 NPIA competition assay with antibody-coated nanoparticles (0.18 nM) binding fluorescently labelled estradiol (5 nM). As a linker system, protein A was coupled covalently onto the beads. Competitor (unlabelled estradiol) was added at concentrations indicated in the graph. Fixed values were $q_1 = 42$ kHz and $c_2 = 0.18$ particles per confocal volume. The intensities were calculated according to $I_x = q_x.c_x$. The **dotted lines** show hyperbolic fits to determine IC_{50} values, which were (a) 11.9 ± 2.4 nM; (b) 13.3 ± 1.3 nM; and (c) 8.8 ± 1.1 nM. Samples were measured 10 times for 2 s each in experimental buffer at room temperature. Reprinted with permission from Schaertl *et al.*, A novel and robust homogenous fluorescence-based assay using nano particles for pharmaceutical screening and diagnostics. *Journal of Biomolecular Screening* 5 (2000) 227–237. Copyright 2000 Sage Publications, Inc.

antibody on the nanocarrier recognizes the β subunit of hCG whilst an Alexa546 labelled anti-α-hCG antibody binds to the other subunit of the hormone. Figure 5.13 shows the calculated value of I_2 (corresponding to the nanocarrier intensity) from the FIDA experiments as a function of different target and detection antibody concentrations as hCG is titrated into the assay. Clearly, a range of antibody concentrations provides comparable results but at high analyte concentrations the bound and free antibodies are saturated with hCG preventing the sandwich being formed. This overtitration is more of an issue at lower nanocarrier concentrations (Figure 13(c)) as might be expected.

This type of nanoparticle carrier assay is suitable for protein–protein, receptor–ligand, DNA–DNA and DNA–protein interactions, which makes it a versatile tool for drug discovery, cross reactivity screening, clinical diagnostics and many other applications. Schaertl and co-workers have also demonstrated

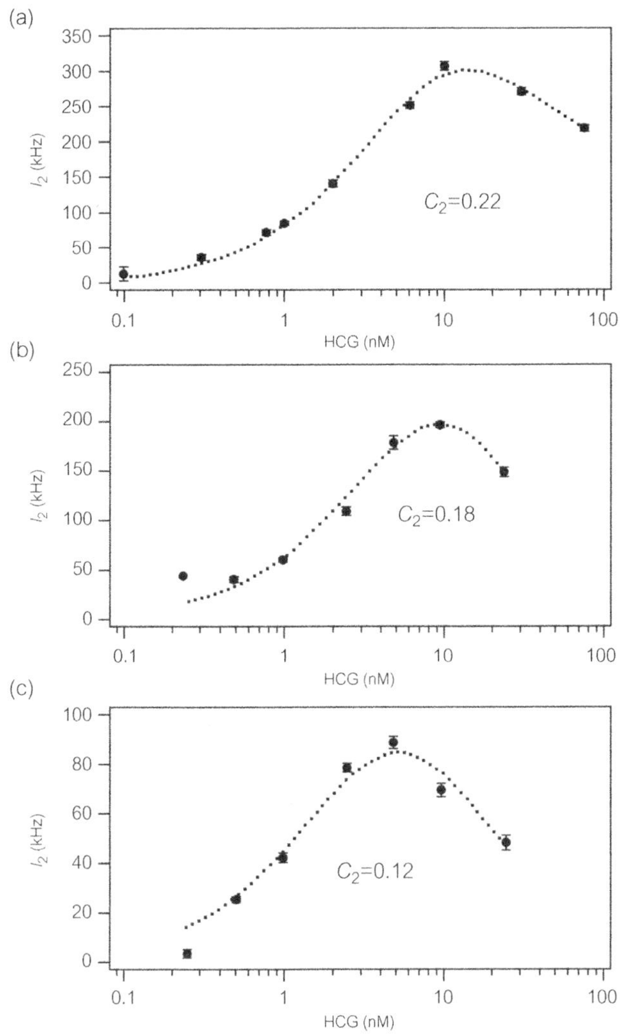

Figure 5.13 NPIA sandwich assay using antibody-coated nanoparticles binding hCG. As a linker system, the streptavidin-biotin interaction was used. The second (detection) antibody was labelled and accumulated on the bead when sandwiching hCG, which was added in increasing concentrations. In a two-component fit, the fixed concentrations were $q_1 = 50$ kHz and $c_2 = 0.22$ (a), 0.18 (b) and 0.12 (c). Antibody concentrations were 35 nM coating antibody and 40 nM detection antibody (a), 20 nM coating antibody and 25 nM detection antibody (b), 10 nm coating antibody and 12.5 nM detection antibody (c). Reprinted with permission from Schaertl et al., A novel and robust homogenous fluorescence-based assay using nano particles for pharmaceutical screening and diagnostics. Journal of Biomolecular Screening 5 (2000) 227–237. Copyright 2000 Sage Publications, Inc.

the applicability of the technique in a miniaturized HTS format which requires microlitre sample volumes and measurement times as short as 1 s, making it possible to consider screening large libraries with many tens of thousands of compounds in this way.

References

[1] **Bonnet, G, Krichevsky, O,** and **Libchaber, A,** Kinetics of conformational fluctuations in DNA hairpin-loops. *Proceedings of the National Academy of Sciences of the United States of America* **95** (1998) 8602–8606.

[2] **Dimitriadis, G, Drysdale, A, Myers, JK, Arora, P, Radford, SE, Oas, TG,** *et al.*, Microsecond folding dynamics of the F13W G29A mutant of the B domain of staphylococcal protein A by laser-induced temperature jump. *Proceedings of the National Academy of Sciences of the United States of America* **101** (2004) 3809–3814.

[3] **Friel, CT, Beddard, GS,** and **Radford, SE,** Switching two-state to three-state kinetics in the helical protein Im9 via the optimisation of stabilising non-native interactions by design. *Journal of Molecular Biology* **342** (2004) 261–273.

[4] **Capaldi, AP, Kleanthous, C,** and **Radford, SE,** Im7 folding mechanism: Misfolding on a path to the native state. *Nature Structural Biology* **9** (2002) 209–216.

[5] **Wallace, MI, Ying, LM, Balasubramanian, S,** and **Klenerman, D,** Fret fluctuation spectroscopy: Exploring the conformational dynamics of a DNA hairpin loop. *Journal of Physical Chemistry* B **104** (2000) 11551–11555.

[6] **Deniz, AA, Dahan, M, Grunwell, JR, Ha, T, Faulhaber, AE, Chemla, DS,** *et al.*, Single-pair fluorescence resonance energy transfer on freely diffusing molecules: Observation of Förster distance dependence and subpopulations. *Proceedings of the National Academy of Sciences of the United States of America* **96** (1999) 3670–3675.

[7] **Cheung, HC,** in **JR Lakowicz** (Ed.) Resonance Energy Transfer in Topics in Fluorescence Spectroscopy, volume 2: Principles. Plenum Press, New York, 1991.

[8] **Schuler, B, Lipman, EA,** and **Eaton, WA,** Probing the free-energy surface for protein folding with single-molecule fluorescence spectroscopy. *Nature* **419** (2002) 743–747.

[9] **Schuler, B, Lipman, EA,** and **Eaton, WA,** Probing the free-energy surface for protein folding with single-molecule fluorescence spectroscopy (vol 419, pg 743, 2002). *Nature* **421** (2003) 94–94.

[10] **Lipman, EA, Schuler, B, Bakajin, O,** and **Eaton, WA,** Single-molecule measurement of protein folding kinetics. *Science* **301** (2003) 1233–1235.

[11] **Margittai, M, Widengren, J, Schweinberger, E, Schroder, GF, Felekyan, S, Haustein, E,** *et al.*, Single-molecule fluorescence resonance energy transfer reveals a dynamic equilibrium between closed and open conformations of syntaxin 1. *Proceedings of the National Academy of Sciences of the United States of America* **100** (2003) 15516–15521.

[12] **Kapanidis, AN, Lee, NK, Laurence, TA, Doose, S, Margeat, E,** and **Weiss, S,** Fluorescence-aided molecule sorting: Analysis of structure and interactions by alternating-laser excitation of single molecules. *Proceedings of the National Academy of Sciences of the United States of America* **101** (2004) 8936–8941.

[13] **Lee, NK, Kapanidis, AN, Wang, Y, Michalet, X, Mukhopadhyay, J, Ebright, RH,** *et al.*, Accurate FRET measurements within single diffusing biomolecules using alternating-laser excitation. *Biophysical Journal* **88** (2005) 2939–2953.

[14] **Wolcke, J,** and **Ullmann, D,** Miniaturized HTS Technologies—uHTS. *Drug Discovery Today* **6** (2001) 637–646.

[15] **Schaertl, S, Meyer-Almes, FJ, Lopez-Calle, E, Siemers, A,** and **Kramer, J,** A novel and robust homogeneous fluorescence-based assay using nanoparticles for pharmaceutical screening and diagnostics. *Journal of Biomolecular Screening* **5** (2000) 227–237.

SIX

Fluorescence spectroscopy of immobilized single molecules: examples

6.1 Introduction

The time for which a freely diffusing single molecule can be observed is obviously limited by the dimensions of the confocal (detection) volume and the rate at which molecules of interest pass through it. However, as we have seen even given these restrictions fluctuation spectroscopy allows kinetic processes to be studied using the straightforward free-diffusion approach that is unlikely to perturb the analyte. Longer-term observation of kinetic processes, such as protein folding trajectories and enzyme catalysis however, can only be facilitated by immobilizing the molecules in some way. The critical aspect of these experiments is to immobilize the molecules in a manner that does not perturb their structure or function. This seems to be readily achievable in the case of nucleic acids by attachment of moieties such as thiols or biotin onto, for example, the termini of extended regions, which can then be attached to gold or to a surface incorporating the protein avidin, respectively. In this way the 'functional' region can often be isolated from the nearby surface. Such an approach appears to be more problematical in the case of (small) proteins whose tertiary structure is often easily perturbed by minor sequence changes and which show more cooperative folding behaviour. Thus nearby surfaces often tend to entirely denature the protein. It is not possible, for example, to extend one termini with a rigid polyproline peptide, to provide a point of attachment distal to the functional protein, and expect the protein to still fold into a functional form. For proteins then, more novel immobilization strategies must be adopted. In this chapter we will explore some examples of single molecule studies of conformational dynamics, binding and function that address the issues of immobilization and provide excellent demonstrations of the strength of the single molecule approach in a number of fields in biophysics.

6.2 **Single molecule studies of immobilized molecules**

6.2.1 **Quantitation of the oligomeric state of a protein complex**

Many proteins are functional in a complexed state, that is, they do not function as monomers but as dimers, trimers or higher order oligomers. Determination of the oligomeric state of some multi-domain protein complexes is not necessarily straightforward [1] especially in the case of membrane proteins, which are insoluble in aqueous solutions. Electron microscopy and gel chromatography [1] are commonly used, but complex, techniques for determining the oligomeric state of protein complexes. This question can, however, simply be addressed by application of single molecule fluorescence spectroscopy in which the photobleaching of fluorescently labelled and immobilized single protein complexes is monitored. If each monomer in the protein complex can be reliably labelled with a single fluorescent label then, since in a single molecule experiment photobleaching of each dye manifests itself as a quantized drop in fluorescence intensity [2], the number of monomers in the complex can be counted by counting the number of bleaching steps.

Abe and co-workers [1] have put this principle to very good use in the study of the oligomeric state of the H/K-ATPase ion channel. A large body of evidence indicates that the functional form of this enzyme in its native membrane is oligomeric and the nature of this oligomer is of importance in understanding the function of the channel. H/K-ATPase is composed of an α and a β chain and is

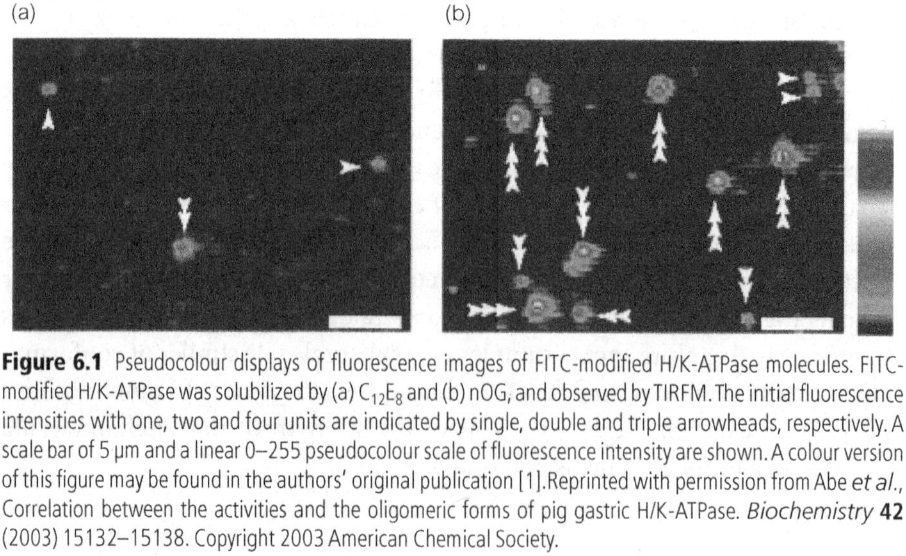

Figure 6.1 Pseudocolour displays of fluorescence images of FITC-modified H/K-ATPase molecules. FITC-modified H/K-ATPase was solubilized by (a) $C_{12}E_8$ and (b) nOG, and observed by TIRFM. The initial fluorescence intensities with one, two and four units are indicated by single, double and triple arrowheads, respectively. A scale bar of 5 μm and a linear 0–255 pseudocolour scale of fluorescence intensity are shown. A colour version of this figure may be found in the authors' original publication [1].Reprinted with permission from Abe *et al.*, Correlation between the activities and the oligomeric forms of pig gastric H/K-ATPase. *Biochemistry* **42** (2003) 15132–15138. Copyright 2003 American Chemical Society.

therefore referred to as an $\alpha\beta$-protomer. The group carried out total internal reflection fluorescence (TIRF) single molecule imaging experiments (see Chapter 3) of a solubilized form of the protein rather than *in situ* in a membrane in order that the results could be compared with electron microscopy and high performance gel chromatography. The protein was labelled with the fluorescent dye FITC at Lys 518 on the alpha chain; although we note the authors do not give much detail about this and do not comment on the absolute requirement in these experiments that there is only one FITC on each protein. The TIRF experiments were carried out using an Olympus objective lens specifically designed for total internal reflection microscopy (PlanApo60xOTIRFM; NA 1.45) and a CCD camera (CCD300RC, DAGE MTI) coupled with an image intensifier (VS-1845, Video Scope International).

Figure 6.1 shows two CCD images of FITC labelled H/K-ATPase molecules on glass prepared by flowing 10 pM concentration solutions over the substrate in two different detergents (octaethylene glycol dodecyl ether ($C_{12}E_8$) and n-octylglucoside (nOG)). The corresponding photobleaching time traces of typical molecules in these images are shown in Figure 6.2. In the case of the $C_{12}E_8$

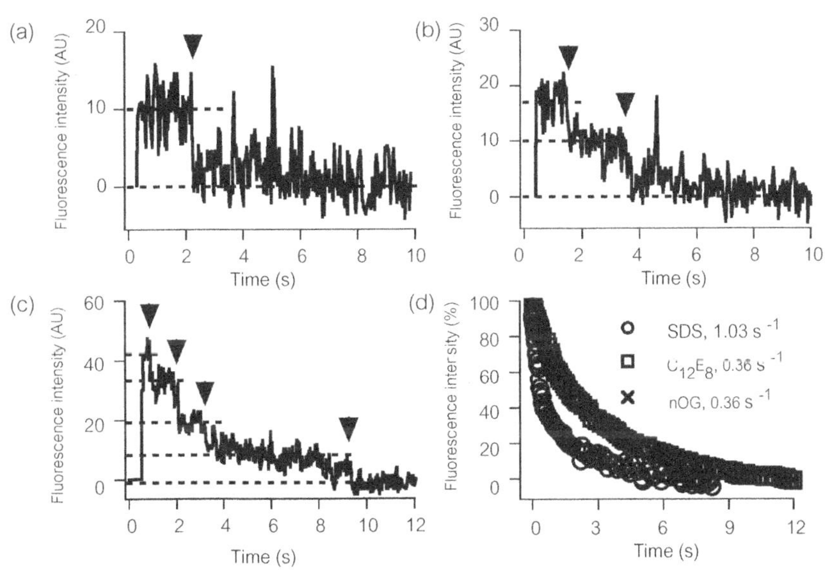

Figure 6.2 Fluorescence trajectories of single particles at the video rate. Typical quantitized photobleaching with one (a) two (b) and four (c) step(s) is shown. The background intensity (the average intensity of 150 video frames 15 s after laser illumination) was set as zero. Photobleaching occurred at the times indicated by the arrowheads. (d) The exposure time for photobleaching was measured for more than 100 single particles, and the data are summarized and plotted as a function of time. The decay could be fitted to a single-exponential function. Reprinted with permission from Abe *et al.*, Correlation between the activities and the oligomeric forms of pig gastric H/K-ATPase. *Biochemistry* **42** (2003) 15132–15138. Copyright 2003 American Chemical Society.

detergent (Figure 6.1(a)) mainly single and double photobleaching steps were observed. However, when nOG was used a number of brighter spots were observed which showed up to four FITC bleaching steps. The statistical analysis of the photobleaching measurements is summarized in the histograms in Figure 6.3. A control sample was studied in a solution of sodium dodecy sulphate (SDS), which is known to reduce the protein to its monomeric state. In this case only single photobleaching steps were observed which are characterized by the single Gaussian distribution in the histogram in Figure 6.3(a). Interestingly, the bleaching rate of the FITC was dependent on the detergent (see Figure 6.2) suggesting that the dyes are in different environments in each case. The FITC is attached to the protein in the ATP binding pocket and therefore the increase in the photobleaching rate in SDS solution (compared with the other two detergents) is assumed to be due to the unfolding of the binding pocket with SDS, resulting in exposure to oxygen radicals in the solution, which promote photobleaching.

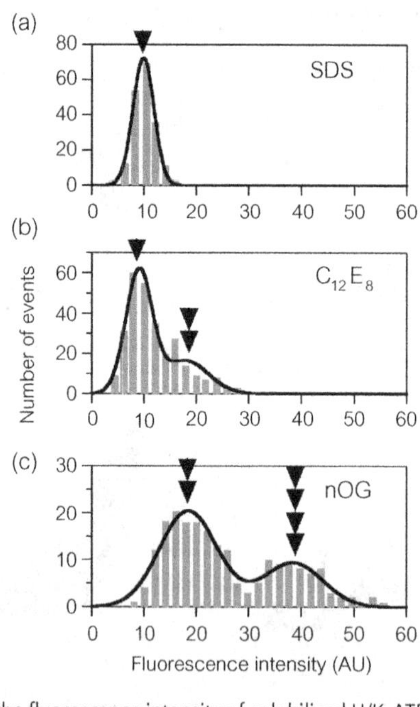

Figure 6.3 Histogram of the fluorescence intensity of solubilized H/K-ATPase. Distributions of fluorescence intensities of FITC- H/K-ATPase solubilized with SDS ((a) $n = 189$), $C_{12}E_8$ ((b) $n = 283$) and nOG ((c) $n = 206$) are shown. The fluorescence intensities of one to four dye molecules were in the linear range of the camera. The solid line indicates the sum of one to four Gaussian components fitted to the data. Single, double and quadruple arrowheads indicate the peak positions of each Gaussian distribution that is responsible for one, two or four fluorescence molecules, respectively. Reprinted with permission from Abe *et al.*, Correlation between the activities and the oligomeric forms of pig gastric H/K-ATPase. *Biochemistry* **42** (2003) 15132–15138. Copyright 2003 American Chemical Society.

From these measurements the authors were able to confirm that the higher order oligomers observed in gel chromatography and electron microscopy experiments were tetraprotomeric and that protomeric and diprotomeric complexes were also formed in nOG. This study benchmarked the photobleaching technique against the results of the other techniques. A natural extension of this study, and a unique strength of the single molecule photobleaching experiments, is that they could be carried out with the channel in its native state in supported lipid membranes on the glass substrates, to remove any influence of the detergent on the oligomeric state.

6.2.2 **Dynamics of proteins at membranes**

Pleckstrin homology (PH) domains play a role in targeting other proteins to cell membranes by binding to specific lipids called phosphoinositols. Little is known about these important protein domains and studies within living cells are challenging not least because of the intrinsic background fluorescence from the large amount of material in cells [3,4]. Mashanov and co-workers [5] have taken a single molecule approach and employed total internal reflection illumination to minimize the volume of the cell that is illuminated. The cells in their experiments sit on a glass coverslip on top of their TIRF microscope which is based on a Zeiss Axiovert inverted optical microscope and uses a through-the-objective TIRF geometry with an intensified CCD detector (see Chapter 3). Binding of proteins labelled with a fluorophore to the plasma membrane appear in the microscope images as points of fluorescence whose time dependence and intensity can be recorded and analysed to gather information about the dynamics of the interaction.

This group of workers is particularly interested in a molecular motor known as myosin X [6] which has a tail region that is thought to bind to cell membranes by its PH domain. Three PH domains were fluorescently labelled by fusing them to green fluorescent protein (GFP) [7] to facilitate single molecule studies. One drawback of this fluorophore system is the relatively short time to photobleaching [8,9] but this was overcome by modulating the excitation light source, using a shutter, such that the sample was repeatedly illuminated for short periods of time. This allowed measurements of up to 900 s to be carried out in which many GFP molecules survived for many seconds before photobleaching. In this way large numbers of observations were obtained giving statistically valid results. In order to ensure that each data set is derived from a true single molecule event in their experiments they apply a set of criteria that they have termed 'DISH'. This stands for (1) the spot must be diffraction limited, (2) the intensity of the emission must be appropriate for a single fluorophore, (3) single step photobleaching must be observed and (4) the half-life before photobleaching of the fluorophore population must be directly proportional to the excitation power. A three pass algorithm is therefore used to ensure the integrity of the data that is retained for analysis: (1) test squares of pixels for single-step photobleaching behaviour, (2) if this is observed then confirm that the point spread function is of the correct size and intensity, (3) then plot trajectories of fluorescence signal versus time with the option of tracking the centroid of the spot to quantify lateral diffusion of the fluorophore.

Initially they carried out a control study of GFP immobilized onto glass using an anti-GFP antibody. This allows the fluorescence behaviour of single GFP molecules to be monitored without the complication of binding and unbinding at the cell membrane when conjugated to a PH domain (and so provides the information required for the DISH criterion—typical size, bleaching half-life and intensity). Figure 6.4(a) shows a typical CCD image of the immobilized GFP molecules, the

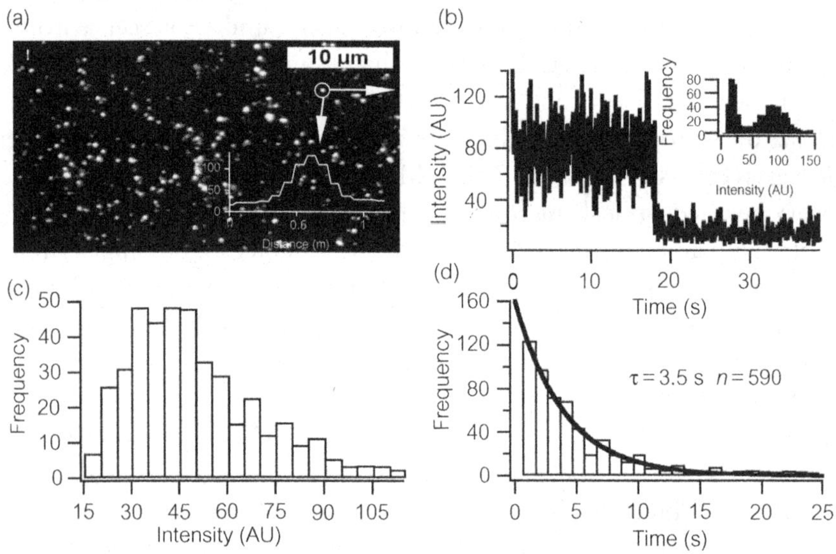

Figure 6.4 Purified eGFP (Clontech) was bound to antibody (Abcam Ltd) that had been adsorbed to a cleaned glass surface. Single fluorophores appear as individual spots of light. (a) Wide-field of view of eGFP stuck to antibody. *Inset*, plot of intensity profile for one spot (width at half-height is 400 nm limited by the resolution of the imaging system). (b) Intensity versus time trajectory for one spot. Inset, histogram of intensity including bright and dark periods and dark counts produced a peak at low values, and the fluorophore produced a signal with mean of 80 units ± S.D. of 20. The noise is due to photon-counting statistics. (c) Intensity distribution of mean intensities of all spots. Note the broad distribution in intensities (much broader than expected from photon shot noise in (b) due to other sources of variance). (d) Distribution of times to photobleaching, average lifetime was 3.5 s. The average time to photobleaching derived from such a plot is inversely proportional to the bulk photobleaching rate measured for an ensemble. Republished with permission of The American Society for Biochemistry and Molecular Biology, Inc., from Mashanov *et al.*, The spatial and temporal dynamics of pleckstrin homology domain binding at the plasma membrane measured by Imaging single molecules in live mouse myoblasts. *Journal of Biological Chemistry* **279** (2004) 15274–15280; permission conveyed through Copyright Clearance Center, Inc.

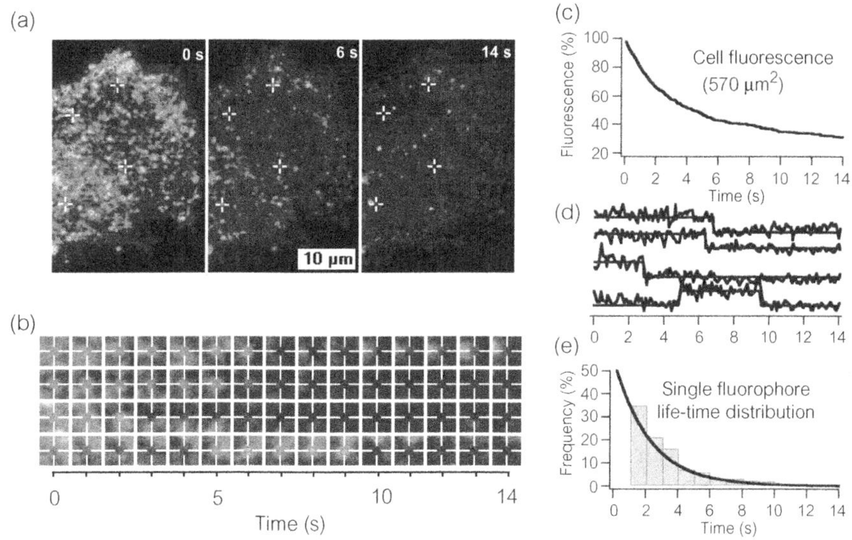

Figure 6.5 Detection of single molecules of eGFP-PH123 in the lamella of a living mouse myoblast under continuous laser illumination. (a) Individual TIRF images taken at 0, 6, and 14 s after the start of recording (average of three 100-ms frames). (b) Representative images of four different individual fluorescent spots marked in (a) by cross-hairs. (c) Variation in the average fluorescence of the total area shown (570 μm^2) during the recording, a rate constant of 0.4 s^{-1}. (d) Records of the fluorescence intensity of the spots shown in (b) measured for every 100-ms frame of the record and plotted against time. Pixel area of 5×5, 0.16 μm^2 (limited by cross-hairs in (b)). (e) Histogram showing the lifetime distribution of 370 individual spots. These were fitted to a single exponential with a rate constant of 0.4 s^{-1}. A colour version of this figure may be found in the author's original publication [5]. Republished with permission of The American Society for Biochemistry and Molecular Biology, Inc., from Mashanov *et al.*, The spatial and temporal dynamics of pleckstrin homology domain binding at the plasma membrane measured by imaging single molecules in live mouse myoblasts. *Journal of Biological Chemistry* **279** (2004) 15274–15280; permission conveyed through Copyright Clearance Center, Inc.

characteristics of these fluorescent spots (Figure 6.4(a), inset) and their single step photobleaching (Figure 6.4(b)). The large variation in individual intensities (Figure 6.4(c)) is attributed to the local environment of each GFP molecule and the random orientation of the GFP molecules on the coverslip, which could have an effect, because polarized excitation light was used (also see Chapter 3 for a discussion of the polarization state of an evanescent wave in TIRF). The angle between the direction of polarization and the absorption dipole moment of the GFP has a strong effect on the absorption probability and therefore on the fluorescence intensity. Figure 6.4(d) shows the mono-exponential distribution of the time before photobleaching observed for GFP molecules. The bleaching lifetime (i.e. inverse of the exponential decay rate), which depends linearly on excitation power as required by the DISH criteria, for this GFP ensemble was determined to be 3.5 s.

Once the behaviour of the fluorescent label was understood the authors moved to study GFP labelled PH within a living cell—in this case a mouse myoblast. Figure 6.5(a) shows the individual spots of light that were observed at the plasma membrane in contact with the glass coverslip under constant laser illumination. Under these conditions, these spots appeared to be photobleaching during the period that they were attached to the membrane. However, under shuttered illumination conditions in which the cell was illuminated for 350 ms in every 5 s, fluorescent spots could be observed for tens or even hundreds of seconds (Figure 6.6). Because of the very rapid diffusion of the GFP labelled PH domain away from the membrane when it undocks, it was not possible to distinguish between photobleaching and dissociation by simple observations. However, since photobleaching depends linearly on excitation power (or in a shuttered illumination experiment it depends linearly on the duty cycle) but the dissociation rate does not, a plot of the rate constant for disappearance of the spots versus average laser power results in a non-zero intercept with the y-axis (see Figure 6.7) which gives an estimate of the dissociation rate constant of 0.05 s^{-1}. A comparison with data from the anti-body immobilized GFP molecules indicates that, as expected, this system has an extremely low off-rate and a near-zero intercept with the y-axis. This analysis

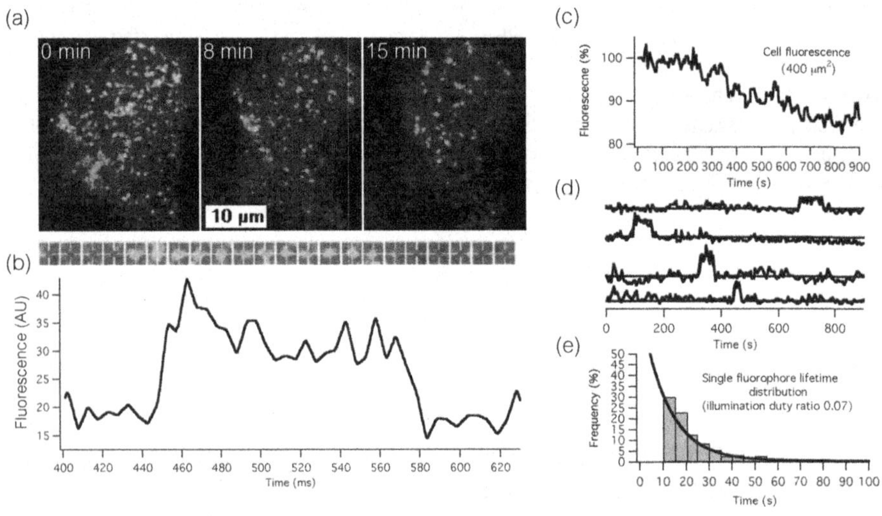

Figure 6.6 Detection of single molecules of eGFP-PH123 in the lamella of a living mouse myoblast under time-lapse recording. An example of one cell that was illuminated for 350 ms in every 5-s interval, giving an illumination duty ratio of 0.07. (a) Individual TIRF images taken at 0, 8, and 15 min during a time-lapse recording. The majority of the fluorescent spots appeared on the membrane after the beginning of the record. (b) Image sequence of a single fluorophore that landed on the cell membrane and stayed attached for over 140 s together with the fluorescence intensity track of this spot measured at 5-s intervals. (c) The average cell fluorescence decreased slightly during recording (by ~15%). (d) Four representative fluorescence intensity tracks of the diffraction-limited areas (5×5 pixel area, 0.16 μm^2) for single fluorophores detected during the record in (a). (e) Histogram showing the lifetime distribution of 775 individual spots. The distribution was fitted by a single exponential with a rate constant of 0.07 s^{-1}. A colour version of this figure may be found in the authors' original publication [5]. Republished with permission of The American Society for Biochemistry and Molecular Biology, Inc., from Mashanov *et al.*, The spatial and temporal dynamics of pleckstrin homology domain binding at the plasma membrane measured by imaging single molecules in live mouse myoblasts. *Journal of Biological Chemistry* **279** (2004) 15274–15280; permission conveyed through Copyright Clearance Center, Inc.

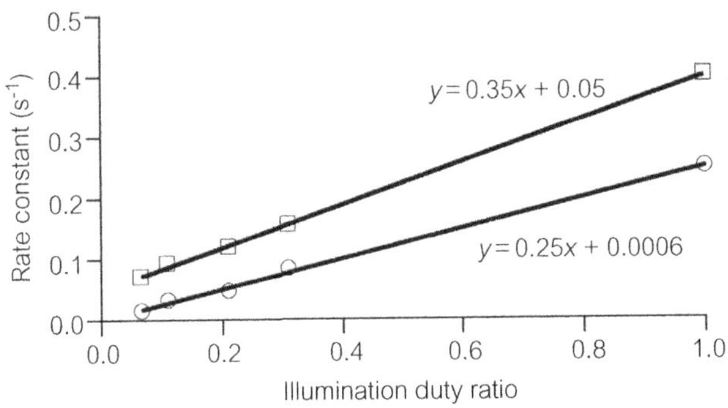

Figure 6.7 Calculation of eGFP-PH123 domain dissociation rate. Lifetime of single eGFP-PH123 molecules on cell membrane depends on the photobleaching and dissociation rate ($k'_{off} = \xi \times k_{pb} + k_d$). k'_{off} was linearly dependent on the average laser illumination, ξ (illumination duty ratio) (open circles). A linear regression fit to the data gives the intercept, k_d, as 0.05 s^{-1} for eGFP-PH123 (squares). By comparison, eGFP molecules attached to GFP antibodies on the glass surface (circles) show that k_d is close to zero (i.e. eGFP is very tightly bound to antibody, and antibody is tightly bound to the glass). Republished with permission of The American Society for Biochemistry and Molecular Biology, Inc., from Mashanov *et al.* The spatial and temporal dynamics of pleckstrin homology domain binding at the plasma membrane measured by imaging single molecules in live mouse myoblasts. *Journal of Biological Chemistry* **279** (2004) 15274–15280; permission conveyed through Copyright Clearance Center, Inc.

does rely on the assumption that the photobleaching and dissociation are both Poissonian processes, but the data presented support this. The on-rate of the PH domain is easily obtained by counting the rate at which fluorescence spots appear in the image per unit area. When the concentration of GFP labelled PH domain in the cell is taken into account (~10 nM) the on-rate is estimated to be 0.028 $\mu M^{-1}\ \mu m^{-2}\ s^{-1}$. The lateral diffusion of bound PH domains was studied as well as the dynamics of binding and dissociation. Interestingly, the GFP fluorescence spots did not diffuse laterally suggesting that the PH domain may be immobilized by secondary binding to the cytoskeleton, or that there is heterogeneity in the distribution of lipids in the membrane forming relatively large and immobile rafts to which the PH domains bind, which opens up interesting questions for further study.

6.2.3 **DNA unwinding by molecular motors**

A very elegant demonstration of the functional detail that can be obtained from single molecule studies of immobilized molecules was given by Ha *et al.* [10]. Their system of interest involves a helicase—a protein which binds to double-stranded DNA and unwinds the helical structure. Helicases are protein motors that are driven by the binding and hydrolysis of ATP. Some helicases function as hexameric rings but the fundamental form of other helicases is a subject of debate, which is addressed in this paper. Previously, bulk studies had been able to reveal some information about the procession of the Rep helicase from *Escherichiaia coli* but the resolution (i.e. in terms of the number of base pairs travelled by the molecular motor) was quite coarse—around 100 bp [11]. In principle single molecule studies of immobilized DNA molecules being unwound by a helicase could provide much greater base pair resolution and begin to unpick the complex and, as yet, only partially understood mechanism.

Several DNA systems were synthesized by the authors to study the processivity of Rep helicase. Figure 6.8 shows data for the first pair of oligonucleotides studied.

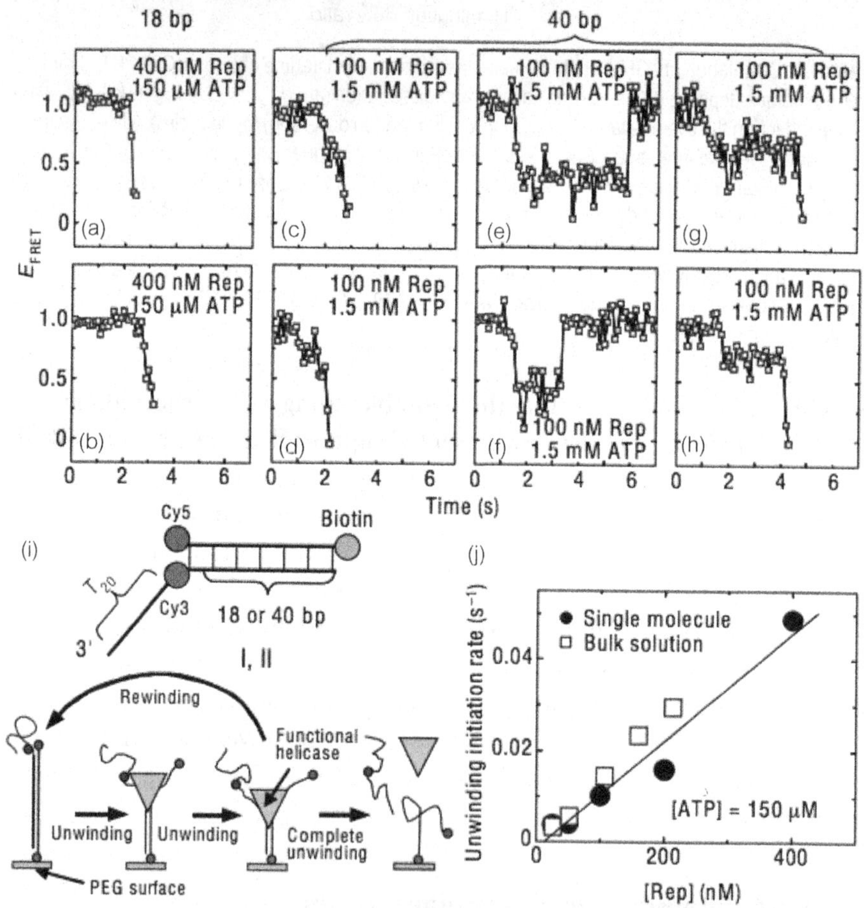

Figure 6.8 FRET assay for single DNA unwinding. (a–h) Single DNA unwinding time records (100-ms bins) typical of four distinct patterns: complete unwinding of 18 bp (a, b) complete unwinding (c, d) stall and DNA rewinding (e, f) and stall and unwinding re-initiation (g, h) of 40 bp. When unwinding is completed (a–d, g–h) the donor strand quickly diffuses away, abruptly terminating the signal. (i) The experimental scheme. (j) The rate of unwinding initiation (the inverse of time between the delivery of enzyme with ATP and the unwinding initiation, (averaged over >50 events for each [Rep]) versus [Rep]). A linear fit and bulk-solution values are shown. A colour version of this figure may be found in the authors' original publication [10]. Reproduced from Ha *et al.*, Initiation and re-initiation of DNA unwinding by the *Escherichia coli* Rep helicase. *Nature* **419** (2002) 638–641 with permission from Nature Publishing Group (Copyright 2002)

The unwinding and rewinding of the DNA is reported by the change in FRET efficiency of the pair of dyes (Cy3 and Cy5 in this case) attached to the end of the double-stranded section of the oliognucleotide (of lengths 18 or 40 bp) which was immobilized to the substrate through a biotin–streptavidin linkage (at the duplex end distal from the junction). This is a common immobilization strategy when studying nucleic acid or nucleic acid/protein systems as oligonucleotides that are derivatized with biotin groups can be readily purchased and such a system ensures that the point of attachment, and so the surface, is well separated from the region of interest (in this case the single–double stranded junction). In this system the surface of a glass slide was covered with polyethylene glycol (PEG) and a biotinylated PEG to permit a streptavidin sandwich to be formed to bind the DNA. The 'PEGylated' glass surface appeared to reduce the amount of non-specific DNA binding by 1000 fold. Two different lengths of double-stranded DNA (18 and 40 bp) were initially studied by total internal reflection illumination using either 33 or 100 ms integration times in the presence of an oxygen scavenger to minimize photobleaching. Figure 6.8(a–h) show typical time traces of the FRET signal when Rep protein and ATP were added to the sample volume. In the case of both DNA constructs a characteristic time elapsed before unwinding was observed. This time was inversely proportional to the concentration of Rep protein added. The length of this initiation step correlated well with data obtained from bulk measurements in solution, which gave confidence that the immobilization technique was not affecting the protein–DNA interaction. Figure 6.8(a–h) shows that a rapid fall in FRET efficiency occurs as the donor and acceptor dyes move further apart when the double helix is unwound. In addition, the time trajectories of the FRET signal reveal periods during which the process stalls which is then followed by rewinding (Figure 6.8(e and f)) or continued unwinding (Figure 6.8(g and h)). This is a beautiful demonstration of how measurement of biomolecular processing at the single molecule level can reveal the inherent heterogeneity of the process. The question that the authors then address is what is the origin of this heterogeneous behaviour.

In order to investigate this question, the authors used a 3rd DNA construct in which the donor was attached at the 3′ end of a $(T)_{19}$ extension on the reverse strand and the acceptor at the single–double strand junction (Figure 6.9). In this construct the FRET signal was sensitive to the conformational fluctuations of the single-stranded section when the Rep protein binds (Figure 6.9). By analysing these fluctuations as a function of [Rep], the authors suggest that after Rep has bound as a monomer it uses ATP to position itself at the junction of the single- and double-stranded sections of the DNA where the complex undergoes conformational fluctuations. These conformational fluctuations continue until at least one more Rep monomer binds forming a complex that is active and can proceed

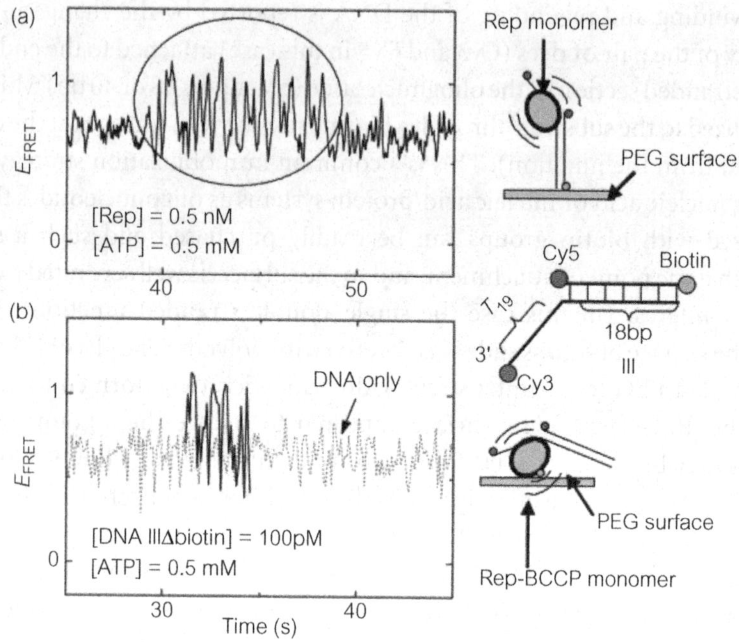

Figure 6.9 ATP-dependent, junction-specific fluctuations of Rep monomer. (a) A burst of FRET efficiency fluctuations (circled) is caused by enzyme binding to surface-immobilized DNA III and conformational fluctuations (33-ms bins). (b) A burst is observed when DNA III (minus biotin) molecules bind to a surface-immobilized, biotinylated Rep (Rep-BCCP) monomer. Fluorescence signal is detectable only during the burst (100-ms bins). Also shown is a typical trace obtained from immobilized DNA III only (grey) for comparison. Accompanying cartoons illustrate Rep monomer conformational fluctuations with either DNA or protein immobilized. A colour version of this figure may be found in the authors' original publication [10].Reproduced from Ha *et al.*, Initiation and re-initiation of DNA unwinding by the Escherichia coli Rep helicase. *Nature* **419** (2002) 638–641 with permission from Nature Publishing Group (Copyright 2002).

to unwind the DNA. The stalls in the process are suggested to arise from either the partial dissociation of the Rep complex, which can recover its functional form if binding of more Rep occurs or the entire complex could dissociate leading to the rewinding of the DNA that they observe in some cases.

6.2.4 **Single molecule protein folding observations**

Immobilization presents an opportunity to observe the folding and unfolding trajectories of single proteins over a long period of time by using single-pair FRET. However, there is a significant body of evidence that indicates that the tethering of a protein to a surface can perturb the native or denatured states or both. Various surface modifications have been attempted to attach various molecules [12–16]. However, tethering proteins to surfaces is widely regarded as problematic and remains a challenge in many areas of biotechnology. Boukobza [17], Rhoades

[18] and colleagues in Gilad Haran's lab in Israel have, however, extended a novel method of restricting a protein in space so that measurements over a longer period of time can be obtained whilst avoiding unwanted interactions. To do this they have encapsulated a protein within a liposome (a spherical self-assembled phospholipid bilayer structure) and tethered the liposome to a lipid-coated glass surface via a streptavidin—biotinylated lipid complex. They used this method to study the folding kinetics of adenylate kinase [18], a 214 amino acid protein that has been well characterized in ensemble measurements and shown to fold in a multi-state manner. This protein therefore provides an ideal opportunity for single molecule studies to dissect a heterogeneous folding reaction.

The protein was mutated to contain two cysteines at positions 73 and 204 and the sulphydryl groups of the residues used to conjugate an Alexa Fluor 488 donor and a Texas Red C2 acceptor. Liposomes were formed in the presence of 3 μM protein ensuring that on average any vesicle encapsulated only a single protein. Single molecule FRET measurements were then carried out using a home built scanning confocal microscope with 488 nm excitation and FRET efficiencies were corrected for the difference in quantum yield and collection efficiency of the optical system at the donor and acceptor wavelengths by monitoring the donor intensities before and after acceptor photobleaching. Polarized light single molecule studies were also carried out. Fluorescence was excited with circularly polarized light and split into its two orthogonal components using a cube polarizer.

The polarized light measurements of donor only labelled protein were used to determine whether the protein molecules were freely diffusing in the vesicles and not interacting with the lipid walls of the container. The distribution of fluorescence polarization for single protein molecules in liposomes and also immobilized directly onto glass are shown in Figure 6.10(a and b). The narrow distribution of the donor fluorescence emission in the liposome in comparison with the protein on glass and the absence of any long-term jumps in the time trace of polarization information (Figure 6.10 (c and d)) is indicative of unhindered free diffusion of the protein suggesting that it is not interacting with the lipid, as desired.

By addition of a suitable amount of a chemical denaturant (in this case guanidinium hydrochloride GdnHCl) the protein can be made to fully unfold or can be poised at the mid-point of a transition determined by an ensemble equilibrium denaturation experiment. Rhoades recorded the histograms of FRET efficiency for the protein under native, denaturing and mid-point conditions (this was determined to be 0.55 M GdnHCl for adenylate kinase under these conditions). These data indicate a FRET efficiency of 0.8 in the folded state and 0.14 in the fully denatured state (2 M GdnHCl). At a GdnHCl concentration of 0.4 M the presence of two sub-populations was detected. At this concentration of denaturant the protein will spend an almost equal amount of time in the folded and unfolded

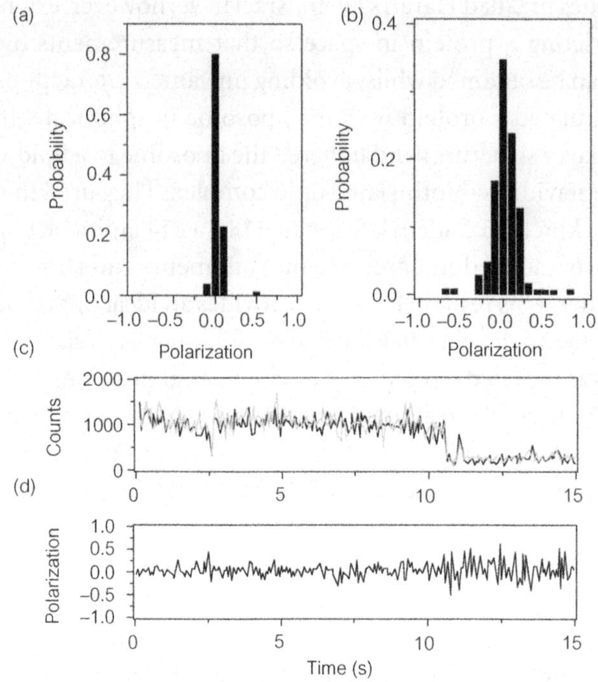

Figure 6.10 (a and b) Distributions of average fluorescence polarization values of single AK molecules labelled with the donor only obtained after excitation with circularly polarized light. The distribution shown in (a) is for molecules trapped in vesicles at 0.4 M GdnHCl, (b) is for molecules adsorbed directly on glass. The very narrow width of the polarization distributions of vesicle-trapped molecules, compared with the width of the distribution of glass-adsorbed molecules, indicates freedom of rotation of the trapped molecules, as discussed in detail by [17]. A similar experiment performed with acceptor-labelled AK molecules and showing analogous results was presented in that article. The polarization distribution of molecules trapped in vesicles under native conditions (data not shown) is indistinguishable from the one shown in (a). (c) A typical time-dependent fluorescence polarization trajectory of a single AK molecule labelled with the donor only and trapped in a lipid vesicle. The vertically polarized (I_V) and horizontally polarized (I_H) components of the fluorescence are shown in gray and black, respectively. (d) The fluorescence polarization calculated from the data in (c). The lack of any long-term jumps in the polarization indicates that this protein molecule does not become static (e.g. due to adsorption on the vesicle wall) for any considerable amount of time. Very similar time traces were obtained from many individual molecules. Reproduced from Rhoades *et al.*, Watching proteins fold one molecule at a time. *Proceedings of the National Academy of Sciences of the United States of America* **100** (2003) 3197–3202 with permission from National Academy of Sciences, USA (Copyright 2003).

conformations and thus, in principle, single molecule experiments carried out under these conditions permit the folding and unfolding pathways to be studied. The authors carried out this experiment and the results for two molecules are shown in Figure 6.11. These data, recorded with a 20 ms integration time, were smoothed using modified versions of algorithms developed by researchers who study the conduction of ion channels in membranes. These algorithms avoid the

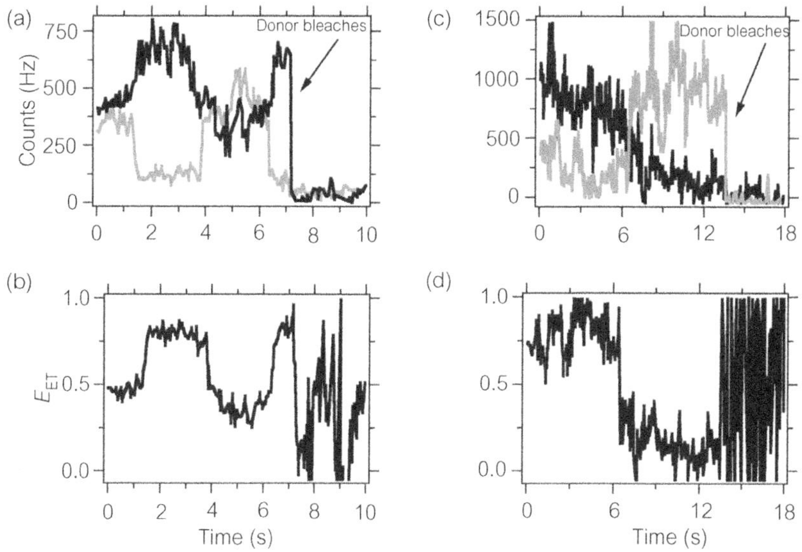

Figure 6.11 (a and c) Time traces of individual vesicle-trapped AK molecules under midtransition conditions with the acceptor signal in black and the donor in grey. The traces were collected with 20-msec time bins. They then were smoothed by using the forward–backward nonlinear filter developed by Chung and Kennedy [19] for ion-channel current analysis. In this filter, predictors derived from the data are adaptively weighted to ensure that fast intensity jumps will not be smeared, as happens when standard rolling-average procedures are used. The nonlinear filter as used here reduces the noise in the trajectories by a factor of ~4 while correctly preserving intensity transitions. (b and d) E_{ET} trajectories calculated from the signals in (a) and (c), respectively. In (a) and (b) several transitions occur between states that are essentially within the 'folded' ensemble, whereas in (c) and (d) a single transition takes the molecule from the folded to the 'denatured' ensemble. Note that transitions can be strictly recognized by an anticorrelated change in the donor and acceptor fluorescence intensities as opposed to uncorrelated fluctuations sometimes appearing in one of the signals. Reproduced from Rhoades *et al.*, Watching proteins fold one molecule at a time. *Proceedings of the National Academy of Sciences of the United States of America* 100 (2003) 3197–3202 with permission from National Academy of Sciences, USA (Copyright 2003).

smearing and distortion of rapid steps that can occur in time traces like these when a simple running average is calculated. Anti-correlated changes in donor and acceptor intensity are clearly observed (Figure 6.11 (a and c)) which correspond to changes in protein conformation that are reflected in the calculated FRET efficiency time traces for these two experiments (Figure 6.11 (b and d)). To further elucidate the folding pathways the data obtained from many single protein trajectories were transformed into a two-dimensional plot with the initial FRET efficiency before a transition on the ordinate and the final value on the abscissa (Figure 6.12(a)). The large amount of scatter in this plot indicates a wide variation in the start and end points of the folding/unfolding pathways of the molecules. Furthermore, a histogram of the difference in FRET efficiency before and after a transition revealed two distributions; one for folding (centred at around +0.2 as

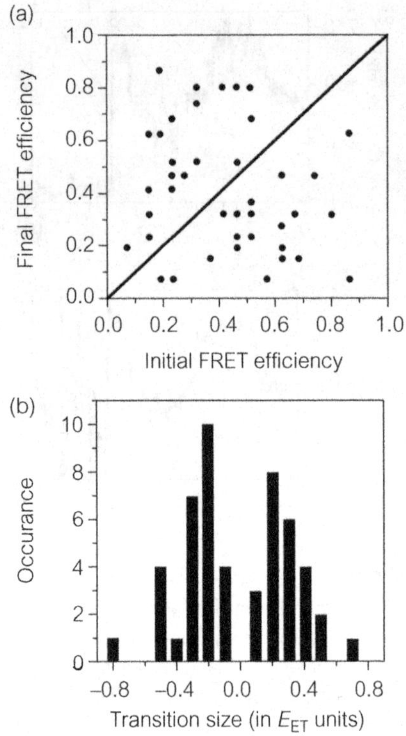

Figure 6.12 (a) Map of folding/unfolding transitions obtained from single-molecule trajectories. Each point represents the final versus initial FRET efficiency for one transition. The line is drawn to distinguish folding and unfolding transitions; above the line are folding transitions (efficiency increases), and below the line are unfolding transitions (efficiency decreases). (b) Distribution of transition sizes (i.e. final minus initial efficiencies) as obtained from the map in a. The two branches of the distribution represent unfolding and folding transitions, respectively. The overall similarity of the shape of the two branches indicates uniform sampling of the energy landscape. They both peak at a low efficiency value, signifying a preference for small-step transitions. Reproduced from Rhoades, *et al.*, Watching proteins fold one molecule at a time. *Proceedings of the National Academy of Sciences of the United States of America* **100** (2003) 3197–3202 with permission from National Academy of Sciences, USA (Copyright 2003).

the final efficiency is larger than the initial) and one for unfolding (centred at around -0.2) thus indicating a preference for step sizes of around 0.2–0.3 FRET units. This is not equivalent to the difference between the fully folded and fully unfolded FRET efficiencies (0.8 and 0.14) identified earlier and the authors interpret this as implying that the proteins do not change from fully folded to fully unfolded states but undergo a series of smaller intermediate steps. This reveals the heterogeneity of the kinetic pathways and the complexity of the so-called energy landscape over which the protein travels.

The slow folding of adenylate kinase prevents the authors observing many transitions between structurally persistent states because photobleaching typically occurs first, preventing the kinetic rate constants being measured with ease. Many of the steps that are observed in the time traces are very rapid (i.e. faster than 20 ms) and detail within these steps cannot therefore be resolved. However, some transitions are much slower taking >1 s (see Figure 6.13 (a–c)). The authors discuss these slow transitions in terms of motion over the energy landscape that is hindered by shallow traps. Slow changes in overall protein dimensions reflected in FRET efficiency suggest that these conformational changes are entropically driven (i.e. only small changes in entropy are allowed for each step)

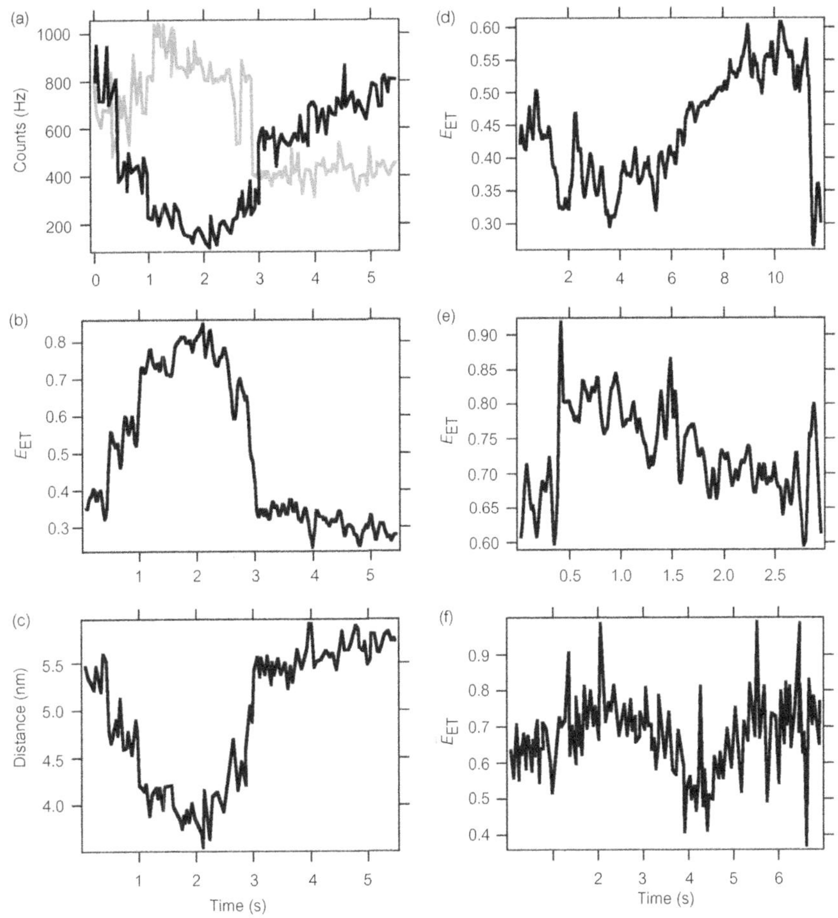

Figure 6.13 Time-dependent signals from single molecules showing slow folding or unfolding transitions. (a) Signals showing a slow folding transition starting at ~0.5 s and ending at ~2 s. The same signals display a fast unfolding transition as well (at ~3 s). The acceptor signal is shown in black and the donor is shown in grey. (b) E_{ET} trajectory calculated from the signals in (a). (c) The interprobe distance trajectory showing that the slow transition involves a chain compaction by only 20%. The distance was computed from the curve in (b) by using a Forster distance (R_0) of 49 Å. This Forster distance was calculated by assuming an orientation factor (κ^2) of 2/3. However, the point discussed here does not depend on the exact value of these parameters. (d–f) Additional E_{ET} trajectories demonstrating slow transitions. These transitions were identified, as already noted, by anticorrelated donor–acceptor intensity changes. Reproduced from Rhoades, *et al.*, Watching proteins fold one molecule of a time. *Proceedings of the National Academy of Sciences of the United States of America* **100** (2003) 3197–3202 with permission from National Academy of Sciences, USA (Copyright 2003).

and that the large abrupt changes in conformation represent enthalpically driven process (i.e. large changes in entropy appear to be allowed).

This sophisticated discussion of protein pathways over a complex energy landscape does rely on some assumptions, such as there being no interaction at all between the protein and the liposome and that the dyes themselves cannot adopt certain fixed conformations with respect to the folded or unfolded protein. A natural question is whether similar slow conformational changes and intermediate FRET signals are observed in well-characterized two-state folding proteins? The same group addressed this very question using the cold shock protein from *Thermotoga maritime* [20], which is known to fold in a two-state fashion. In this case, no such slow conformational changes were observed, only abrupt jumps in FRET efficiency that occurred on a 100–200 μs timescale. The folding rate constant obtained by these single molecule observations was in good agreement with that obtained from bulk experiments (~0.4 s^{-1}). Thus it appears that the slow changes in FRET efficiency observed for the much larger adenylate kinase molecule may well reflect meandering through many local traps on the energy landscape and are not an artefact of the protein interacting with the liposome or dyes. Such an observation provides a powerful incentive for further similar studies with a range of proteins and their mutants.

6.2.5 **Conformational dynamics of single ribozyme molecules**

Steven Chu's [21–23] group have generated an impressive body of work around single molecule studies of the folding and catalytic activity of the ribozymes. Ribozymes are protein-independent RNA enzymes. Like their protein-based counterparts the catalytic function of ribozymes is highly dependent on their fold and therefore are strongly influenced by conformational dynamics. It is this aspect of the system that Chu and co-workers have studied so effectively using single molecule techniques.

Figure 6.14(a) shows the structure of a hairpin ribozyme which is a minimal active form of a large RNA structure comprised of two independent helix-loop-helix domains labelled loop A and loop B. This molecule can exist in a linear extended form referred to as the undocked state and a bent form in which loop A docks with loop B. This hairpin ribozyme reversibly cleaves its substrate (another piece of RNA labelled S in Figure 6.14). Previous studies have shown that this process has several stages: (1) the substrate binds to the ribozyme in its undocked state, (2) the complex forms into the docked state, (3) the substrate is cleaved into two shorter pieces and (4) the complex undocks and the products are released. In order to study this system with single molecule sensitivity the A strand of the RNA was labelled at the 3′ and 5′ ends with Cy3 and Cy5 respectively and loop B was functionalized with biotin in order that the ribozyme could be immobilized on a glass substrate. Single molecule measurements of FRET were made using both a total internal reflection geometry and a scanning confocal setup (see Chapter 3). Figure 6.14(b) shows the complex catalytic pathway of this hairpin ribozyme with rate constants derived by these single molecule measurements as described later.

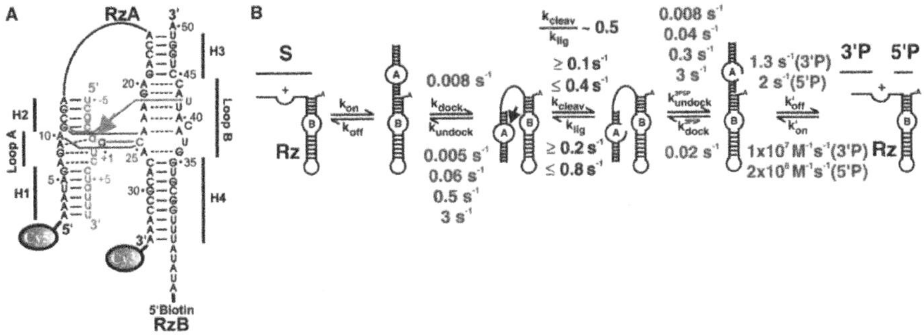

Figure 6.14 Structural dynamics and function of the hairpin ribozyme. (a) The two-strand (RzA, RzB) hairpin ribozyme (SV5 EH4) used in this study binds substrate to form domain A, comprising helices H1 and H2 (short lines, Watson-Crick base pairs) and the symmetric internal loop A. Domain A is connected by a flexible hinge to domain B of the ribozyme, comprising helices H3 and H4 and the asymmetric internal loop B. Noncanonical base pairs are indicated as dashed lines. Biotin and the fluorophores Cy3 and Cy5 were attached as indicated. (b) The putative reaction pathway of the hairpin ribozyme. The rate constants measured are given. A colour version of this figure may be found in the authors' original publication [23]. Reprinted with permission from Zhuang *et al.*, Correlating structural dynamics and funetion in single ribozyme molecules. *Science* **296** (2002) 1473–1476. Copyright 2002 AAAS.

Figure 6.15 shows the histrogram of FRET efficiencies after addition of the substrate to the system to create bound complexes, then removal to permit the cleavage process to be observed. Three distinct states can be identified via their different FRET efficiencies corresponding to the undocked state to which substrate is bound (i.e. longest distance between donor and acceptor when the ribozyme is in its fully linearized state), the docked state in which the bent conformation brings the donor and acceptor together to their closest separation and what the authors term the 'S-free' state which is the intermediate FRET efficiency corresponding to the ribozyme in an intermediate undocked state when the substrate is not bound. Measurement of the kinetics of formation of the S-free state shows that the population of this state increases with time as more substrate is cleaved and product is released. This reaction time course was found to be identical to that measured by bulk ensemble methods. Neither time course could be fit by a single exponential function, indicating heterogeneous reaction kinetics.

Figure 6.16 reveals more detail of the cleavage process. When substrate binds, 98–99% of the molecules studied exhibit a fall in FRET efficiency, indicating that the molecules move into the undocked state when substrate binds. Then the molecules are seen to switch stochastically between the docked and undocked states (Figure 6.16(a)). Evidence that cleavage only occurs when the system is in the docked state comes from the observation that for 90–95% of molecules studied the intermediate FRET efficiency is adopted after the ribozyme has been in the docked state (Figure 6.16(b) gives a typical example). The remaining molecules,

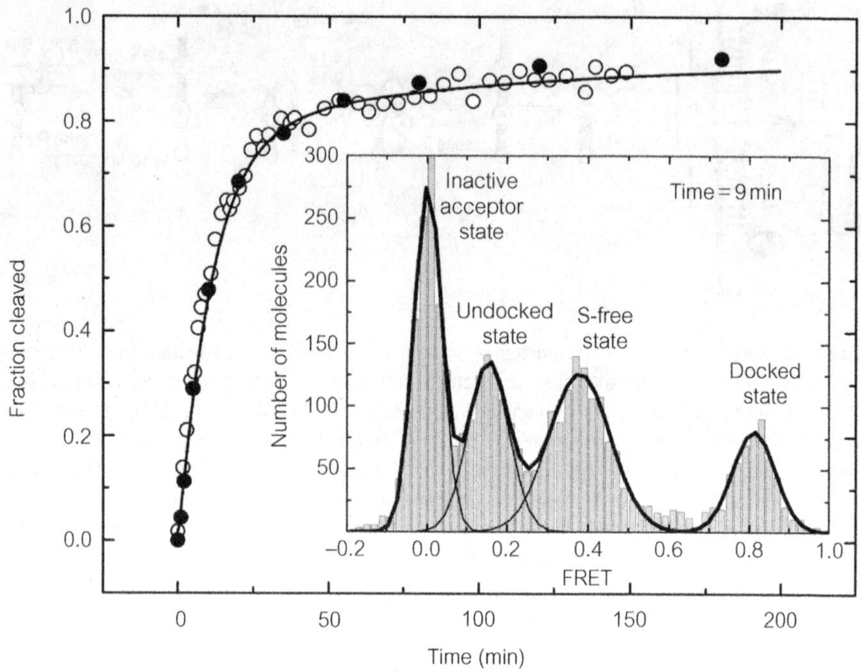

Figure 6.15 Single-molecule and bulk solution measurements of enzymatic activities. The open symbols show the reaction time course of surface-immobilized ribozyme. For initiation of cleavage, a buffer containing 200 nM substrate was added to the immobilized ribozymes for 30 s to allow substrate binding; free substrate was then removed from the buffer. During the reaction, the FRET distribution showed three distinct ribozyme populations: undocked (FRET ~0.15), docked (~0.81), and substrate-free ribozymes (~0.38). The peak at FRET ~0 was due to inactive Cy5. The substrate-free fraction is plotted against time. In a control experiment with noncleavable substrate the substrate-free fraction accumulated with a rate constant slower than $4 \times 10^{-5}\ s^{-1}$, indicating that substrate dissociation is much slower than cleavage. The solid symbols show the reaction time course for the same ribozyme free in solution, as determined by gel electrophoresis and autoradiography. The data cannot be fit by a single-exponential function, indicating heterogeneous reaction kinetics. The solid curve is a numerical fit using the rate equations. Reprinted with permission from Zhuang *et al.*, Correlating structural dynamics and function in single ribozyme molecules. *Science* **296** (2003) 1473–1476. Copyright 2002 AAAS.

which show a transition from the undocked state to the S-free state, probably result when occupancy of the docked state is too brief to observe in the experiments (which have a 2 s time resolution) or from slow substrate dissociation without cleavage.

The kinetics of docking and undocking can simply be obtained from observation of the dwell times in each state. This analysis reveals that whilst there is a single rate ($0.008\ s^{-1}$) that describes docking, in agreement with ensemble experiments, there are four observed undocking rates where in ensemble experiments only one is observed ($0.005\ s^{-1}$, $0.06\ s^{-1}$, $0.5\ s^{-1}$, and $3\ s^{-1}$). Only the

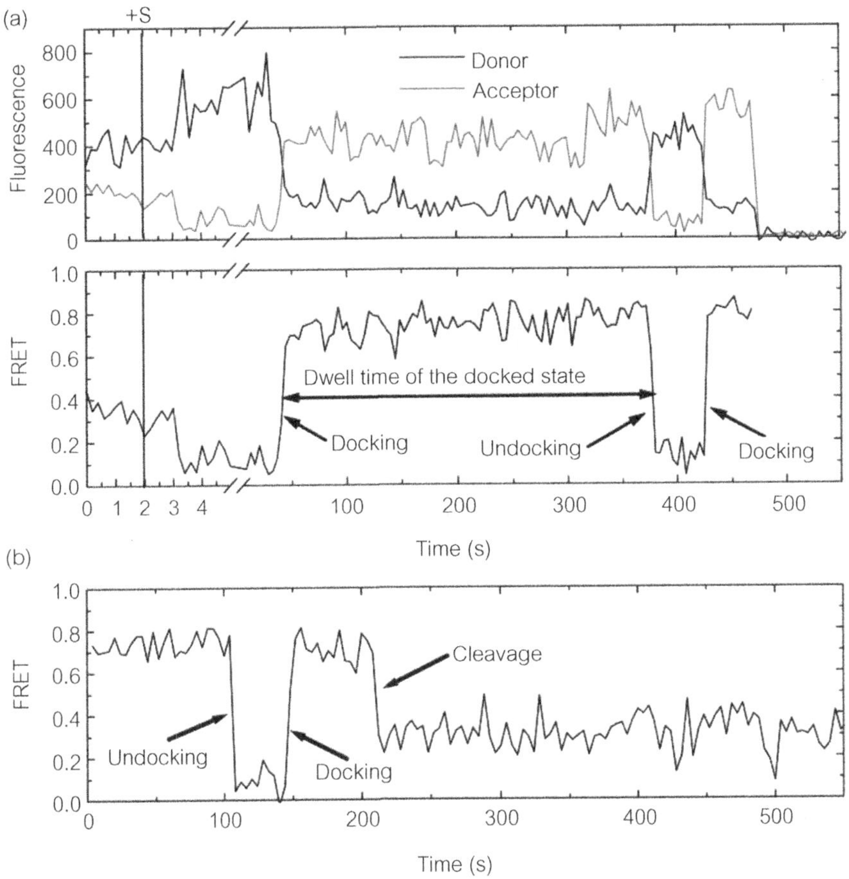

Figure 6.16 Following the structural dynamics and function of single ribozyme molecules. (a) Typical fluorescence time trace of a single ribozyme upon substrate binding. Standard buffer containing 200 nM substrate was added to the sample at 2 s. The delay between substrate arrival and FRET signal change is consistent with the binding rate of substrate. Fluorescence signals from the donor and acceptor are indicated by black and gray lines, respectively. (b) FRET time trace of a single ribozyme-cleavable substrate complex, showing docking, undocking and cleavage as indicated. Reprinted with permission from Zhuang *et al.*, Correlating structural dynamics and function in single ribozyme molecules. *Science* **296** (2002) 1473–1476. Copyright 2002 AAAS.

slowest undocking rate had previously been observed because the fast undocking of the others implies that the states are insufficiently populated to be observed in ensemble experiments—a crucial advantage of studying heterogeneity of structure and function by single molecule methods.

Interestingly, the time trajectories of individual molecules reveal what the authors term a 'memory effect' in which individual molecules repeat similar dwell times in the docked state suggesting that structural features are somehow

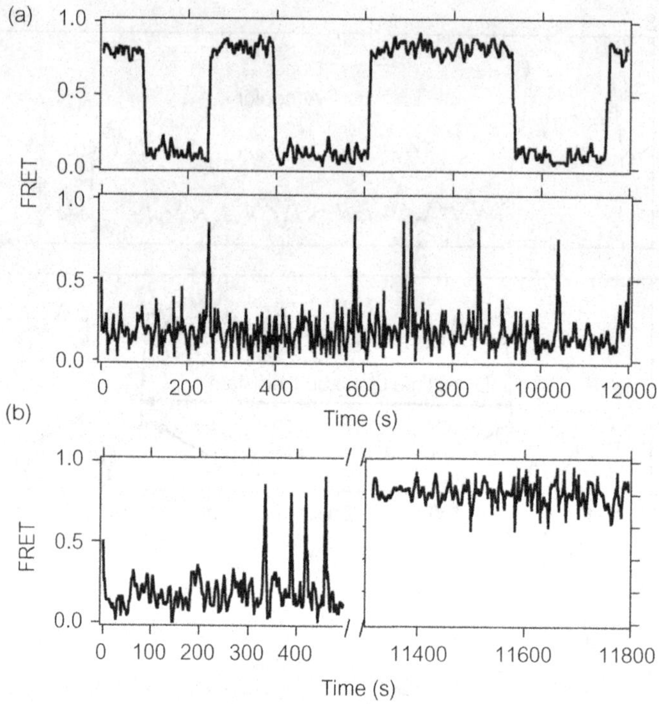

Figure 6.17 Docked states have a strong memory effect, as indicated by FRET time traces of single ribozyme–substrate complexes. (a) Memory effect of the undocking kinetics where a molecule rarely switches between different docked states. (b) An example of memory loss after 3 h. The excitation was shut off after ~ 500 s for 3 h to prevent dye photobleaching. Reprinted with permission from Zhuang *et al.*, Correlating structural dynamics and function in single ribozyme molecules. *Science* **296** (2003) 1473–1476. Copyright 2002 AAAS.

'remembered' after undocking (Figure 6.17). This memory effect appears to be lost over long periods of time (e.g. memory loss occurred after 3 h in Figure 6.17(b)) suggesting that there are some slow conformational dynamics that control the undocking rate.

The authors go on in this work to study a range of mutants to further unpick the nature of these heterogeneous kinetics and most recently have reported a detailed phi analysis of the P1 duplex docking to the pre-folded core of a ribozyme [21] derived from the self-splicing group I intron of *Tetrahymena thermophila.* In such a study the effects of a series of mutations on the kinetic rate constants are measured in order to infer the importance of certain interactions in the transition state. These experiments are very powerful demonstrations of how single molecule methods can contribute significantly to our understanding of the function of complex structures such as these large RNA enzymes and the heterogeneous conformational dynamics that play a key role in determining their function.

References

[1] **Abe, K, Kaya, S, Hayashi, Y, Imagawa, T, Kikumoto, M, Oiwa, K,** *et al.*, Correlation between the activities and the oligomeric forms of pig gastric H/K-ATPase. *Biochemistry* **42** (2003) 15132–15138.

[2] **Ying, LM** and **Xie, XS**, Fluorescence spectroscopy, exciton dynamics, and photochemistry of single allophycocyanin trimers. *Journal of Physical Chemistry B* **102** (1998) 10399–10409.

[3] **Schwille, P, Haupts, U, Maiti, S,** and **Webb, WW**, Molecular dynamics in living cells observed by fluorescence correlation spectroscopy with one- and two- photon excitation. *Biophysical Journal* **77** (1999) 2251–2265.

[4] **Mortelmaier, M, Kogler, EJ, Hesse, J, Sonnleitner, M, Huber, LA,** and **Schutz, GJ**, Single molecule microscopy in living cells: Subtraction of autofluorescence based on two color recording. *Single Molecules* **3** (2002) 225–231.

[5] **Mashanov, GI, Tacon, D, Peckham, M,** and **Molloy, JE**, The spatial and temporal dynamics of pleckstrin homology domain binding at the plasma membrane measured by Imaging single molecules in live mouse myoblasts. *Journal of Biological Chemistry* **279** (2004) 15274–15280.

[6] **Kovacs, M, Wang, F,** and **Sellers, JR**, Mechanism of action of myosin X, a membrane-associated molecular motor. *Journal of Biological Chemistry* **280** (2005) 15071–15083.

[7] **Kubitscheck, U, Kuckmann, O, Kues, T,** and **Peters, R**, Imaging and tracking of single GFP molecules in solution. *Biophysical Journal* **78** (2000) 2170–2179.

[8] **Schwille, P, Kummer, S, Moerner, WE,** and **Webb, WW**, Fluorescence correlation spectroscopy (PCS) of different GFP mutants reveals fast light-driven intramolecular dynamics. *Biophysical Journal* **76** (1999) A260.

[9] **Haupts, U, Maiti, S, Schwille, P,** and **Webb, WW**, Dynamics of fluorescence fluctuations in green fluorescent protein observed by fluorescence correlation spectroscopy. *Proceedings of the National Academy of Sciences of the United States of America* **95** (1998) 13573–13578.

[10] **Ha, T, Rasnik, I, Cheng, W, Babcock, HP, Gauss, GH, Lohman, TM,** et al., Initiation and re-initiation of DNA unwinding by the Escherichia coli Rep helicase. *Nature* **419** (2002) 638–641.

[11] **Dohoney, KM** and **Gelles, J**, Chi-sequence recognition and DNA translocation by single RecBCD helicase/nuclease molecules. *Nature* **409** (2001) 370–374.

[12] **Dickson, RM, Cubitt, AB, Tsien, RY,** and **Moerner, WE**, On/off blinking and switching behaviour of single molecules of green fluorescent protein. *Nature* **388** (1997) 355–358.

[13] **Ha, T, Glass, J, Enderle, T, Chemla, DS,** and **Weiss, S**, Hindered rotational diffusion and rotational jumps of single molecules. *Physical Review Letters* **80** (1998) 2093–2096.

[14] **Jia, YWTalaga, DS, Lau, WL, Lu, HSM, DeGrado, WF,** and **Hochstrasser, RM**, Folding dynamics of single GCN4 peptides by fluorescence resonant energy transfer confocal microscopy. *Chemical Physics* **247** (1999) 69–83.

[15] **Lu, HP, Xun, L,** and **Xie, XS**, Single-molecule enzymatic dynamics. *Science* **282** (1998) 1877.

[16] **Wazawa, T, Ishii, Y, Funatsu, T,** and **Yanagida, T**, Spectral fluctuation of a single fluorophore conjugated to a protein molecule. *Biophysical Journal* **78** (2000) 1561–1569.

[17] **Boukobza, E, Sonnenfeld, A,** and **Haran, G**, Immobilization in surface-tethered lipid vesicles as a new tool for single biomolecule spectroscopy. *Journal of Physical Chemistry B* **105** (2001) 12165–12170.

[18] **Rhoades, E, Gussakovsky, E,** and **Haran, G**, Watching proteins fold one molecule at a time. *Proceedings of the National Academy of Sciences of the United States of America* **100** (2003) 3197–3202.

[19] **Chung, SH** and **Kennedy, RA**, Forward-backward non-linear filtering technique for extracting small biological signals from noise. *Journal of Neuroscience Methods* **40** (1991) 71–86.

[20] **Rhoades, E, Cohen, M, Schuler, B,** and **Haran, G**, Two-state folding observed in individual protein molecules. *Journal of the American Chemical Society* **126** (2004) 14686–14687.

[21] **Bartley, LE, Zhuang, XW, Das, R, Chu, S, and Herschlag, D**, Exploration of the transition state for tertiary structure formation between an RNA helix and a large structured RNA. *Journal of Molecular Biology* **328** (2003) 1011–1026.

[22] **Zhuang, XW, Bartley, LE, Babcock, HP, Russell, R, Ha, TJ, and Herschlag, D**, et al., A single-molecule study of RNA catalysis and folding. *Science* **288** (2000) 2048–2051.

[23] **Zhuang, XW, Kim, H, Pereira, MJB, Babcock, HP, Walter, NG, and Chu, S**, Correlating structural dynamics and function in single ribozyme molecules. *Science* **296** (2002) 1473–1476.

SEVEN
The outlook for single molecule fluorescence measurements

7.1 Outlook

Single molecule sensitivity provides a platform for fundamental and applied studies that are increasing in their complexity and ambition at a fast pace. There are undoubtedly exciting discoveries to be made in many areas, particularly in biophysics. Even in the few brief examples that have been given in this book we have had a glimpse of the insights that single molecule measurements have already provided. The roles which structural and functional heterogeneity play in many biological processes, revealed by single molecule methods, will be a topic of intense study in the coming years.

One exciting area for future development is the combination of single molecule manipulation techniques with fluorescence measurements [1,2]. There are a number of tools that can be used to manipulate single biological molecules such as proteins and nucleic acids for example, the atomic force microscope [3], optical tweezers [4] and the patch clamp [5]. By combining single molecule fluorescence measurements with one of these manipulation techniques it will be possible to obtain multiple experimental observables of conformational changes or biochemical reactions [6–8]. This will allow more rigorous testing of theoretical models and molecular dynamics simulations and will allow distance measurements of conformational change by fluorescence to be correlated with force, for example, in the power stroke of a molecular motor or during DNA twisting. One key challenge in this area is the short lifetime of the common fluorescent dyes which photobleach in seconds. This is in contrast to the many hours during which a single molecule can, in principle, be manipulated. Improved dye characteristics [9] and the use of quantum dots as fluorophores [10] in single molecule experiments are essential if such studies are to be successful.

Perhaps the most exciting prospect for single molecule fluorescence will be the translation of the technique to operation within a living cell [11–14]. The enormous complexity of the cell and the inherently low concentration of any particular molecule of interest within each cell strongly indicate that single molecule studies will have a major role to play in cell biology in the future. Of course, total internal reflection microscopy has already been used to study intracellular single molecule events (see Chapter 6). However, in order to perform single molecule spectroscopy throughout the volume of the cell and routinely and rapidly track biomoelcules as they move around the cell [15], it will be necessary to make improvements in fluorescent labelling methodologies in order to move away from green fluorescent protein and its analogues which do not have ideal characteristics for single molecule experiments [16] and also make improvements in confocal imaging technology. Such developments will undoubtedly open a new era in cell biology.

The challenge presented by the relatively poor characteristics of the available dyes and the difficulties in the labelling of specific sites within a protein or other large molecule with perfect homogeneity and high success rate is significant. Whilst quantum dots provide a solution to the problem of photobleaching [10] they are relatively large and one cannot envisage experiments in which such structures are successfully internalized within large proteins without affecting structure. One possible solution would be to develop single molecule sensitive experiments capable of operating in the near UV so that intrinsic protein fluorophores (tryptophan and tyrosine) could be used as the probes of structure and dynamics [17]. This is a significant challenge not only because of the worsening detector efficiency towards the UV but also due to the low quantum yield (tryptophan has a protein-incorporated quantum yield of about 10%) and poor photostability of these amino acids. However, improvements in detector technology, careful design of the collection optics and possibly the use of artificial fluorescent amino acids [18,19] to replace tryptophan in proteins could make intrinsic fluorescence measurements of protein structure a reality which would open a huge range of potential experiments to test in detail the current theories of folding.

Time resolution is also an important issue in the study of biomolecular dynamics and reactions. There is a fundamental limit to how many photons per unit time can be expected from a single fluorophore that is reporting on protein folding, for example, even if the overall detection efficiency of the optical instrumentation is near 100%, which is of course not the case. The excitation/relaxation cycle of a fluorophore may take several nanoseconds and, given the excitation photon fluxes used and the quantum yield of the dyes, a reasonable estimate is that one can expect a maximum of a single fluorescence photon per 100 ns. Perhaps in the ideal case with detection efficiency near to 100% there will be sufficient time resolution in single molecule experiments to characterize some biological processes but investigation of the pathway between states, in protein folding for example, may well take place on the sub-microsecond timescale, leaving little detectable signal. An interesting approach would be to employ flash freezing of the sample using liquid nitrogen, similar to existing electron cryomicroscopy methods [20] for example, after a mixing process so that the system is frozen far from equilibrium in glassy water. Cryo-single molecule fluorescence measurements to ensure that the system remains frozen might allow some of the most transiently populated states in protein folding to be seen for the first time, although an immediately obvious problem of randomly fixing the transition dipoles of the fluorescent probe molecules must be overcome.

Nanotechnology is clearly an area in which single molecule sensitivity has high potential impact [21,22]. Regardless of their purpose, nanoscale machines will require a connection to the macroscopic world. Instructions and data may need to be communicated in both directions and the delivery and collection of single photons from organic and inorganic structures is likely to play a key role in this interface. A single fluorophore represents the tightest possible focusing of an optical field and a single fluorescing molecule is a point dipole source in which the energy of the photon is concentrated in the near-field in approximately one cubic nanometer. Data may clearly be stored on a nanometer length scale as excitons in single molecules and further levels of information may be stored in polarization states or via electronic or photo-switchable conformations [23]. Photons may also provide a convenient method of putting energy into nanoscale machines to drive them to perform mechanical or chemical functions. One aspect of single molecule spectroscopy that we have not discussed in any detail in this book is the quantum nature of the process but we must not forget that single photon experiments provide a platform for quantum computation/cryptography processes [24].

In Chapter 5 we discussed one example of the use of single molecule fluctuation spectroscopy as a potentially high throughput tool for biological assays [25]. Another driving force in drug discovery is the need to minimize the amount of sample required simply because the amounts that are available in the early stage of pharmaceutical development are tiny. Clearly single molecule spectroscopy provides a high throughput tool that in principle can operate with the absolute minimum of sample. Quantitative analysis is the major challenge in high throughput screening [26,27] both with ensemble and single molecule approaches. However, in principle FCS and FIDA both provide quantitative measurements of biomolecular interactions and it seems very likely that single molecule spectroscopy will become more and more widely used in the early stages of drug development.

Single molecule spectroscopy is a technology that underpins a range of fundamental and applied fields of study. There is no doubt that there are challenging and exciting prospects for those choosing to work in this area.

References

[1] **Harada, Y, Funatsu, T, Tokunaga, M, Saito, K, Higuchi, H, Ishii, Y,** *et al.*, Single Molecule Imaging and Nanomanipulation of biomolecules. *Methods in Cell Biology*, 55 (1998) 117–128.

[2] **Yanagida, T,** Single molecule nano-bioscience. *Journal of Pharmacological Sciences* **91** (2003) 2P–2P.

[3] **Brockwell, DJ, Smith, DA, Radford, SE,** Protein folding mechanisms: New methods and emerging ideas. *Current Opinion in Structural Biology* **10** (2000) 16–25.

[4] **Ashkin, A,** Optical trapping and manipulation of neutral particles using lasers. *Proceedings of the National Academy of Sciences of the United States Of America* **94** (1997) 4853–4860.

[5] **Yang, S, Zhou, WX,** and **Zhang, YX,** New advance in in vivo patch clamp technique. *Progress in Biochemistry and Biophysics* **31** (2004) 870–873.

[6] **Heinz, WF, Weston, KD, Jolivet, V, Navarro, B, Bernardi, P,** and **Goldner, LS,** A combined atomic force, confocal, and total internal reflection microscope for single molecule microscopy. *Biophysical Journal* **82** (2002) 201.

[7] **Sarkar, A, Robertson, RB,** and **Fernandez, JM,** Simultaneous atomic force microscope and fluorescence measurements of protein unfolding using a calibrated evanescent wave. *Proceedings of the National Academy of Sciences of the United States of America* **101** (2004) 12882–12886.

[8] **van Dijk, MA, Kapitein, LC, van Mameren, J, Schmidt, CF,** and **Peterman, EJG,** Combining optical trapping and single-molecule fluorescence spectroscopy: Enhanced photobleaching of fluorophores. *Journal of Physical Chemistry* B **108** (2004) 6479–6484.

[9] **Willets, KA, Callis, PR,** and **Moerner, WE,** Experimental and theoretical investigations of environmentally sensitive single-molecule fluorophores. *Journal of Physical Chemistry B* **108** (2004) 10465–10473.

[10] **Hohng, S** and **Ha, T,** Single-molecule quantum-dot fluorescence resonance energy transfer. *Chemphyschem* **6** (2005) 956–960.

[11] **Ichinose, J** and **Sako, Y,** Single-molecule measurement in living cells. *Trends in Analytical Chemistry: TRAC* **23** (2004) 587–594.

[12] **Mashanov, GI, Tacon, D, Knight, AE, Peckham, M,** and **Molloy, JE,** Visualizing single molecules inside living cells using total internal reflection fluorescence microscopy. *Methods* **29** (2003) 142–152.

[13] **Mashanov, GI, Tacon, D, Peckham, M,** and **Molloy, JE,** The spatial and temporal dynamics of pleckstrin homology domain binding at the plasma membrane measured by imaging single molecules in live mouse myoblasts. *Journal of Biological Chemistry* **279** (2004) 15274–15280.

[14] **Pramanik, A,** Ligand-receptor interactions in live cells by fluorescence correlation spectroscopy. *Current Pharmaceutical Biotechnology* **5** (2004) 205–212.

[15] **Levi, V, Ruan, QQ,** and **Gratton, E,** 3-D particle tracking in a two-photon microscope: Application to the study of molecular dynamics in cells. *Biophysical Journal* **88** (2005) 2919–2928.

[16] **Chirico, G, Cannone, F, Beretta, S, Diaspro, A, Campanini, B, Bettati, S,** *et al.*, Dynamics of green fluorescent protein mutant2 in solution, on spin-coated glasses, and encapsulated in wet silica gels. *Protein Science* **11** (2002) 1152–1161.

[17] **Lippitz, M, Erker, W, Decker, H, van Holde, KE,** and **Basche, T,** Two-photon excitation microscopy of tryptophan-containing proteins. *Proceedings of the National Academy of Sciences of the United States of America* **99** (2002) 2772–2777.

[18] **Acchione, M**, **Guillemette, JG**, **Twine, SM**, **Hogue, CWV**, **Rajendran, B**, and **Szabo, AG**, Fluorescence based structural analysis of tryptophan analogue—AMP formation in single tryptophan mutants of Bacillus stearothermophilus tryptophanyl-tRNA synthetase. *Biochemistry* **42** (2003) 14994–15002.

[19] **Broos, J**, **ter Veld, F**, and **Robillard, GT**, Membrane protein-ligand interactions in Escherichia coli vesicles and living cells monitored via a biosynthetically incorporated tryptophan analogue. *Biochemistry* **38** (1999) 9798–9803.

[20] **Unger, VM**, Electron cryomicroscopy methods. *Current Opinion in Structural Biology* **11** (2001) 548–554.

[21] **Doty, RC**, **Fernig, DG**, and **Levy, R**, Nanoscale science: a big step towards the Holy Grail of single molecule biochemistry and molecular biology. *Cellular and Molecular Life Sciences* **61** (2004) 1843–1849.

[22] **Whitesides, GM**, Nanoscience, nanotechnology, and chemistry. *Small* **1** (2005) 172–179.

[23] **White, SS**, **Ying, LM**, **Balasubramanian, S**, and **Klenerman, D**, Individual molecules of dye-labeled DNA act as a reversible two-color switch upon application of an electric field. *Angewandte Chemie-International Edition* **43** (2004) 5926–5930.

[24] **Moerner, WE**, Single-photon sources based on single molecules in solids. *New Journal of Physics* **6** (2004) art. no.-88.

[25] **Schaertl, S**, **Meyer-Almes, FJ**, **Lopez-Calle, E**, **Siemers, A**, and **Kramer, J**, A novel and robust homogeneous fluorescence-based assay using nanoparticles for pharmaceutical screening and diagnostics. *Journal of Biomolecular Screening* **5** (2000) 227–237.

[26] **Rudiger, M**, **Haupts, U**, **Moore, KJ**, and **Pope, AJ**, Single-molecule detection technologies in miniaturized high throughput screening: Binding assays for G protein-coupled receptors using fluorescence intensity distribution analysis and fluorescence anisotropy. *Journal of Biomolecular Screening* **6** (2001) 29–37.

[27] **Wolcke, J** and **Ullmann, D**, Miniaturized HTS Technologies—UHTS. *Drug Discovery Today* **6** (2001) 637–646.

Index

Note: *f* following a page number indicates a figure and *t* indicates a table.

Made in the USA
Las Vegas, NV
14 September 2026

54613748R00155